Removing Barriers

EDITED BY
JILL M. BYSTYDZIENSKI
AND
SHARON R. BIRD

Removing Barriers

Women in Academic Science, Technology, Engineering, and Mathematics

INDIANA UNIVERSITY PRESS

Bloomington and Indianapolis

This book is a publication of

Indiana University Press
601 North Morton Street
Bloomington, IN 47404-3797 USA

http://iupress.indiana.edu

Telephone orders 800-842-6796
Fax orders 812-855-7931
Orders by e-mail iuporder@indiana.edu

Library of Congress Cataloging-in-Publication Data

Removing barriers : women in academic science, technology, engineering, and mathematics / edited by Jill M. Bystydzienski and Sharon R. Bird.
p. cm.
Papers originally presented at a 2002 conference at Iowa State University, "Retaining Women in Early Academic Science, Technology, Engineering and Mathematics (STEM) Careers."
Includes bibliographical references and index.
ISBN 0-253-34679-7 (cloth : alk. paper) — ISBN 0-253-21817-9 (pbk. : alk. paper)
1. Women in science—Congresses. 2. Women scientists—Employment—Congresses. 3. Feminism and science—Congresses. I. Bystydzienski, Jill M., date II. Bird, Sharon R., date
Q130.R46 2005
305.43′5—dc22

2005017212

1 2 3 4 5 11 10 09 08 07 06

For Nora Gresch

To my sister, Rebecca D. Lemcke
srb

Contents

Acknowledgments

This edited volume would not have been possible without the hard work and support of numerous people. The 2002 conference at Iowa State University, "Retaining Women in Early Academic Science, Technology, Engineering and Mathematics (STEM) Careers," yielded the keynote addresses and papers which constitute the chapters of this book. The organization of the conference, and a great deal of the subsequent effort of getting the contributions peer-reviewed, was largely done by Nora Gresch, 2002–2003 assistant to the director of Women's Studies at Iowa State. It is, therefore, to Nora Gresch that this book is dedicated. We are grateful for her enthusiasm and optimism, and for the time and energy she contributed to the project.

In addition to the generous support of the 2002 conference by the National Science Foundation, we also greatly appreciate additional financial support received from several units at Iowa State University: the office of the Dean of the College of Liberal Arts and Sciences, the Dean of the Engineering College, and the Provost's office.

We want to thank Dee Mortensen, sponsoring editor, and Elizabeth Marsh, assistant sponsoring editor, at Indiana University Press for their excellent work in shepherding the manuscript through the production process and Shoshanna Green for her expert copyediting of the text.

Finally, we want to acknowledge the important work of all the contributors to this volume, whose research and dedication to improving the status of women and other underrepresented groups in the sciences, engineering, and mathematics are helping to create more equitable structures and a more inclusive climate for all people in these academic fields.

This material is based upon work supported by the National Science Foundation under Grant HRD-094556. Any opinions, findings, and conclusions or recommendations expressed herein are those of the authors and do not necessarily reflect the views of the National Science Foundation.

Introduction

Jill M. Bystydzienski and Sharon R. Bird

Despite several decades of research on and interventions to benefit women in science, technology, engineering, and mathematics (STEM) fields, the progress in effectively changing the representation of women in many STEM careers has been slow. Thus, while women are currently slightly overrepresented among all high school and college graduates, and constitute the majority of undergraduates and master's-level graduates in the biological sciences, in areas such as physics, computer science, and engineering women continue to obtain only a small proportion of college degrees (see table I.1). What is most discouraging is that, in the last decade, relatively little progress has been made in recruiting and retaining more women in these fields. And while African American, Hispanic, and American Indian women earn a higher proportion of bachelor's and graduate degrees in STEM than their ethnic group male counterparts (NSF 2003), they nevertheless still obtain only tiny fractions of all college degrees in these fields. For example, of the 14,313 Ph.D.s awarded to United States citizens in 2002 in STEM fields, only 353 were obtained by African American women, 103 by Latinas, and 32 by Native American women (Hill 2003).

In some science fields, the seeming increase in the proportion of women is a statistical artifact, especially at the bachelor's level, as the actual numbers of women have remained steady or have dropped, while the total numbers of all graduates have declined (see table I.2). The decreasing white male enrollment in science and engineering accounts for much of this decline (NSF 2003, 24).

The progressively decreasing numbers of women in STEM from undergraduate to graduate levels is further reflected in their even smaller proportions among the faculty. In four-year colleges and universities, women constitute 36 percent of the overall faculty, but only 22 percent of professors of physical and related sciences and 9 percent of engineering faculty (NSF 2003, 266). A recent study of science and engineering departments in Association of American Universities institutions showed that only about 5 percent of the full professors and 4.2 percent of the chairs in engineering, mathematics and statistics, earth sciences, chemistry, and physics and astronomy combined were female (Niemeier and Gonzales 2004). It is still the rule that the higher the educational level, the fewer the women, and especially so in STEM fields.

Another problem documented by recent studies (Etzkowitz, Kemelgor, and Uzzi 2000, and chapters 3, 7, 8, and 12 of this volume) is that even the women who are "successful" in these fields, that is, who persevere and do not drop out,

Table I.1. Women in Science and Engineering by Field
and Percent of Degrees Awarded, 1990/1998

Field	Bachelor's (1990/98)	Master's (1990/98)	Ph.D. (1990/98)
Astronomy	—/34.5	21.0/29.7	15.6/20.0
Chemistry	—/45.7	39.3/45.9	24.0/29.7
Physics	12.4/19.3	16.3/18.2	10.3/12.6
Atmospheric Sciences	28.1/23.5	21.1/26.1	16.9/17.7
Mathematics & Statistics	46.0/46.8	40.1/41.7	17.7/25.5
Computer Science	30.2/26.9	27.7/29.1	15.6/18.4
Biological Sciences	41.8/55.3	51.1/52.7	37.3/42.9
Engineering	15.4/18.6	13.6/19.8	8.5/14.8

Source: National Science Foundation. Division of Science Resources Statistics. 2003. *Women, Minorities, and Persons with Disabilities in Science and Engineering: 2002.* Arlington, Va.: National Science Foundation (NSF 03-312).

Table I.2. Bachelor's Degrees Awarded to Women in Selected Fields, 1990/1998

	Physics	Mathematics	Computer Science	Engineering
Total Men & Women	3,740/3,455	14,674/12,094	27,695/27,674	64,705/60,914
Women	651/666	6,811/5,659	8,374/7,439	9,973/11,339
Percent Women	17.0/19.3	46.4/46.8	30.2/26.9	15.4/18.6

Source: National Science Foundation. Division of Science Resources Statistics. 2003. *Women, Minorities, and Persons with Disabilities in Science and Engineering: 2002.* Arlington, Va.: National Science Foundation (NSF 03-312).

experience frustrations and dilemmas that negatively affect the quality of their careers. The numbers do not convey the real lives of women scientists who have had to make enormous personal sacrifices and often work much harder than their male counterparts in order to prove themselves in difficult circumstances.

What are the reasons for the continuing underrepresentation of women in STEM as well as the problems experienced by those who persevere? What barriers continue to thwart progress for women in these fields? What strategies can be used to illuminate the practices that construct sexist formal knowledge, and that essentialize socially created notions of "femininity" and "masculinity"? And what remedies exist or can be developed to overcome the barriers? These are the questions addressed in this collection.

The chapters in this book consist of several keynote addresses previously published in a special issue of the *NWSA Journal* (Bystydzienski 2004) and selected papers presented at the October 2002 conference on Retaining Women in Early Academic Science, Technology, Engineering and Mathematics (STEM) Careers at Iowa State University.[1] Taken together, these articles examine the status of women in STEM fields, both past and present; the barriers that keep women from full participation in science, technology, and engineering; and the particular obstacles that exist in these fields for women of color, especially African American (and African) women.

This collection highlights, as well, gender biases in the *practice of* science (chapters 9, 10, and 11). Feminist study of scientific practice emphasizes that merely increasing the number of women who "do science" does not ensure that women's perspectives and positions *as women* will influence the process of constructing formal knowledge. Masculinist interpretations[2] of asexual reproduction in living organisms, for example, are commonly embraced in biology as gender-neutral facts. Critiques of such interpretations, however, are more widely acknowledged and accepted by scholars outside STEM disciplines (e.g., in women's studies, the humanities, and the social sciences) than within them (Hammonds and Subramaniam 2003).

Perhaps the most important aspect of this volume is that, in addition to a thorough examination of barriers to women's equal participation in STEM fields, it offers pragmatic, concrete suggestions for transforming academic science, technology, engineering, and mathematics. Taken together, this collection argues that in order for STEM fields to be equally open to women and men from all racial and ethnic groups, systemic change has to take place. Women need to be represented more equally, but institutional structures and disciplinary cultures must also be changed. Strategies for transforming institutions range from organization-wide hiring and family policies (the focus of the "women in science" approach) to changing the practices of science and engineering (the concern of the "feminist science" approach).

Reasons for Low Numbers of Women in STEM

Until recently, a great deal of the research and most of the intervention programs developed on college and university campuses emphasized women's perceptions, expectations, and choices and focused on fitting women into existing science and engineering departments, programs, and laboratories. Since earlier research found sex differences in learning and achievement (Benbow and Stanley 1980; Berryman 1983; Maccoby and Jacklin 1966), it was assumed that women were "deficient" in math and science and lacked motivation to participate. Hence, they had to be individually encouraged, mentored, and appropriately socialized to enter and remain in the science, engineering, and technology fields. However, more recent research has found that the gaps between the sexes in course enrollment and achievement in math and the sciences have become statistically insignificant (Clewell and Campbell 2002; NSF 2003, 5), and that performance levels of women and men in STEM are not significantly different.

For example, girls and boys have the same aptitude for mathematics and science, as demonstrated by class grades, achievement tests (Catsambis 1995), and enrollment in most advanced high school science and mathematics classes (Greenfield 1997), while the performance of women faculty and faculty of color, as measured by number of publications and other criteria, does not differ from that of white men (Jackson 2004).

Nevertheless, despite evidence that women and men are equally capable of careers in STEM fields, percentages of women and other underrepresented groups in most science and engineering fields remain small. Hence, research and interventions that construe women as "the problem" in need of change have increasingly been shown to be an inappropriate approach to increasing the numbers of women in STEM.

Another approach to explaining and improving the lagging participation of women in science, engineering, and technology fields is known as the "pipeline" theory. This theory holds that if sufficient numbers of women are encouraged to enter STEM fields, the gender gap in science and technology eventually will disappear. A "pipeline," like that for transporting water or natural gas, symbolizes the career trajectory from elementary school to initial employment. The rate of flow into scientific careers is measured by passage through transition points such as graduation and continuation to the next educational level. The pipeline begins with a relatively large pool of females in elementary school (where there are no gender differences in achievement and interest in science) and gradually leaks potential STEM recruits at every joint, beginning in eighth grade, where girls' interest in math and science wanes (Catsambis 1994), then in college, where over half of all women who start out majoring in science, math, or engineering switch to other majors (Seymour and Hewitt 1997). An even smaller number of women go on to graduate study, and still fewer obtain graduate degrees and become STEM faculty, full professors, and department heads.

While the analogy of the leaky pipeline is apt, it nevertheless suggests an approach to solving the problem of low numbers of women in STEM that is not that different from the approach implied by the "women as 'deficient'" model. It assumes that if girls are encouraged to study science, particularly at the middle school level, then women's and men's participation in STEM will become more equal over time, and more women will attain higher levels of education and leadership positions. However, this approach, by focusing on the "supply side," neglects an analysis of the "demand side," in particular organizational resistance to change and the persistence of systemic barriers to women's entry into and full participation in STEM professions (Etzkowitz, Kemelgor, and Uzzi 2000). Moreover, those who use the pipeline metaphor do not question why the pipeline works the way it does, nor the context within which the pipeline is situated (Hammonds and Subramaniam 2003, 940).

In recent years, activists, social science researchers, and increasing numbers of policymakers have begun to recognize what many women's and gender studies scholars have been advocating for some time now: that the (remaining) barriers to women's progress in academia are systemic. Thus, rather than trying to change women to fit the sciences and engineering, these fields need to be changed in order to fully embrace women.[3]

Many institutional leaders are shifting their approaches to the problems of women's inclusion in STEM. For example, as Sue Rosser points out in chapter 3, in a special meeting at MIT in 2003, high-level administrators from the most prestigious U.S. universities acknowledged that barriers still exist for women and that abolishing them will require significant changes in institutional procedures as well as within the scientific and engineering fields. The ADVANCE program, initiated by the National Science Foundation in 2001, encourages institutional, rather than individual, solutions to empower women to participate fully in STEM (Georgia Institute of Technology 2003). In 2003, the American Institute of Physics issued a report advocating changing the culture of science and math education to encourage greater interest in and motivation for physics among underrepresented groups (AIP 2003a).[4]

It is also more widely recognized today that women and people of color face barriers in STEM fields that are the product of both historical and contemporary practices of science. Feminist study of scientific practice seeks to reveal sexist, heterosexist, and racist (as well as classist, ableist, and other) biases in the practice of science, and rejects the claim that theories and hypotheses are formulated in a manner that is completely "value-free." Acknowledging that values influence what researchers choose to study and how they collect evidence enables more accurate assessment of the limits of research. This does not mean, however, that feminist critiques of science advocate the relativist notion that all methods for conducting science, no matter how careless or rigorous, are equally useful and valid. Acknowledging that the conclusions drawn from scientific research depend on the social and material circumstances under which questions of science are formulated, data are gathered, and theories are tested promotes better science through the inclusion of many different standpoints (Hammonds and Subramaniam 2003; Mayberry, Subramaniam, and Weasel 2001; Rosser 1992; Tuana 1989).

The focus on external barriers for women in science, technology, engineering, and mathematics, as well as on obstacles created by prevailing practices in these fields, has great promise, because it can lead to a transformation of these fields to the benefit of both women and men. Given the recent change among academic leaders in thinking about and dealing with the problems of inclusion of women in STEM, the studies and analyses presented, as well as the recommendations proposed, by the contributors to this book are timely and offer a real opportunity for more widespread implementation.

Barriers for Women in STEM in Higher Education

Historically, women's participation in science, technology, and engineering in the United States has been subject to numerous barriers outside their control. As the first two chapters in this volume show, while women made considerable contributions to these fields over the last century, and even achieved some degree of acceptance, their progress has been thwarted by limited opportunities in male-dominated structures. Sally Gregory Kohlstedt demonstrates in chapter 1 that the entry of women into science and technology fields was not a linear process and that, despite considerable activism, patterns of discrimination have continued to circumscribe women's achievements into the current period. In chapter 2, Amy Bix focuses on the history of women in engineering, emphasizing the strategies female engineers employed to improve conditions for women in education and employment. Though positive changes have taken place as a result, many challenges still remain as engineering continues to be one of the fields least open to women.

Not unlike the military and politics, academic science is a gendered institution.[5] This means that the practices and procedures through which students and faculty are recruited, processed, and deemed successful (or not) tend to advantage men (and constructions of masculinity) over women (and constructions of femininity). While the entire system of higher education was established initially by and for elite men (Noble 1992), the traditionally masculine values of hierarchy, challenge, competition, and independence have persisted especially in fields like physics, computer science, and engineering, which remain largely filled by men. Hence, when women enter these male spaces they are confronted with a culture and structures that many of them (and some men also) find to be at best bothersome and at worst hostile and excluding. Despite the fact that women (undergraduate and graduate students and faculty) today in STEM departments in U.S. colleges and universities are most often academically as well prepared as, or better prepared than, men (Clewell and Campbell 2002; Seymour and Hewitt 1997), many more of them than their male counterparts drop out of these fields or do not go on to higher levels. And among those who persevere, many experience serious difficulties and obstacles as they pursue science careers.

What sort of systemic barriers does the male-dominated and masculinist culture pose for women in science? An implicit male standard permeates science, according to which researchers are most productive when their time is devoted to scientific investigation to the exclusion of all other aspects of life (Noble 1992). It is assumed that those who are really serious about a career in science will make their work the highest priority of all. Thus women who have or may want to have children, or even just a social life with family and friends, are made to feel that their desire to have a life outside of science renders them unfit for scientific careers. Most male scientists, at least until recently, had wives who took care of their children and homes and thus the men were able to devote most of their time and energy to their work. Women, however, typically have not had that type of support from spouses. As Sally Gregory Kohlstedt points out in chapter 1, many of the women pioneers in science were unattached and childless, which meant that they could devote themselves exclusively to scientific inquiry, while those who did have positive and supportive spousal relationships were able to do outstanding work.

Sue Rosser demonstrates in chapter 3 that the most significant obstacle for many women in science and engineering today is the need to balance career demands and family and personal issues. The respondents in Rosser's study were successfully employed female scientists, mainly at research universities, and thus the work and family challenges they identified can be assumed to be even more problematic for women in relatively less favorable situations.

What is interesting today, as more women enter the sciences, is that many are unwilling to give up motherhood for a scientific career and want to be able to have both. Increasingly, too, younger men pursuing science also want time for family life. These women and men, however, continue to confront academic norms that demand total focus on their work and especially extreme pressure in early career stages, which coincide with women's child-bearing years. Hence, many women, after discovering these expectations, do not even consider scientific careers, while others leave after attempting unsuccessfully to balance graduate school or faculty demands and motherhood (Wasserman 2000).

Although research has demonstrated that the notion that advancing age inhibits high-quality scientific work is unfounded (Merton and Zuckerman 1973), the assumption that youth is associated with high scientific achievement still pervades academic science and is reflected in how scientific careers are structured. The normative career path proceeds from undergraduate studies to graduate school, with a doctoral dissertation completed within seven years, and a typically short postdoctoral position; tenure must be achieved within six years. Once this has been done, a slower pace is allowed. Most scientists go through this very intense and demanding trajectory between their early twenties and their early or midthirties, which is also the optimal time for women to bear children and establish families. Thus the structure of academic careers poses difficulties for women in particular, but also for those men who aspire to being scientists but also want to be involved with their partners or spouses and their children.

Another barrier for women in science is what is commonly known as the "weed-out" practice (Etzkowitz, Kemelgor, and Uzzi 2000). In large universities at the bachelor's level, women encounter a system of typically large introductory science courses which is based on a competitive, male model designed to eliminate an unwanted excess of prospective students. To some extent, it continues into graduate school, where only the "best and brightest" who have proven themselves get through. This practice has more negative effects on women than it does on men. The science education system, of which weeding out is a central component, tests for characteristics traditionally associated with masculinity in Western societies and is based on notions such as "the challenge," understood by young men who have encountered it in various rites of passage into manhood (e.g., in sports, fraternities, and the military). This system is intended to promote the ability of young men to tolerate stress, pain, and humiliation by developing their "character," or personal strength and self-control. Nurturing is deliberately denied, because it is assumed that it will make men weak and vulnerable, and thus young men are driven to look inward for sources of fortitude and outward to bond with other men in adversity. While some men do not feel comfortable in and resist this system, most are at least familiar with its expectations (Connell 2000).

Although most faculty members in the sciences intend to treat their male and female students alike, this seeming equality actually works against women because they are expected to behave in ways that are contrary to the socially constructed stereotypes and imperatives of femininity. Thus, by challenging all students in a science class or a graduate program to prove themselves as they are faced with a rapid curriculum pace and rigid evaluation procedures, the faculty send messages to the female minority that are often at odds with the women's experiences and self-identity. Many women switch their majors from the sciences to other fields because they get discouraged and do not wish to continue having to prove themselves (Seymour and Hewitt 1997).

Related to the weed-out system is a teaching and pedagogical approach prevalent in the sciences that emphasizes traditional techniques, especially lectures and quantitative (as opposed to conceptual) laboratory activities. As Heidi Fencl and Karen Scheel point out in chapter 15, while the traditional teaching style is not appreciated by many men, it is especially contrary to the learning environment in which most women feel comfortable—an environment that emphasizes cooperation, interaction, and experience.

Teaching methods, according to Fencl and Scheel, affect the retention of students, especially women, in introductory science courses. Despite the growing recognition that scientists need to work more collaboratively in academia and beyond, and thus that it is advantageous to cultivate the teaching styles preferred by women (and many men) in the sciences, there has been very little change toward the use of nontraditional teaching models in science classrooms, and especially at the undergraduate introductory level.

Molly Dingel explains in chapter 8 that the institutionalized culture and dominant teaching practices in first-year college courses result in a classroom atmosphere that women experience as unsupportive. Women students are perceived by peers and teachers in the science classroom as being out of place and lacking knowledge. Female students also feel pressured to achieve unrealistic levels of perfection and thus are discouraged by what they assume to be their own failures.

Dingel draws attention to the overall climate of classrooms, departments, and campuses, and to the attitudes, behaviors, and standards that tend to validate and embrace the needs, abilities, and potentials of men over those of women, and of white Anglo-Europeans over racial and ethnic minority group members. In many of the science fields where women are in the minority, they frequently encounter an unwelcoming or even hostile climate. Aspects of a chilly climate may include women's being ignored, discriminated against, or sexually harassed. Often, male peers manifest unwelcoming attitudes and behaviors, making rude and insensitive comments about and to their female counterparts (Crawford and MacLeod 1990; Seymour 1995). Frequently women, who are socialized to tolerate inappropriate and even abusive behavior, make excuses for men ("boys will be boys") and respond by ignoring the behavior ("reacting just makes it worse"). However, if they are exposed to many such instances, over time, the accumulation of negative experiences leads many women to drop out (Pascarella et al. 1997).

Faculty and staff in science departments also frequently engage in behavior that contributes to a chilly climate for women. While some of this is overt, like sexist jokes or deliberate exclusion, other behavior is more subtle. For example, as Cynthia Burack and Suzanne Franks indicate in chapter 4, the way in which science and engineering faculty tell the story of their own programs often unconsciously equates these fields with masculinity and whiteness and contributes to resistance to diversity. For instance, statements such as "Some of our best students are women and minorities" or "To succeed in science you need hard skills, but soft skills are helpful also" inadvertently put women and other underrepresented groups in a subordinate relationship to majority men. In contrast, stating instead that "Women and men from all ethnic groups succeed in our program" or "Successful scientists have strong technical and communication skills" embraces a more inclusive view of the benefits of science and engineering to all.

The climate for women of color is frequently even more damaging than for majority group women. As several chapters in this book indicate, African American and other minority women have to deal with both sexism and racism as they pursue careers in science and technology fields. In chapter 6, Sandra Hanson analyzes data from a longitudinal survey that indicate higher persistence rates for young African American women than for white women in the sciences. Despite these women's strong interest and involvement in science fields, supported by unique resources in their communities of origin, their interest in science nevertheless lessens over time because of such aspects of the chilly climate as white teachers' low expectations, lack of African American female mentors, and lack of recognition of African Americans' contributions to STEM fields. In chapter 7, Josephine Beoku-Betts reports on her study of African women who pursued graduate science work at Western universities. She documents the remarkable determination and resilience of these women in the face of multiple barriers to them as Third World women of color. In chapter 12, Anne MacLachlan similarly shows that women of color in graduate science and technology programs in the University of California system experienced even greater obstacles than majority women. They reported more faculty doubt about their abilities than did white women; they commented on subtle changes in colleagues' behavior suggesting they did not belong; and they indicated that they were seen not as full persons or future scientists, but as representatives of their ethnic groups and were scrutinized and judged on that basis. Both racism and sexism situate women of color in less powerful positions relative to their male and white counterparts, which affects how they acquire scientific knowledge and how they experience their education or faculty status in the context of their lives in and outside of academia. Their structural position and the exclusionary practices they encounter in the scientific community are shaped by gendered and racialized institutional systems and processes.

Contrary to the popular view of the scientist as a lone practitioner spending countless hours in the lab conducting experiments, the practice of science today increasingly involves teams of academic scientists who apply for grants and work together on research and publications. As Sue Rosser in chapter 3 and Anne MacLachlan in chapter 12 demonstrate, despite this reality many women scientists do not benefit from such collaborative efforts, as they are often excluded from departmental networks and denied important information. Academic female scientists tend to be more isolated than their male peers and often pursue funding and research on their own. While this may be an asset in some respects (see chapter 3), a recent comparison of collegial relationships and networks of men and women scientists who possessed the same education and work experience showed that the women had smaller networks of departmental and external ties, which prevented them from collaborating productively on research projects. Women's smaller and less effective networks were linked to token overload (shouldering responsibilities that their male peers did not) and power imbalances (being viewed by colleagues as professional unequals and having to work harder to prove their worth) (Etzkowitz, Kemelgor, and Uzzi 2000).

Although most of the barriers discussed by the contributing authors affect women across the STEM fields, there are also more specific discipline-based obstacles that limit the advancement of women. For instance, Amy Bix in chapter 2 and Cynthia Burack and Suzanne Franks in chapter 4 focus on engineering, indicating that the particular historical and cultural development of engineering makes it especially difficult for women to navigate this most traditionally male profession. While Bix shows that despite a century of women's activism to change engineering education there has been relatively little progress for women, Burack and Franks point out how the normative assumption of the superiority of the white male engineer leads to resistance to diversity in engineering culture.

Similarly, Mo-Yin Tam and Gilbert Bassett in chapter 5 and Anna Martinson in chapter 14 focus on information (computer) technology (IT), a field in which women are even more severely underrepresented today than twenty years ago. Tam and Bassett report that the gender gap in college students who major in information technology remains wide, even though the difference between women and men in math performance has declined significantly in the past decade. The gender gap in IT majors, in turn, means that women are not as well poised as men to take advantage of job opportunities in a rapidly growing "high IT core" of occupations. Martinson suggests, along similar lines, that a very pronounced male bias in computer games (violence and sexist representations) is related to gender differences in the use and enjoyment of electronic games, and women's lower interest in computer science. And Abbe Herzig, in chapter 13, indicates that the isolation of doctoral students from authentic practice in mathematics is an obstacle they have to overcome to develop a sense of belonging in that particular field.

Feminist Study of Scientific Practice

In recent years, an increasing number of feminist scholars have focused on the study of gender in scientific fields, taking an approach distinct from that of "women in science." As Evelyn Hammonds and Banu Subramaniam (2003) explain, feminist study of science differs from the women in science project in that the latter focuses on "the structures within scientific communities that have led to the marginalization of women . . . while scholars working on gender and science focus on the gendered nature of science itself" (928). The dominant and pervasive practices of science, as feminist theorists point out, including its language, logic, methods, and interpretations, are masculinist (Bleier 1986; Haraway 1991; Harding 1991; Hubbard 1990; Keller 1982; Kourany 2002; Rose 1994; Tuana 1989). Masculinist science practices are characterized by sexist and racist metaphors, exclusionist methodologies, the depersonalization of natural knowledge, and the barring of human passion and interest from the production of knowledge (Lederman and Bartsch 2001, 3–4).

The sexist and heterosexist biases in the language and metaphors of biology are by now well known to those who are familiar with feminist analyses of science. In this volume, Dana Dudle and Meryl Altman (chapter 11) show, for example, that Carolus Linnaeus's sexual system of plant classification was based on what he called "the marriages of the plants," according to which plants were placed in a hierarchy of *classes* based on the number, position, and length of their stamens (male organs) and into *orders* within classes based on the number of their pistils (female organs). As Dudle and Altman point out, this system was sexist and inaccurate in that it assumed that the "hermaphroditic" flowers (having both male and female organs) could only be fertilized by male flowers, in a "public marriage where husband and wife have separate beds and husbands live with wives and concubines." While the use of such metaphors seems laughable today, remnants of sexism still creep into reproductive biology.

Kirsten Smilla Ebeling explains in chapter 9 that the binary classification system of sex and reproduction that prevails in biological texts and academic research on animal species continues to shape the production of biological knowledge. Heterosexual reproduction and sexual dimorphism are constructed as the norm against which animal species are evaluated. Ebeling explains that binary sexes in the animal kingdom, though constructed as predominant, are but one of many forms. Some species, for example, alter their methods of reproduction depending on circumstances, while in other species, individuals can change their sex. Biological theories and explanations that abnormalize species that do not conform to binary concepts of sex and reproduction, in turn, help reinforce social constructions of human values that sustain human sex and gender power asymmetries.

Feminist critiques of science have also questioned the methodological bases of scientific inquiry. As Carla Fehr demonstrates in chapter 10, scientists are fond of using the methodology of reductionism, which pares down objects of scientific investigation to their smallest parts or finer and finer levels. This approach creates a hierarchy of natural phenomena as well as of the disciplines that study the different levels of organization (thus biological phenomena are explained by their chemistry, chemistry is explained by physics, and physics by mathematics); decontextualizes the phenomena under investigation (seldom is an attempt made to relate or reintegrate the knowledge obtained from studying natural phenomena at the different levels); and encourages determinism (the smaller parts are assumed to cause the higher-level phenomena). It also privileges reductive methods over all others, and does not allow intellectual space for a plurality of methods that may engage the world at a variety of levels of organization. This rigid methodology rejects alternative, more pluralistic approaches to research that are more inclusive of women and others whose perspectives have been marginalized in science.

Such critiques contribute to the emerging field of feminist science studies, which aims to illuminate the relationship between science and gender (as well as race, class, and sexuality), recognizing that science has been a central force in constructing human differences and inequalities (Hammonds and Subramaniam 2003). As a number of feminist scholars who have contributed to this area suggest, gendered practices in the sciences constitute significant barriers for women (e.g., Keller 1977; Rosser 1992; Tuana 1989).[6]

Removing Barriers to Women in STEM

What solutions exist or can be developed to the problems for women in science, technology, engineering, and mathematics? It is clear that the obstacles for women are embedded in the way that academic science is organized and practiced, so that the remedies have to be systemic rather than individual, though individual hearts and minds have to change as well (as the collectivity of individual faculty, students, staff, and administrators makes up the system of departments, programs, and universities).

Since scientific fields in academia are organized as departments and many of the barriers for women are found at this level, one way to model systemic change is to envision forms of departmental organization and culture that are without the common barriers women face. A useful model comes from the work of Henry Etzkowitz, Carol Kemelgor, and Brian Uzzi, who in their book *Athena Unbound: The Advancement of Women in Science and Technology* (2000) present "the relational department." What is encouraging is that this model is not just hypothetical; it is based on cases identified by the authors from among the many they examined.

The relational department is characterized by a collegial and collaborative climate that is relatively free of the stress and defensiveness women and members of other underrepresented groups often experience in traditional departments. Women students and faculty in relational departments report that their colleagues show personal concern for one another. This helps create emotional closeness and shared experiences, leading to a greater sense of personal well-being. In these departments students and faculty work together under the leadership of a chair who is committed to facilitating continual interaction among all members and to open communication. The goal of the department is to instill confidence in undergraduate and graduate students and in untenured faculty, and to mentor rather than weed out those who do not "fit in." Such departments typically have a larger-than-usual proportion of women and minority group members and also are sometimes leaders in developing campus programs for improving the climate for women and other underrepresented groups.

One example is the School of Industrial Engineering at the University of Oklahoma, where Betty Harris and colleagues found that 50 percent of the majors and 45 percent of faculty were female. Faculty and students in the department worked collaboratively on research projects, faculty were highly accessible to students, and there were many opportunities for work-related and social interaction. Women reported very positive experiences in this setting (Harris et al. 2004).

How can a department that is traditional and not particularly welcoming to women be transformed to become more relational? Change can begin at different points and levels in the system, depending on the specific context, because the context into which reforms are introduced is critical to their acceptance—the specific history, culture, and organization of a department or campus will determine whether proposed changes will be accepted. Sometimes a particularly skillful chair can mobilize the faculty to begin to change the departmental culture. Other times, faculty and student grassroots efforts can start the transformation. Occasionally, a higher administrator such as a dean or provost will initiate changes and provide financial and other incentives to departments that are willing to implement them. Usually change does not take place all at once but rather proceeds piecemeal, focusing on removing one or two specific barriers for women.

Sue Rosser in chapter 3 indicates that some departments and universities have started to implement more family-friendly policies and practices. Stopping the tenure clock, providing on-site day care, and facilitating dual-career employment can help both women and men faculty. Day care and accommodations in coursework can also be implemented for graduate and nontraditional undergraduate students.

More recently, the Georgia Institute of Technology, under the auspices of an NSF ADVANCE grant, developed and implemented an "active service–modified duties" procedure that has enabled faculty members with family responsibilities to carry more flexible workloads. Faculty members who take this modified leave are continuing to work on their research, but teach less and are excused from service. They are also eligible for extension of the tenure clock. Georgia Tech also established a child-care center with 120 places. These are not the only initiatives that this university has put into practice. The ADVANCE grant also enabled the development of research teams comprising undergraduate and graduate students and faculty who collect and analyze institutional data on barriers for women in the sciences and engineering. The grant also led to the establishment of numerous activities to foster interaction among the faculty, and the organization of national conferences where best practices can be shared and opportunities for networking provided for women in the sciences (Georgia Institute of Technology 2003).

Other departments and campuses have focused on science education, making attempts to move away from the traditional introductory-level class based on lectures and intended to weed out the majority of students. As research has shown that students, especially women, are more likely to drop out of courses with the traditional format (see Fencl and Scheel, chapter 15), some science departments at large research universities have begun to teach introductory courses in physics and chemistry using a mix of lectures and interactive pedagogies, which has been linked to retention of women students.

Neal Abraham argues in chapter 16 that much can be learned from the four-year liberal arts colleges about teaching and nurturing undergraduate women students in the sciences, as these colleges have a long track record of producing a disproportionate number of undergraduate science majors who go on to complete their Ph.D.s. Many of these colleges have nurtured a pedagogical approach focused on providing learning and challenging experiences for all rather than filtering and testing to find the elite few who survive. Liberal arts colleges such as Bryn Mawr and DePauw have a long-standing tradition of recruiting and retaining women in the sciences, and many of their programs and approaches can be modified to suit research universities. These colleges also have formed partnerships and collaborations with research universities to provide them with graduate students and programmatic ideas. Liberal arts colleges also typically have more women faculty in the sciences than do research universities and are able to attract and retain them because they offer more collaborative and supportive cultures.

There are also a number of excellent initiatives to transform graduate education. As Anne MacLachlan shows in chapter 12, science and engineering departments at several institutions have established orientation programs, support groups, mentoring programs, and research and teaching seminars to improve the graduate experience for women and students from historically underrepresented groups. Activities such as periodic meetings of each cohort of students with faculty to discuss issues pertinent to the students' stage of work as they are going through the graduate program, effective advising and mentoring programs, and the building of collaborative research teams and networks in and outside departments have served to reduce the stress associated with graduate work and to develop a more welcoming, collegial climate.

Abbe Herzig points out in her examination of graduate studies in mathematics (chapter 13) that if the recruitment and retention of women students and students of color is the goal, policymakers and administrators must develop strategies for creating educational structures that enable all students to participate fully in the discipline and develop a sense of belonging. Herzig recommends greater flexibility in programs to help students balance the competing demands of school, family, and other commitments, hands-on learning that leads to a deeper understanding of core theories and concepts, and formally designated opportunities for women and members of other underrepresented groups to develop lasting, meaningful, and positive relationships with faculty and with other students. Molly Dingel (chapter 8) adds that creating a learning environment in which women students feel as entitled, legitimate, and authoritative as men will require conscious and sustained efforts to deconstruct the dominant masculinist culture and ideology that prevails in college science classrooms.

And, as Anna Martinson argues in chapter 14, offering academic courses in computer game design that integrate an understanding of gender issues with the knowledge of design skills can lead to the development of computer games that are more female-friendly. She indicates that several universities are beginning to offer such courses, as well as degrees in computer game design, and that this provides opportunities for women to become designers (and to develop games "in their own image") rather than only users of the technology.

Sandra Hanson (chapter 6), Josephine Beoku-Betts (chapter 7), Anne MacLachlan (chapter 12), and Abbe Herzig (chapter 13) indicate that initiatives and programs aimed at removing the barriers for women in STEM have to be sensitive to the needs of diverse women and must contain components aimed specifically at improving the climate for women of color. These authors agree that an understanding of different positionalities and experiences among women in STEM is crucial to the development of effective institutional strategies for increasing and retaining a larger number of women in scientific careers. MacLachlan shows how graduate education can be transformed to include orientation programs and support groups for all women students as well as women and men of color. And Herzig explains that programs that replace the prevailing "sink-or-swim" ideology of graduate education with one that focuses on cultivating students' potential through mentoring relationships and systemic socialization into the profession will not only increase retention of students from underrepresented groups, but will also enhance the potential for creating diversified approaches to the production of knowledge.

Finally, as chapters 9, 10, and 11 indicate, feminist critiques of science have sought to change how scientists "do science" by addressing the gender biases in scientific epistemology (how and why scientists see and understand the world the way they do) and in the actual practice of science. Although the dissemination of this critique is still limited, a growing number of courses, seminars, and conferences offered for those in the sciences are beginning to increase awareness among students and faculty about gender biases in STEM fields. For example, at Texas A&M University, a recent project titled "Changing Faculty through Learning Communities," comprising a series of workshops and seminars, succeeded in making engineering faculty more aware of gender schemas (see chapter 17 in this volume) and differences in ways that women and men communicate and approach science (Covington and Froyd 2004).

Women's and gender studies programs are increasingly offering courses on women in STEM, and collaborations between women scientists and feminist theorists are becoming more frequent. One such attempt at collaboration is described by Dana Dudle and Meryl Altman in chapter 11. An evolutionary plant biologist and a literary critic working with feminist theory discuss how the concepts of sex and gender have been put to use in their respective fields and how gendered metaphors have both obscured and helped to clarify biological knowledge and feminist theories. The collaborators in this project come to better understand that uses of sex and gender in their respective areas of knowledge can be mutually informative and that plant biology and feminist theory have had a parallel but shared goal: to better explain and account for the diversity among plants and humans, respectively.

We end this collection with a chapter by Virginia Valian, whose work on gender schemas and women's cumulative disadvantage has had considerable impact in academia. As the author of a landmark book, *Why So Slow? The Advancement of Women* (1998), and a featured speaker at many universities in North America and Europe, Valian has carried her message about the disadvantages for women in the professions, especially in the STEM fields, to many academic audiences. Here, she summarizes the research related to gender schemas and offers strategies and remedies for gender equity in higher education institutions.

Given the current shift among researchers, activists, and institutional leaders in approaches to the problems of women's inclusion in academia, the time is ripe for the development and implementation of strategies that will transform the academy, making it more inclusive of women and other underrepresented groups in science and related fields. However, we still need more research to document how institutions operate to both exclude and include women in the sciences, engineering, and technology. We also need to disseminate more positive examples of institutions and departments where inclusion has been achieved, and of practices that are transforming the sciences themselves. This book contributes to these aims by providing an account of the institutional barriers for women in STEM fields as well as accounts of some of the best practices and possible means of removing these obstacles. It is our hope that this collection will be useful to those who are looking for ways to change academic institutions so that they can fully embrace the talent and contributions of diverse women and men.

Notes

1. The event, sponsored by the Women's Studies Program and funded largely by National Science Foundation Grant HRD-0094556, brought together faculty, graduate students, and administrators in women's studies and STEM fields from fifty-five universities, mainly in the United States.

2.   Masculinist interpretations, as well as masculinist values and structures, are those that institutionalize and privilege ideas and patterns of behavior that are associated with hegemonic masculinity; in science, for example, such ideas include notions of "objective" science and of an "ideal worker" whose personal life does not interfere with professional life.

3.   This is not to say that theories like that of the pipeline are no longer utilized. Indeed, the pipeline model still provides much of the rationale for examining institutional and cultural barriers for women in STEM disciplines.

4.   Physics remains one of the fields in which women least participate—in 2002 only 10 percent of faculty and 17 percent of graduate students were women (AIP 2003b).

5.   For a discussion of the concept of gendered institutions, see, for example, Acker 1990, 1992.

6.   It should be pointed out, however, that some of these scholars see the feminist science studies project to be unrelated to the concerns of the women in science approach. For a discussion of this position see Hammonds and Subramaniam 2003.

References

Acker, Joan. 1990. "Hierarchies, Jobs, Bodies: A Theory of Gendered Organizations." *Gender and Society* 4(1): 139–58.

———. 1992. "From Sex Roles to Gendered Institutions." *Contemporary Sociology* 21(5): 565–69.

AIP (American Institute of Physics). 2003a. "Report Advocates Changing the Culture of Science and Math Education." *AIP Bulletin of Science Policy News* 63(May 14): 1–3.

———. 2003b. *2002 Workforce Report.* College Park, Md.: AIP Statistical Research Center.

Benbow, Camilla P., and Julian C. Stanley. 1980. "Sex Difference in Mathematical Ability: Fact or Artifact?" *Science* 210: 1262–64.

Berryman, Sue E. 1983. *Who Will Do Science? Minority and Female Attainment of Science and Mathematics Degrees: Trends and Causes.* New York: The Rockefeller Foundation.

Bleier, Ruth, ed. 1986. *Feminist Approaches to Science.* Oxford: Pergamon.

Bystydzienski, Jill M., ed. 2004. "(Re)Gendering Science Fields." Special issue, *NWSA Journal* 16(1).

Catsambis, Sophia. 1994. "The Path to Math: Gender and Racial-Ethnic Differences in Mathematics Participation from Middle School to High School." *Sociology of Education* 67(3): 199–215.

———. 1995. "Gender, Race, Ethnicity, and Science Education in the Middle Grades." *Journal of Research in Science Teaching* 32(3): 243–57.

Clewell, Beatriz Chu, and Patricia B. Campbell. 2002. "Taking Stock: Where We've Been, Where We Are, Where We're Going." *Journal of Women and Minorities in Science and Engineering* 8(3–4): 255–84.

Connell, R. W. 2000. *The Men and the Boys.* Berkeley: University of California Press.

Covington, Kimberly, and Jeff Froyd. 2004. "Challenges of Changing Faculty Attitudes about the Underlying Nature of Gender Inequities." In *Proceedings of the 2004 American Society for Engineering Education Annual Conference and Exposition.* Salt Lake City, Utah.

Crawford, Mary, and Margo MacLeod. 1990. "Gender in the College Classroom: An Assessment of the 'Chilly Climate' for Women." *Sex Roles* 23(3–4): 101–22.

Etzkowitz, Henry, Carol Kemelgor, and Brian Uzzi. 2000. *Athena Unbound: The Advancement of Women in Science and Technology.* New York: Cambridge University Press.

Georgia Institute of Technology. 2003. *NSF Advance Program for Institutional Transformation Annual Report.* http://www.advance.gatech.edu/AnnualReport03.pdf. Accessed April 1, 2005.

Greenfield, Teresa Arambula. 1997. "Gender and Grade Level Differences in Science Interest and Participation." *Science* 81(3): 259–75.

Hammonds, Evelynn, and Banu Subramaniam. 2003. "A Conversation on Feminist Science Studies." *Signs: Journal of Women in Culture and Society* 28(3): 923–44.

Haraway, Donna. 1991. *Simians, Cyborgs, and Women: The Reinvention of Nature.* New York: Routledge.

Harding, Sandra. 1991. *Whose Science? Whose Knowledge? Thinking from Women's Lives.* Ithaca, N.Y.: Cornell University Press.

Harris, Betty, et al. 2004. "Gender Equity in Industrial Engineering: A Pilot Study." *NWSA Journal* 16(1): 186–93.

Hill, Susan. 2003. *Science and Engineering Doctorate Awards: 2002.* Arlington, Va.: National Science Foundation, Division of Science Resource Statistics.

Hubbard, Ruth. 1990. *The Politics of Women's Biology.* New Brunswick, N.J.: Rutgers University Press.

Jackson, Judy. 2004. "The Story Is Not in the Numbers: Academic Socialization and Diversifying the Faculty." *NWSA Journal* 16(1): 172–85.

Keller, Evelyn Fox. 1977. "The Anomaly of a Woman in Physics." In *Working It Out: 23 Women Writers, Artists, Scientists, and Scholars Talk about Their Lives and Work,* ed. Sara Rudick and Pamela Daniels, 77–91. New York: Pantheon.

———. 1982. "Feminism and Science." *Signs: Journal of Women in Culture and Society* 7(3): 589–602.

Kourany, Janet A. 2002. *The Gender of Science.* Upper Saddle River, N.J.: Prentice Hall.

Lederman, Muriel, and Ingrid Bartsch, eds. 2001. *The Gender and Science Reader.* New York: Routledge.

Maccoby, Eleanor E., and Carol N. Jacklin. 1966. *The Psychology of Sex Differences.* Stanford, Calif.: Stanford University Press.

Mayberry, Maralee, Banu Subramaniam, and Lisa Weasel, eds. 2001. *Feminist Science Studies: A New Generation.* New York: Routledge.

Merton, Robert K., and Harriet Zuckerman. 1973. "Age and Scientific Productivity." In *The Sociology of Science: Theoretical and Empirical Investigations,* ed. Robert K. Merton, 323–45. Chicago: University of Chicago Press.

Niemeier, Debbie, and Cristina Gonzales. 2004. "Breaking into the Guildmasters' Club: What We Know about Women Science and Engineering Department Chairs at AAU Universities." *NWSA Journal* 16(1): 157–71.

Noble, David. 1992. *A World without Women: The Christian Clerical Culture of Western Science.* Oxford: Oxford University Press.

NSF (National Science Foundation). Division of Science Resources Statistics. 2003.

Women, Minorities, and Persons with Disabilities in Science and Engineering: 2002. Arlington, Va.: National Science Foundation (NSF 03-312). Accessed February 16, 2005.

Pascarella, Elizabeth, et al. 1997. "Women's Perception of a 'Chilly Climate' and Their Cognitive Outcomes during the First Year of College." *Journal of College Student Development* 38(2): 109–20.

Rose, Hillary. 1994. *Love, Power, and Knowledge: Towards a Feminist Transformation of the Sciences*. Bloomington: Indiana University Press.

Rosser, Sue. 1992. *Biology and Feminism: A Dynamic Interaction*. New York: Twayne.

Seymour, Elaine. 1995. "The Loss of Women from Science, Mathematics, and Engineering Undergraduate Majors: An Explanatory Account." *Science Education* 79: 437–73.

Seymour, Elaine, and Nancy M. Hewitt. 1997. *Talking about Leaving: Why Undergraduates Leave the Sciences*. Boulder, Colo.: Westview.

Tuana, Nancy, ed. 1989. *Feminism and Science*. Bloomington: Indiana University Press.

Valian, Virginia. 1998. *Why So Slow? The Advancement of Women*. Cambridge, Mass.: MIT Press.

Wasserman, Elga R. 2000. *The Door in the Dream: Conversations with Eminent Women in Science*. Washington, D.C.: Joseph Henry.

Part One.   *History of Women*
             in STEM Fields

1 Sustaining Gains: Reflections on Women in Science and Technology in the Twentieth-Century United States

Sally Gregory Kohlstedt

Advances in science and technology played key roles in the two world wars and in a space exploration program for which the twentieth century will undoubtedly be remembered. In that century, too, industrialized nations gave unprecedented support to systematic inquiry and, in many cases, privileged experimental and quantitative research. We may or may not agree with the lists of outstanding discoveries of the twentieth century—including radio and television, laser and fiber optics, nuclear technologies, genetic research, geological plate tectonics, and a myriad of others that made it onto lists published at the end of 2000 (e.g., National Academy of Engineering et al. 2000). But we do need to understand where women and gender fit into these twentieth-century stories of scientific and technological accomplishments and to produce new, inclusive, and more complex accounts that acknowledge both women's achievements and the discrimination that they faced. It is critical to understand that gains may be eroded and that sustaining women's opportunities still requires attentive activism.

In the past three decades historians have provided us new stories, data, and analytical perspectives that reveal both the accomplishments and the disparities that accompanied technological change in women's lives (Ogilvie and Harvey 2000). Some undercut conventional wisdom. For example, while many electrical appliances reduced the actual physical labor required, they also often brought, as the title of Ruth Schwartz Cowan's book suggests, *More Work for Mother* (1983). The so-called labor-saving devices carried expectations that women would use these tools to produce more complex and sophisticated menus, take advantage of the ease of electrical irons to create more elaborately ruffled dresses, and maintain cleaner and more attractively decorated homes. Margaret Rossiter (1982, 1995) has closely examined the women who were involved in science in the United States in two volumes that will be classics and a resource for scholars to mine for years to come. Here, too, her subtitle *Struggles and Strategies* (1982) suggests that there was nothing automatic about women's participation in science. Rossiter conceptualized her project in terms of employment issues (from education and training to promotion and recognition) in an account that balances the extraordinary stories of individual survival and accomplishment with the often challenging, even hostile, collective circumstances of women in science and engineering that kept the number of such stories relatively small.

A brief history may reveal something of the patterns of women's participation in science and technology, as well as the particularities of individual women's experiences as moderated by time, place, discipline, and personal circumstances. This history raises multiple issues relating to the gender identities of, and the gender inequalities faced by, academic women. Knowing how women in mathematics, science, and technology have historically positioned themselves in relation to the larger cultural setting should help us as we chart our own way. The specific questions we ask and the analytical approaches that we take to current issues are significantly different from those taken by reform-minded activists a century ago. But in many ways the concerns that they had about recruiting and retaining girls and women in science and technology also involved, in a fundamental way, the problems that we still seek to resolve. The fact that their solutions produced significant successes but also contained an undertow (Evans 2003) that sometimes constrained them and their successors is instructive as we measure our own sense of achievement and lay out short- and long-term goals and plans.

This essay selectively dips into history, allowing us to see the "struggles and strategies" that frame our own experiences a century later. What should emerge as particularly compelling is the extent to which women in scientific fields consciously forged a path that they hoped would bring advantages to the women who came after them. Since I am a historian of science, I am better at reviewing the past than forecasting the future—but, as an activist, I am convinced that we must create personal and collective visions of what we want to accomplish in our own immediate environments and then raise our sights to think ahead about what might, indeed must, happen locally, nationally, and globally.

Women of Distinction

While women have been part of science and technology throughout history, in the late nineteenth century professionalizing practices circumscribed previous amateur activities even as they rationalized meritocracy and certification as the criteria for participation. A women's rights movement already had fostered single-sex colleges alongside coeducational opportunities in all but the most elite or military institutions, and science had flourished in many of them. By the early twentieth century, a number of American women had earned Ph.D.s in the natural and social sciences. Moreover, an exceptional European woman provided a symbol of achievement that could not be ignored by scientists in Europe or North America.

The most famous woman in science in the twentieth century by nearly any measure was Marie Curie, the Polish-French physical scientist whose work on radium and related topics in France earned her two Nobel Prizes. Her work with Pierre Curie led the public to view her as an "eminent collaborator," and a 1903 illustration in *Vanity Fair* presented her with her husband Pierre as the little woman behind the great man (Pycior 1993). Marie Curie understood well that their close collaboration might be an issue, and she published a single-authored paper quite early in her career, unlike Pierre, all of whose publications were coauthored (Quinn 1995; Pycior 1996). Pierre died within a decade after their marriage, so much of Marie's life and work were carried on more independently. She established a public image as a serious-minded scientist as well as a devoted single mother of two daughters, the eldest of whom also later earned a Nobel Prize with her husband. What is less known is that Marie Curie also taught in a local normal (teacher training) school, apparently wanting to encourage women in science; she occasionally recruited students from there to work in her laboratory, and was interested in the networks that they formed to work with one another.

Curie, of course, was exceptional, and became an icon in the United States after a triumphant tour in the 1920s that netted the chemist a gram of radium. Yet, as Margaret Rossiter (1982, 122–28) has pointed out, Curie may have inadvertently created a "Curie effect" in which her brilliance in research distanced her as a role model—after all, who could aspire to match her accomplishments? Indeed, her achievements led some North American academic men to suggest that of course they would hire a woman, if only they could find a Marie Curie.

Less daunting as a role model, perhaps, was the woman scientist and educator who was named one of twelve "outstanding women in America" in 1925, Anna Botsford Comstock. One of the early female graduates of Cornell University, Anna Botsford married a professor of entomology, John Comstock, and became a skillful illustrator for his pioneering text that applied evolutionary principles in that field; he was among the rare husbands who gave their wives full credit on the title page. In the 1890s Anna Comstock became involved in an effort to introduce more systematic nature study into the schools of New York State, and eventually she became a leading figure in the national curriculum movement known as "nature study." Although Cornell's board of trustees turned down her proposed appointment as an assistant professor, she stayed on in the lesser position of instructor to teach classes on nature study and to conduct field trips to demonstrate educational techniques. She published an authoritative textbook, edited *Nature-Study Review* for nearly a decade, and served as president of the Nature-Study Society. Popular as a lecturer on nature study, Comstock talked often with teachers in New York and neighboring states and also taught occasionally at Stanford and Berkeley at the turn of the century. Comstock carved out a niche for her own interest in natural history and biology and simultaneously created a new site where women's traditional prospects might intersect with exciting opportunities that involved genuine career advancement. Opening this area led some women to become nature study supervisors for entire school systems, from New York City to Oakland, California, and others to teach the subject at normal schools (Comstock 1953; Henson 1996). Such specialized niches, however, could also be both isolating and difficult to sustain (Appel 1999).

These issues are particularly evident in the work of Ellen Swallow Richards, who created a field of home economics that built on advances in chemical testing, breakthroughs in understanding the germ origins of some diseases, and a growing demand for professional arenas in which to use the talented women earning scientific degrees (Lippincott 2003). Home economists took their professional lead from Richards, whose position at the Massachusetts Institute of Technology (MIT) gave her the stature and colleagues for a movement to apply science to domestic and institutional "household" issues. Richards drew as well on the domestic science programs being offered in land grant colleges as complements to the agricultural curriculum for men (Stage and Vincenti 1997). Richards's connections with wealthy women in Boston provided her with financial resources to create model programs that dealt with individual nutrition and standards of cleanliness even as they lobbied for legislation to establish and regulate community standards (Tomes 1998). Both Comstock and Richards expanded opportunities for women interested in science through their very specific programs that did not directly challenge male domains; they created places of high credibility where collaboration with other women was not only possible but probable. Their niches, however, also became in certain ways limiting spaces where women were segregated from other male and female scientists.

In the first half of the twentieth century, women with advanced degrees were predominantly found on the campuses of women's colleges (where they sometimes constituted half or more of the faculty). They also began to take positions at coeducational land grant schools, at a few major universities, and in the expanding normal school systems (gradually merged into state colleges) which supplied teachers to the public schools. Some women scientists never married, like the well-known geneticist and Nobel Prize winner Barbara McClintock and the paleontologist Winifred Goldring, who had "met no man she liked so well as her work" (Sicherman and Green 1980, 282–83), while others, like Annie Alexander, established long-term partnerships with women (Stein 2001). A significant number married and had children even as they conducted academic careers. Dual-career couples have demonstrated that partnerships can be highly creative and complementary, as these men and women shared experiences and commitments (Rossiter 1995, 115–21). Such was the case for the Curies, the Comstocks, ecologists Frederick and Edith Clements, Alva and Gunnar Myrdal (sociologist and economist, respectively), biochemists Gerty and Carl Cory (who together won a Nobel Prize in 1947), and anthropologists Margaret Mead and Gregory Bateson while they did their research in Polynesia (Abir-Am and Outram 1987). However, the situation might be or turn negative if the two were in the same field and the husband's career flourished through the collaboration while the wife's contribution was masked, as apparently happened for the promising young physicist Mileva Maric, who married Albert Einstein (Renn and Schulmann 1992). Margaret Rossiter (1980) also has noted how often the men featured in *American Men of Science* were, in fact, married to women in their fields and thus able to have informal and sometimes unacknowledged assistance in their work.

Historians and sociologists reaffirm the observation that personal relationships, as well as intellectual capacities, are fundamental in framing careers (Zuckerman and Cole 1991). Research and personal memoirs also suggest how positive situations provided support and established safe havens during the sometimes tumultuous early years when a career is established, and, alternatively, how the family and community obligations that women are expected to assume could compromise or even curtail a promising career (Ajzenberg-Selove 1994; Cassell 1998; Laslett and Thorne 1997). Studying the biographies of women scientists reduces the contrast between personal and professional and reinforces our postmodern understanding that each life is, indeed, heterogeneous in its construction and multiple in its identities. These individual women's stories reveal the capacity to do outstanding work and to pursue science in education that enables others to work independently and creatively, and also collaboratively.

Expansion of Opportunities for Women

Historical circumstances of time and place play a critical role in the lives of individuals. Curie, Comstock, and Richards had the advantage of coming of age in an era when opportunities for women were expanding and when new patterns in higher education established stronger certification requirements and an academic standard of research productivity that made merit part of evaluation. Historians have documented expanding educational opportunities for women in single-sex and coeducational colleges, increasing opportunities to study for advanced degrees in European universities and some North American ones, and the expansion of areas of scientific inquiry, in some of which women scientists concentrated (Rossiter 1982). These included astronomy (women astronomers worked on star clusters under Wilhelmina Fleming's supervision at the Harvard Observatory [Lankford 1997; Mack 1990]) and botany (many women found positions in the U.S. Department of Agriculture, because they could be paid less and were often more educated than their male competitors for these posts) (Rossiter 1982, 228–35). However, not all fields opened to women. Very few women were able to study and then to practice engineering in newly professionalizing specializations in civil, electrical, mechanical, and chemical engineering. Certification here was a handicap, as Amy Bix shows in chapter 2 of this volume, because engineering schools were remarkably resistant to women in the early decades of the twentieth century and, even after their doors opened, their structural and cultural frameworks could be explicitly unwelcoming. At the other end of the spectrum, some women went into the social sciences precisely because here they could investigate questions that related to women: questions on the psychology of sex differences, the intellectual and physical capacities of women, the significance of gendered identities, and even women's sexuality (Deacon 1997; Rosenberg 1982).

Success for women in science came from personal perseverance and creative work (characteristics that were fundamental to surviving). However, what facilitated opportunities for these women was often activism that was smart, strategic, and persistent. At least a few of those who had found a happy career niche or pushed their way into reluctant institutions became activists themselves, working for the advancement of women in science. Their strategies were several.

One approach, as Margaret Rossiter (1982) has documented well, was to produce and publicize data that revealed patterns of discrimination and thwarted ambition for women with advanced degrees seeking and even earning degrees in science. The Association of Collegiate Alumnae, which became the American Association of University Women (AAUW) and is probably best known for its fellowship program, had been formed to consider women's opportunities in higher education. Its leadership watched as the increasing number of women taking advanced degrees at the turn of the century failed to be appointed to professorships or higher-status positions in industry and government. Studies of individual institutions and specific fields were published in the *AAUW Journal* and thus documented the continuing inferior position of women on all but women's college faculties as measured by their numbers, their rank (disproportionately at the instructor level even at women's colleges), and their low salaries (Levine 1995). Women faculty also urged the creation of Committee W of the American Association of University Professors (AAUP), which had been founded in the 1910s to deal with faculty issues, particularly academic freedom. The Committee W report in 1921 made clear, for example, that there were no women at all on faculty at twenty-seven of the hundred major coeducational colleges studied and that at the others women held only 4 percent of the full professorships, and less than that if highly feminized fields like home economics and physical education were omitted. A report eight years later found that little had changed and that women's low status in academia could not be correlated with degrees and length of experience (Rossiter 1982, 160–66). Nonetheless, home economics and its leaders had created a space for at least some of the women aspiring to careers in nutrition and related fields (Nerad 1999).

Another strategy was a reaction to the underrecognition of women's achievements. While individual women relied on sympathetic male mentors for their entrance into education and employment, these supporters had limited capacity to influence the larger profession. Women understood, even if they were not often privy to the correspondence and commentary of their male colleagues, that they simply were not recognized as eligible by many men scientists. This is shown in a letter, quoted by Rossiter, from Robert A. Milliken of the California Institute of Technology to the president of Duke University in 1936 in response to information that Duke had hired a woman physicist:

> I should feel that my chance of building a very strong department would be better if I made my choices among the most outstanding of the National Research Fellows or other equally outstanding young men. . . . Women have done altogether

outstanding work and are now in the front rank of scientists in the fields of biology and somewhat in the fields of chemistry and even astronomy, but we have developed in this country as yet no outstanding women physicists. In Europe Fraulein [Lisa] Meitner of Berlin and Madam [Marie] Curie of Paris are in the front rank of the world's recognized physicists. I should, therefore, expect to go farther in influence and get more for my expenditure if in introducing young blood into a department of physics I picked one or two of the most outstanding younger men, rather than if I filled one of my openings with a woman. I might change this opinion if I knew of other women who had the accomplishments and attained to the eminence of Fraulein Meitner. (1982, 192–93)

The standards for women were thus not the same as for men; very few, if any, of those women scientists could be expected to match Curie or Meitner (Sime 1996). Thus, even as women were gaining more advanced degrees and membership in professional societies, Margaret Rossiter notes, "New barriers and forms of stratification arose almost as rapidly as others fell" (1982, 305).

It was this exceptionally high standard that discouraged some women but that led others to overcredential themselves (multiple degrees were common). Some also found ways to work collectively, showing the achievements of women by creating special prizes, collecting and publicizing data on women's achievements, and challenging those gatekeepers who, like James McKeen Cattell (who persisted for decades in calling his dictionary *American Men of Science* even while including women), had the power to recognize and give status to women's accomplishments.

In response to the argument that women did not fulfill their early promise and become outstanding, women began to establish prizes for achievement rather than simply fellowships for promising younger women. Thus in 1937 women funded the Annie Jump Cannon Prize for women in astronomy and the Francis P. Garvan Medal in chemistry, awards to be given at the annual meeting of the related professional associations. More women scientists were also promoted and recognized by the annual Achievement Award of the AAUW, the Woman of the Year Award, sponsored by the *Ladies' Home Journal,* and similar honors (Rossiter 1982, 305–12).

The problem of underrecognition also led women to create their own organizations—a third strategy—that would ensure that their résumés would also highlight scientific leadership. Membership in male-dominated institutions, when it was gained, did not lead to committee appointments, elected office, or awards. Iota Sigma Pi, a national honor society for women chemists with early chapters at the University of California at Berkeley in 1900 and at the University of Washington in 1910, eventually held meetings for women at the American Chemical Society. Sigma Delta Epsilon, founded at Cornell in 1921, later renamed itself Graduate Women in Science. These organizations sought to coordinate women with graduate degrees across the disciplines with the goal of reducing feelings of isolation, providing a forum to air grievances, and even engaging in occasional activism. When constrained in male societies, women also formed specialized groups, like the Women's Anthropological Society of America and the Society of Women Geographers (Kass-Simon and Farnes 1990). The flip side, which caused some prominent women to stay aloof from such organizations, was that this activity, like the career niches, might create a kind of ghetto for women and take them out of the mainstream organizations. Nonetheless, women's organizations remained both a haven and a site for activism for women in the various disciplines well into the twentieth century.

Women's collaboration could help create optimism or at least stem feelings of isolation. Handbooks and other publications on careers for girls and women in the sciences were popular products of women's organizations during the first half of the twentieth century (Lafollette 1988). Indeed the statistics made it clear that more girls than ever were graduating from high schools (sometimes even more frequently than boys in the first decade) and were using that certification to attend business schools, normal schools, and colleges (Tolley 2002). The expanding women's college network, which included more Catholic as well as secular private schools, tended to hire women faculty members, which made advanced degrees more attractive and useful.

Erosion

There were signs, however, that things were not going as they should. Too many women with advanced degrees failed to get jobs and others quickly hit what would later be called the "glass ceiling" that limited their prospects for upward mobility. During the 1920s, women's educational advancement seemed to plateau and in some ways women scientists even began to lose ground.

Territorial marking, backlash, and simply sexist prejudices worked in conjunction with cultural sensibilities to reestablish older boundaries and erect new barriers (Cott 1987). Some historical work on masculinity and science is suggestive. Robert Nye (1997) has pointed out that even as science and medicine became more professional at the end of the nineteenth century, men encoded masculine displays of "honor" by requiring certain rituals for admission that encouraged a degree of gentlemanly debate and contention. Ruth Oldenziel (1997, 1999) has studied how in the twentieth century engineers worked to make technology masculine. Her argument is that shop floor culture, technology-school culture, and even competitions like the Fisher Body Company's miniature car contests of the 1930s, in which young men were encouraged to build models and women were encouraged to admire their handiwork and perhaps contribute by sewing upholstery, sustained male identification with technology. Such projects obscured women's earlier participation in science and technology and created, quite deliberately, men's "romance with technology" during the 1920s and 1930s.

Contemporary feminist scholars have worried about the interwar and post–World War II years, concerned that they provide a warning signal that the pattern of gradual improvement on several fronts could be reversed despite the evident progress made in the 1970s and 1980s. Certainly the post–World War I economy, followed by the depression of the 1930s, did not help, nor did the curious mix of conservative politics and flapper giddiness work to advance women's career prospects (Greenwald 1980; Kessler-Harris 1982; Tyack and Hansot 1992). Some have noted that in the aftermath of finally winning the vote with passage of the Nineteenth Amendment (a multigenerational challenge), the women's movement itself seemed to take a deep breath and relax even as the National Woman's Party addressed broader social and economic issues in the 1920s (Cott 1987) and some former suffragists became peace activists (Alonso 1993). There is also evidence of backlash, demonstrated by the letter of Robert Milliken, as the men who had trouble finding employment felt that they deserved positions more than women and were abetted by well-established men who concurred. Whatever the reasons, the interwar years slowed and, in some cases, reversed women's progress, and even World War II provided expanded opportunities only temporarily (Howes and Herzenberg 1999).

Of course, some women continued to take degrees in science and mathematics and to achieve in a variety of ways during these midcentury years. Barbara McClintock made major contributions to genetics that ultimately earned her a Nobel Prize (Keller 1983; Comfort 2001). Grace Murray Hopper was involved with the major computer science breakthroughs in World War II and shortly thereafter. For her efforts in the development of the UNIVAC, the rear admiral was named "Computer Science Man of the Year" in 1969 (Williams 2001; Reynolds 1999). Biologist Rachel Carson did her most courageous work during this postwar period, including publishing the pathbreaking *Silent Spring*, which blew the cover on pesticides and their impact on air, water, and living things (Lear 1997).

During and after the Second World War, women in the sciences (and most fields) continued to be paid less than men, and few seemed to make the remarkable advances that were more common among the men who were in better-equipped laboratories and had more staff and scientific research support. While the war had raised expectations with its call for "manpower," and companies like Curtiss Wright had recruited women for engineering posts during the war years, within just a few years those women who did work in strategic scientific and technology areas were out of work and back at home (McIntire 1993). The period was also scarred by the suppression of some women's work, most significantly that of Rosalind Franklin, whose image suggesting the double helix was surreptitiously viewed by James B. Watson and then used by him and Crick to produce their Nobel Prize–winning work (Maddox 2002; Sayre 1975).

The most evident constraint on women pursuing science careers in the postwar 1950s was the expectation that they would stay at home and that children were their highest priority (May 1999). The headline of a San Diego newspaper, announcing the awarding of the Nobel Prize in Physics to Maria Goeppart Mayer, crowed, "San Diego Mother Wins Nobel Physics Prize." Within the first two paragraphs, readers also learned that she was a red-haired college professor and mother of two (Rossiter 1995, 332).

The midcentury doldrums affected even home economics, where women had been in charge, maintaining high standards and a scientific outlook. Male deans

Figure 1.1. As part of the effort to increase the number of skilled workers during World War II, these Curtiss-Wright Cadettes at Iowa State University learned aerospace engineering in an intensive two-year class.
Iowa State University Library/Special Collections Department.

and more male faculty (many of them clustered in the higher-status area of child development) came to power in the 1950s in many colleges, which were then forcibly renamed, restaffed, and reoriented away from the programs that had created scientific opportunities for women in nutrition, textiles, and chemistry since the turn of the century (Rossiter 1997a; Stage and Vincenti 1997). Even at women's colleges a higher proportion of men became administrators and faculty. The growth in enrollment of these women's colleges and the establishment of new ones during this period could cause it to be seen as a "golden age" for women's colleges, but the subliminal theme of preparation for family life seemed to women like Betty Friedan (1963) to work against long-term career planning for these college women in the 1950s. Moreover, this growth spurt would be challenged as traditional men's colleges, often under pressure from the women's movement, began to admit women undergraduates in the 1970s and many women's colleges either went coeducational or failed (Solomon 1985).

Although there was less evident activism during the middle decades of the century, the actual number of women in science and technology continued to increase in most fields. The postwar baby boom meant that many women took positions in primary and secondary education, and to a limited extent in the growing number of junior and technical colleges. Many, especially those who had taken advanced degrees in the 1940s and 1950s, continued to push AAUW, as well as the research division of the National Education Association, to produce educational and employment data itemized by sex and discipline so that women's circumstances could be understood in relation to those of men. A survey in 1954–55 showed that there were 3,600 women scientists in higher education, but that they were concentrated at the poorest and least prestigious institutions. At the twenty leading universities in 1960, women were concentrated in home economics, with a few clusters in genetics and anatomy; but representation was negligible or nonexistent in physics, geology, engineering, and some biological sciences like entomology (Rossiter 1995, 186–90). Moreover, the rapidly expanding federal funds for research during this same period were in many ways in inverse ratio to the concentrations of women, since considerable funds were going to engineering and to physics, fields where the numbers of women were lowest. Women were not discouraged from studying science in colleges in the 1950s, but as a 1955 graph projecting demand for labor indicated by putting a skirt on the silhouette of "teaching needs," women earning science degrees were to know their target career (Rossiter 1997b, 178).

Second Wave: Women in Science

Patterns of participation, promising in the early decades of the twentieth century, were more uneven through midcentury when opportunities in science rose significantly in parallel with a reconstituted women's movement. The post-Sputnik support for science through the National Defense Education Act of 1958 provided fellowships for women as well as men in sciences, foreign languages, and other defense-related areas. In just over a decade the political and social movement for the so-called second wave of women's rights was symbolized and made specific by a myriad of activisms that ranged from political initiatives, including Title IX (which forbids sex discrimination at educational institutions that receive federal funds) and the much-discussed Equal Rights Amendment, to women's studies programs on campuses, women's health programs, and initiatives to reconfigure the home lives of working women by providing more child-care and other resources. A number of recent semiautobiographical accounts provide fresh insights into the activism of that generation (Brownmiller 1999; Evans 2003; Freedman 2002; Rosen 2000; Tobias 1997).

One technique was the now well-established tradition of pushing professional societies like the American Physical Society to gather data and then using various publicity tactics to get the findings before the public. The sometimes feisty Women's Bureau of the Department of Labor was an intermittent ally in this effort. One remarkable woman, Betty Vetter, made a significant difference. Vetter had a degree in chemistry and served in the WAVES during World War II, and, after teaching in a number of places and getting married, in 1963 she joined the staff of the nonprofit Commission on Scientific Manpower, later the Commission on Professionals in Science and Technology (CPST), in Washington, D.C. For the next three decades of her life, she was the editor of *Scientific,*

Engineering, Technical Manpower Comments and concentrated on compiling and analyzing data on women and minorities across fields and from undergraduate degrees through faculty appointments.[1] This material demonstrated both where government policies were having some effect and where problems lingered.[2]

In many ways the concerns of women in science reflected those of the earlier activist generation: they wanted to offer girls and women in science encouragement and access, provide opportunities for those who persisted in such studies, illuminate the patterns of discrimination that limited their opportunities, and recognize those who demonstrated what women could accomplish. What was clear in the new movement was that women were, in some instances, positioned better than they had been earlier and that the arguments being formulated for equality were more adamant in the second wave and brought along a more sophisticated set of social science tools (Tharp 2002).

Changes did occur. Among the most striking were in engineering, where the number of women getting an engineering education went up dramatically during the 1970s and most of the 1980s. Unfortunately other indicators, such as ACT (American College Testing) scores, did not show comparable improvement, suggesting that a few of the best and brightest women were doing very well in math and science but, on average, precollege young women were being left behind. A graph created by the National Science Foundation (NSF) of differential persistence in science and engineering in 1977 provided evidence of what became known as the pipeline problem, with women dropping out at virtually every decision point: high school graduation, choosing a major in junior year, B.S. graduation, M.S. graduation, and qualifying exams, until finally very few of the women who had earlier expressed interest in science were still in the pipeline and completing Ph.D.s (Alper 1993). Scholars began to show when and how dramatically women's interest in science declined in the course of their advanced education and used the data provided by federal agencies to demonstrate that the divergence between girls and boys began to occur well before high school (Lafollette 1988; Oldenziel 1999; Rossiter 1982). Women started to gain ground in the 1970s and 1980s as programs of a number of agencies, including the NSF, pushed at the pipeline problem at essentially every level. Even so, by the late 1980s, the data showed that women's participation was plateauing or even declining in computer science. Many activists were also frustrated by the persistence of gender differences across fields, indicating that women continued to cluster in the less prestigious and less well-supported areas. Patterns of placement remained considerably different for men and for women, with men rather evenly spaced among the social science, life science, physical science, and engineering (including computer science) sectors, while women were highly concentrated in psychology and the life sciences and marginally represented in engineering, computer science, and mathematics. Equally discouraging was the probability that the problems at every level contributed to a cumulative disadvantage (Clark and Corcoran 1986). Worried women wondered why gains were not being sustained and whether, in fact, in a more conservative social and political climate women were slipping back into the pattern of the 1920s.

The 1990s and Beyond

In the 1990s, feminist scholars advocated for analysis of the important data that were being gathered by various federal agencies (collected and thus readily available) and compiled by the indefatigable Betty Vetter. The results revealed, once again, that promising gains were not being consolidated and that women's previous levels of participation in science and engineering fields had not been sustained over the course of the century or even over the course of the past decade. A study sponsored by the National Research Council and chaired by Mildred Dresselhaus summarized issues and research and proposed advocacy and monitoring as mechanisms to increase the number of women in science and engineering (Committee on Women in Science and Engineering 1991, 1992). Qualitative assessments pointed to the specific issues faced by female undergraduate students and offered explanations for their high drop-out rate, including Sheila Tobias's widely cited study, *They're Not Dumb, They're Different* (1990). Another study interviewed men and women in an effort to find out *Who Succeeds in Science?* (Sonnert and Holton 1995). Various handbooks also offered specific advice on how to give women with scientific aspirations *A Hand Up* (American Women in Science 1993).

An NSF (2000) report titled *Women, Minorities, and Persons with Disabilities in Science and Engineering* focused on the period from 1980 to 1997. It revealed that while the number of postgraduate men remained considerably higher than that of women in science and engineering, the number of men was beginning to decline in the 1990s while the number of women had stabilized. A slow but steady climb in the overall number of Ph.D.s in science and engineering earned by women between 1966 and 1996 was certainly heartening, but the gap remained nearly ten thousand (see figures 1.2 and 1.3).

Between 1970 and 1999, women had become a larger percentage of graduate students in all categories. However, old patterns persisted, with women gaining well above 50 percent of graduate degrees in psychology and behavioral sciences and reaching 50 percent in the biological and social sciences, but still under 30 percent overall, and well below that in engineering and the physical sciences.

Overall, academic employment was increasing for women in the last two decades of the twentieth century (as it was for men). Although the gender gap was slowly diminishing, the absolute numbers do not tell the entire story. According to a survey completed in 1999 by the NSF and a panel at the National Academy of Sciences (NAS), women continued to be clustered in lower ranks or even outside the ranks of tenure, in positions as instructors, lecturers, and adjunct faculty; relatively few were full professors (Long 2001). When salaries were compared in the aggregate, women did not catch up to men over the course of their careers, but actually seemed to lose ground. Thus, most recent data confirmed

Figure 1.2. Doctoral degrees awarded in science and engineering (S&E) fields and in non-S&E fields by sex, 1966–1997.

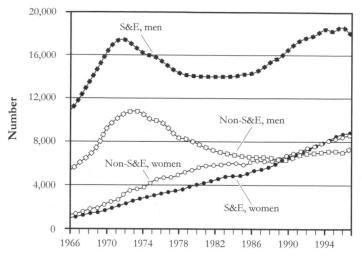

Survey of Earned Doctorates.
Women, Minorities, and Persons with Disabilities in Science and Engineering. Washington, D.C.: National Science Foundation, Division of Science Resources Statistics, 2002.

Figure 1.3. Proportion of science and engineering (S&E) graduate students who are women by field, 1990 and 1999.

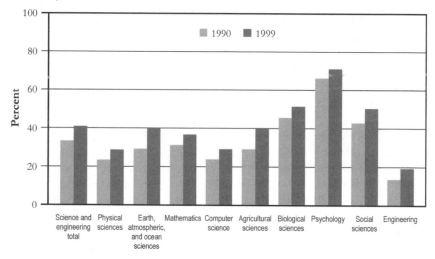

Survey of Graduate Students and Postdoctorates in Science and Engineering, various years.
Women, Minorities, and Persons with Disabilities in Science and Engineering. Washington, D.C.: National Science Foundation, Division of Science Resources Statistics, 2002.

that, despite some occasional and field-specific advances worthy of note, in general women had not made substantial headway in the 1990s. Women actually lost ground in mathematics and in computer science, the "hot field" of that decade, as well as over the course of their working lives (NSF 2002).

This whirlwind tour of statistics is intended to introduce readers to data sources so that they can use the data strategically. We can encourage those who need to be shown that, given opportunities, women do go into science and technology. However, we also need to make clear that there is still a long way to go before gender equity is achieved in science and technology.

Reflections on the Current Condition

Our foremothers presented equity as a primary goal, and many of us continue to think that equity for women is fundamental even as we work to factor gender considerations about real-life circumstances of women (especially family care) into the discussion (Davis et al. 1996). To talk about equity is to require statistical data, and I was surprised when gathering data for this essay to find that disaggregated labor statistics that reveal the sex and minority status of those taking graduate degrees in science and engineering were no longer being compiled and published by federal departments for education and for labor, or by women's organizations such as AAUW, as they had been earlier. The issue is not ignored completely, however, and federal funding supported the work of proactive scholars like Sue Rosser, whose book *Female-Friendly Science* is widely used by educators concerned about gender equity in the classroom (Rosser 1990, 1995, 1997).

In its effort to understand how gender plays a role in career choices, AAUW, for example, has done several important studies that concentrate on the school years and especially on those critical middle school years when girls seem to turn away from some of their science options. AAUW's highly visible publications on sexual harassment, discrimination, and even the issue of girls and computers (AAUW 2000) are very important. But this attention to the early years of education has come at the expense of research on the later patterns in graduate school and employment. We have now less authoritative and comparative data on women at advanced levels of education and those working in academe than we had three decades ago. Some of the best data are provided by the NSF, perhaps because microbiologist Rita R. Colwell heads the foundation and key long-standing staff members attend to these issues. Selected data have also been compiled and put online by the Association for Women in Science (AWIS), and also by the Commission on Professionals in Science and Technology, especially in *Salaries of Scientists, Engineers, and Technicians* (Commission on Professionals in Science and Technology 2002). Useful discussion is provided in *Women, Minorities, and Persons with Disabilities,* a congressionally mandated biennial review by the NSF. We must call for more data that will help us understand what has happened and is happening to women who are already in science and engineering.

Londa Schiebinger's (1999) book *Has Feminism Changed Science?* answers her own rhetorical question with a cautious affirmative because of numerical and statistical increases in women across most fields. A number of scholars have demonstrated that women's involvement in some fields has posed new research questions and that women's concerns have also changed workplace dynamics (Creager, Lunbeck, and Schiebinger 2001; Haraway 1989; Harding and O'Barr 1987; Keller 2001; Rose 1994). These scholars have argued that if science is different in 2000 than it was in 1900 or 1970, the change is primarily due to those women in scientific and technological fields who have not only done good work but have used their energy and ideas to promote change; it's not due to their biology (Schiebinger 1998).

It is important to celebrate women in science and engineering. The models of achieving women are everywhere around us. It remains important to highlight the continuing trickle of women Nobel laureates; they are less celebrated than Marie Curie or Anna Botsford Comstock were among their contemporaries. We need to show achieving women doing all kinds of scientific work in their infinite variety of personal and professional profiles. In the past decade a number of memoirs and collections revealed the life stories of women who have, one way or another, made it in science (Eisenhart and Finkel 1998; Morse 1995; Murray 2000; Pattatucci 1998). Probably more significant than often acknowledged are the fictional characters who attracted us and our children at the turn of the century. Just as Gene Stratton-Porter's series on the adventuresome girl of the Limberlost inspired a generation of young naturalists, young women (and men) can be inspired, perhaps, by Hermione Granger, the strong and intellectual if stereotypically nerdy girl who gets Harry Potter out of trouble, and Dana Scully, the strong, independent, rational co-lead on the television series *The X-Files*.[3] Advocacy, promotion, marketing—we need to identify allies who agree that women are needed in the sciences to influence the scientific and public agenda. Studies have made it clear that rhetoric within science resonates with male imagery and outlooks and only broader female involvement will create a space where there is gender diversity and gender neutrality in language and behavior (Cohn 1987; Traweek 1988). The early-twentieth-century strategies were relatively effective, but their inability to bring sustained change indicates that we must pursue more fine-tuned approaches and build strong infrastructures that can better withstand political and cultural backlash (Fox 1998, 2000).

The most important initiatives are often local and can reverberate, as with the initiative at MIT, where faculty women's complaints prompted an investigation (with data) that has provided publicity and new grounds for individual women there to argue their case not only for salaries but for space and other resources (MIT 1999). It is always hard to measure the impact of such activist projects, which, like Shyamala Rajender's class-action suit in the mid-1970s that resulted in court-supervised scrutiny of hiring at the University of Minnesota, seem to have a ripple effect across the country. Rajender's case certainly brought changes at Minnesota (Saitre 1992).

Individual actions also may have far-reaching effects. For instance, Anne

Hibner Koblitz wrote a biography of Anna Kovalevskaia, a political activist and exceptional mathematician of the mid-nineteenth century. Inspired by the life of this Russian radical who boldly dedicated her life to advancing women, Anne Koblitz used the royalties of the book plus other funds to establish the Kovalevskaia Fund to support women in science, especially in Southeast Asia and Central America (Koblitz 1993; Koblitz and Koblitz 2000). The very process of identifying such women and then offering them modest awards had a significant effect because university administrators and political officials paid attention when an American came to present a plaque and grant. The awardees gained personal recognition, additional research funds, and broader visibility for women in their countries in mathematics and science generally. This global perspective is very important. For example, the Women's International Science Collaboration Program aims to link women scientists around the world with women in the United States for collaborative research.[4] We may learn much from comparative work that will explain how differently women have been received in various sciences at different times and places (Canel, Oldenziel, and Zachmann 2000; Etzkowitz, Kemelgor, and Uzzi 2000; Stolte-Heiskanen 1991).

In October 2002, U.S. Senator Ron Wyden, chair of the Senate's Science, Technology, and Space Subcommittee, held hearings to review what had happened to advance women in mathematics, science, and technology in the thirty years since Title IX had been passed. The Department of Education's representative, W. Todd Jones, touted "good news," highlighting the kinds of data that indeed show increasing numbers. By contrast, the women who testified, including Marcia Greenberger, co-president of the National Women's Law Center, and Geraldine Richmond, professor of chemistry at the University of Oregon, were less complacent. Under questioning, Todd admitted that no recent studies of compliance with Title IX had been done. The women then pressed various issues, pointing out the declining numbers of women in computer science, family issues still confronting women, and the failure of universities to follow through on earlier ambitious plans for diversity. They once again urged the Department of Education to hold universities accountable for their lack of progress in making resources available to women—and made sure that gender issues were put on the table.[5]

History is not tidy, but it is useful to know. A significant reason to tell *herstory* is to keep in view the strong, resilient, and achieving women who worked to advance other women, but that heroic account alone lacks the complementary reality that such strength and resilience are honed in often challenging personal and professional experiences (Kohlstedt 1999). Moreover, history like the one just told here makes it clear that there is no guarantee of progress or even stability in the gains made in policy and practice. That awareness keeps us from being complacent. History also frees us from personal guilt and from arrogance when we understand much broader patterns of cultural conditions, situational challenges, and personal support systems. History has taught me, at least, to modify any undue pride about my achievements (given the fact that I luckily came of age while women were pushing to expand opportunities for women), and to mitigate any self-blame for failures that were structural and not personal.

Looking back a hundred years from now, some historian will again recount that we had women of achievement, women who worked collaboratively in science, and women who left a legacy through their advocacy for opportunity, equity, and social justice. We may also be at a critical juncture where the gains are eroded and research institutions deepen their discipline, or where opportunities expand and diversity pushes science and technology in directions that are even more responsive to the economic and social problems of our generation. Women can be, indeed must be, among the leadership directing the future of science, engineering, and medicine.

Notes

I would like to thank Margaret Rossiter, Mark Jorgensen, two anonymous reviewers, and Jill Bystydzienski for their constructive readings of an earlier version of this paper.

1. The CPST produced a short history as part of its fiftieth anniversary celebrations in 2003. The report indicates how much more active the organization became when Betty Vetter joined its staff, but gives no detail about her initiatives.
2. The Betty McGee Vetter Papers are housed in the Special Collections Library at Iowa State University, Ames.
3. The models will change, but youngsters at the turn of the twentieth century are swept up in J. K. Rowling's Harry Potter series and many college students watch *The X-Files*.
4. For more information, see the program's Web site at http://www.aaas.org/programs/international/wisc/.
5. A report of the committee hearing was published in the *Chronicle of Higher Education*, October 4, 2002, on the Web at http://chronicle.com/daily/2002/10/2002100403n.htm (accessed March 21, 2005).

References

AAUW (American Association of University Women). 2000. *Tech-Savvy: Educating Girls in the New Computer Age*. Washington, D.C.: American Association of University Women Educational Foundation.

Abir-Am, Pnina G., and Dorinda Outram, eds. 1987. *Uneasy Careers and Intimate Lives: Women in Science, 1789–1979*. New Brunswick, N.J.: Rutgers University Press.

Ajzenberg-Selove, Fay. 1994. *A Matter of Choices: Memoirs of a Female Physicist*. New Brunswick, N.J.: Rutgers University Press.

Alonso, Harriet Hyman. 1993. *Peace as a Women's Issue: A History of the U.S. Move-

ment for World Peace and Women's Rights. Syracuse, N.Y.: Syracuse University Press.

Alper, J. 1993. "The Pipeline Is Leaking Women All the Way Along." *Science* 260:409–11.

American Women in Science. 1993. *A Hand Up: Women Mentoring Women in Science*. Washington, D.C.: American Women in Science.

Appel, Toby A. 1999. "Physiology in American Women's Colleges: The Rise and Decline of a Female Subculture." In *History of Women in the Sciences: Readings from Isis*, ed. Sally Gregory Kohlstedt, 305–25. Chicago: University of Chicago Press.

Brownmiller, Susan. 1999. *In Our Time: Memoir of a Revolution*. New York: Dial.

Canel, Annie, Ruth Oldenziel, and Karin Zachmann, eds. 2000. *Crossing Boundaries, Building Bridges: Comparing the History of Women Engineers, 1870s–1990s*. Amsterdam: Harwood Academic.

Cassell, Joan. 1998. *The Woman in the Surgeon's Body*. Cambridge, Mass.: Harvard University Press.

Cattell, James McKeen. 1933. *American Men of Science: A Biographical Dictionary*. New York: Science.

Clark, Shirley M., and Mary Corcoran. 1986. "Perspectives on the Professional Socialization of Women Faculty: A Case of Accumulative Disadvantage." *Journal of Higher Education* 57: 20–43.

Cohn, Carol. 1987. "Sex and Death in the Rational World of Defense Intellectuals." *Signs: Journal of Women in Culture and Society* 12(3): 687–719.

Comfort, Nathaniel. 2001. *The Tangled Field: Barbara McClintock's Search for the Patterns of Genetic Control*. Cambridge, Mass.: Harvard University Press.

Commission on Professionals in Science and Technology. 2002. "Salaries of Scientists, Engineers, and Technicians." http://www.cpst.org. Accessed December 24, 2002.

Committee on Women in Science and Engineering. 1991. *Women in Science and Engineering: Increasing Their Numbers in the 1990s*. Washington, D.C.: National Academy Press.

———. 1992. *Women Scientists and Engineers Employed in Industry: Why So Few?* Washington, D.C.: National Academy Press.

Comstock, Anna Botsford. 1953. *The Comstocks of Cornell: John Henry Comstock and Anna Botsford Comstock*. New York: Comstock Publishing Associates.

Cott, Nancy F. 1987. *The Grounding of Modern Feminism*. New Haven, Conn.: Yale University Press.

Cowan, Ruth Schwartz. 1983. *More Work for Mother: The Ironies of Household Technology from the Open Hearth to the Microwave*. New York: Basic.

Creager, Angela N. H., Elizabeth Lunbeck, and Londa Schiebinger, eds. 2001. *Feminism in Twentieth-Century Science, Technology, and Medicine*. Chicago: University of Chicago Press.

Davis, Cinda-Sue, et al., eds. 1996. *The Equity Equation: Fostering the Advancement of Women in the Sciences, Mathematics, and Engineering*. San Francisco: Jossey-Bass.

Deacon, Desley. 1997. *Elsie Clews Parsons: Inventing Modern Life*. Chicago: University of Chicago Press.

Eisenhart, Margaret A., and Elizabeth Finkel. 1998. *Women's Science: Learning and Succeeding from the Margins*. Chicago: University of Chicago Press.

Etzkowitz, Henry, Carol Kemelgor, and Brian Uzzi. 2000. *Athena Unbound: The Ad-*

vancement of Women in Science and Technology. New York: Cambridge University Press.

Evans, Sara. 2003. *Tidal Wave: How Women Changed America at Century's End.* New York: Free Press.

Fox, Mary Frank. 1998. "Women in Science and Engineering: Theory, Practice, and Policy in Programs." *Signs: Journal of Women in Culture and Society* 24(1): 201–23.

———. 2000. "Organizational Environments and Doctoral Degrees Awarded to Women in Science and Engineering Departments." *Women's Studies Quarterly* 28: 47–61.

Freedman, Estelle. 2002. *No Turning Back: The History of Feminism and the Future of Women.* New York: Ballantine.

Friedan, Betty. 1963. *The Feminine Mystique.* New York: Dell.

Greenwald, Maurine Weiner. 1980. *Women, War, and Work: The Impact of World War I on Women Workers in the United States.* Westport, Conn.: Greenwood.

Haraway, Donna. 1989. *Primate Visions: Gender, Race, and Nature in the World of Science.* New York: Routledge.

Harding, Sandra G., and Jean F. O'Barr, eds. 1987. *Sex and Scientific Inquiry.* Chicago: University of Chicago Press.

Henson, Pamela M. 1996. "The Comstocks of Cornell: A Marriage of Interests." In *Creative Couples in the Sciences,* ed. Helena M. Pycior, Nancy Slack, and Pnina Abir-Am, 112–25. New Brunswick, N.J.: Rutgers University Press.

Howes, Ruth H., and Caroline L. Herzenberg. 1999. *Their Day in the Sun: Women of the Manhattan Project.* Philadelphia: Temple University Press.

Kass-Simon, G., and Patricia Farnes, eds. 1990. *Women of Science: Righting the Record.* Bloomington: Indiana University Press.

Keller, Evelyn Fox. 1983. *A Feeling for the Organism: The Life and Work of Barbara McClintock.* New York: W. H. Freeman.

———. 2001. "Making a Difference: Feminist Movement and Feminist Critiques of Science." In *Feminism in Twentieth-Century Science, Technology, and Medicine,* ed. Angela N. H. Creager, Elizabeth Lunbeck, and Londa Schiebinger, 98–109. Chicago: University of Chicago Press.

Kessler-Harris, Alice. 1982. *Out to Work: A History of Wage-Earning Women in the United States.* New York: Oxford University Press.

Koblitz, Ann Hibner. 1993. *A Convergence of Lives: Sofia Kovalevskaia, Scientist, Writer, Revolutionary.* New Brunswick, N.J.: Rutgers University Press.

Koblitz, Ann Hibner, and Neal Koblitz. 2000. "The Kovalevskaia Fund." *Mathematical Intelligencer* 22: 62–65.

Kohlstedt, Sally Gregory, ed. 1999. *History of Women in the Sciences: Readings from Isis.* Chicago: University of Chicago Press.

Lafollette, Marcel C. 1988. "Eyes on the Stars: Images of Women Scientists in Popular Magazines." *Science, Technology, and Human Values* 13: 262–75.

Lankford, John. 1997. *American Astronomy: Community, Careers, and Power, 1859–1940.* Chicago: University of Chicago Press.

Laslett, Barbara, and Barrie Thorne, eds. 1997. *Feminist Sociology: Life Histories of a Movement.* New Brunswick, N.J.: Rutgers University Press.

Lear, Linda J. 1997. *Rachel Carson: Witness for Nature.* New York: Henry Holt.

Levine, Susan. 1995. *Degrees of Equality: The American Association of University*

Women and the Challenge of Twentieth-Century Feminism. Philadelphia: Temple University Press.

Lippincott, Gail. 2003. "Rhetorical Chemistry: Negotiating Gendered Audiences in Nineteenth-Century Nutrition Studies." *Journal of Business and Technical Communication* 17: 10–49.

Long, J. Scott, ed. 2001. *From Scarcity to Visibility: Gender Differences in the Careers of Doctoral Scientists and Engineers.* Washington, D.C.: National Academy Press.

Mack, Pamela. 1990. "Strategies and Compromises: Women in Astronomy at Harvard College Observatory, 1870–1920." *Journal for the History of Astronomy* 21: 65–76.

Maddox, Brenda. 2002. *Rosalind Franklin: The Dark Lady of DNA.* New York: Harper-Collins.

May, Elaine Tyler. 1999. *Homeward Bound: American Families in the Cold War Era.* New York: Basic Books.

McIntire, Natalie M. 1993. "Curtiss Wright Cadettes: A Case Study of the Effect of World War II Labor Shortage on Women in Engineering." M.A. thesis, University of Minnesota.

MIT (Massachusetts Institute of Technology). 1999. *A Study of the Status of Women Faculty in Science at MIT.* Special issue of the *MIT Faculty Newsletter* 11(4). http://web.mit.edu/fnl/women/women.html. Accessed February 16, 2005.

Morse, Mary. 1995. *Women Changing Science: Voices from a Field in Transition.* New York: Insight.

Murray, Margaret A. M. 2000. *Women Becoming Mathematicians: Creating Professional Identity in Post–World War II America.* Cambridge, Mass.: MIT Press.

National Academy of Engineering et al. 2000. "Greatest Engineering Achievements of the 20th Century." http://www.greatachievements.org/. Accessed February 16, 2005.

Nerad, Maresi. 1999. *The Academic Kitchen: A Social History of Gender Stratification at the University of California, Berkeley.* Albany: State University of New York Press.

NSF (National Science Foundation). 2000. *Women, Minorities, and Persons with Disabilities in Science and Engineering: 2000.* Arlington, Va.: National Science Foundation (NSF 00-327). http://www.nsf.gov/sbe/srs/nsf00327. Accessed February 16, 2005.

———. 2002. *Science and Engineering Indicators, 2002.* http://www.nsf.gov/sbe/srs/seind02/start.htm. Accessed February 16, 2005.

Nye, Robert A. 1997. "Medicine and Science as Masculine 'Fields of Honor.'" *Osiris* 12: 60–79.

Ogilvie, Marilyn B., and Joy D. Harvey, eds. 2000. *The Biographical Dictionary of Women in Science: Pioneering Lives from Ancient Times to the Mid-twentieth Century.* New York: Routledge.

Oldenziel, Ruth. 1997. "Boys and Their Toys: The Fisher Body Craftsman's Guild, 1930–1968, and the Making of a Male Technical Domain." *Technology and Culture* 38: 60–98.

———. 1999. *Making Technology Masculine: Men, Women, and Modern Machines in America, 1870–1945.* Amsterdam: Amsterdam University Press.

Pattatucci, Angela M., ed. 1998. *Women in Science: Meeting Career Challenges.* Thousand Oaks, Calif.: Sage.

Pycior, Helena M. 1993. "Reaping the Benefits of Collaboration while Avoiding Its Pit-falls: Marie Curie's Rise to Scientific Prominence." *Social Studies of Science* 23: 301–23.

———. 1996. "Pierre Curie and 'His Eminent Collaborator Mme. Curie': Complementary Partners." In *Creative Couples in the Sciences,* ed. Helena M. Pycior, Nancy Slack, and Pnina Abir-Am, 39–56. New Brunswick, N.J.: Rutgers University Press.

Quinn, Susan. 1995. *Marie Curie: A Life.* New York: Simon and Schuster.

Renn, Jurgen, and Robert Schulmann, eds. 1992. *Albert Einstein/Mileva Maric: The Love Letters.* Princeton, N.J.: Princeton University Press.

Reynolds, Moira Davison. 1999. *American Women Scientists: Twenty-Three Inspiring Biographies, 1900–2000.* Jefferson, N.C.: McFarland.

Rose, Hilary. 1994. *Love, Power, and Knowledge: Towards a Feminist Transformation of the Sciences.* Bloomington: Indiana University Press.

Rosen, Ruth. 2000. *The World Split Open: How the Modern Women's Movement Changed America.* New York: Viking.

Rosenberg, Rosalind. 1982. *Beyond Separate Spheres: Intellectual Roots of American Feminism.* New Haven, Conn.: Yale University Press.

Rosser, Sue V. 1990. *Female-Friendly Science: Applying Women's Studies Methods and Theories to Attract Students.* New York: Pergamon.

———, ed. 1995. *Teaching the Majority: Breaking the Gender Barrier in Science, Mathematics, and Engineering.* New York: Teachers College Press.

———. 1997. *Re-engineering Female Friendly Science.* New York: Teachers College Press.

Rossiter, Margaret W. 1980. "Women's Work in Science, 1880–1910." *Isis* 71: 381–98.

———. 1982. *Women Scientists in America: Struggles and Strategies to 1940.* Baltimore, Md.: Johns Hopkins University Press.

———. 1995. *Women Scientists in America: Before Affirmative Action, 1940–1972.* Baltimore, Md.: Johns Hopkins University Press.

———. 1997a. "The Men Move In: Home Economics in Higher Education, 1850–1970." In *Rethinking Home Economics: Women and the History of a Profession,* ed. Sarah Stage and Virginia B. Vincenti, 17–33. Ithaca, N.Y.: Cornell University Press.

———. 1997b. "Which Science? Which Women?" In *Women, Gender, and Science: New Directions,* ed. Sally Gregory Kohlstedt and Helen Longino, 169–85. Chicago: University of Chicago Press.

Saitre, Sara. 1992. "The Rajender Report on the Status of Women." *Minnesota Alumni Association Magazine,* July–August: 11–25.

Sayre, Anne. 1975. *Rosalind Franklin and DNA.* New York: W. W. Norton.

Schiebinger, Londa. 1998. "Creating Sustainable Science." *Osiris* 12: 201–16.

———. 1999. *Has Feminism Changed Science?* Cambridge, Mass.: Harvard University Press.

Sicherman, Barbara, and Carol Hurd Green, eds. 1980. *Notable American Women: The Modern Period.* Cambridge, Mass.: Harvard University Press.

Sime, Ruth Lewin. 1996. *Lise Meitner: A Life in Physics.* Berkeley: University of California Press.

Solomon, Barbara Miller. 1985. *In the Company of Educated Women: A History of Women and Higher Education in America.* New Haven, Conn.: Yale University Press.

Sonnert, Gerhard, and Gerald Holton. 1995. *Who Succeeds in Science? The Gender Dimension.* New Brunswick, N.J.: Rutgers University Press.

Stage, Sarah, and Virginia B. Vincenti, eds. 1997. *Rethinking Home Economics: Women and the History of a Profession.* Ithaca, N.Y.: Cornell University Press.

Stein, Barbara R. 2001. *On Her Own Terms: Annie Montague Alexander and the Rise of Science in the American West.* Berkeley: University of California Press.

Stolte-Heiskanen, Veronica, ed. 1991. *Women in Science: Token Women or Gender Equality?* New York: Oxford University Press.

Tharp, Ann Marie. 2002. "Career Development of Women Engineers: The Role of Self-Efficacy and Supports-Barriers." Ph.D. dissertation, University of Minnesota.

Tobias, Sheila. 1990. *They're Not Dumb, They're Different: Stalking the Second Tier.* Tucson, Ariz.: Research Corporation.

———. 1997. *Faces of Feminism: An Activist's Reflections on the Women's Movement.* Boulder, Colo.: Westview.

Tolley, Kimberley. 2002. *The Science Education of American Girls: A Historical Perspective.* London: Routledge-Falmer.

Tomes, Nancy. 1998. *The Gospel of Germs: Men, Women, and the Microbe in American Life.* Cambridge, Mass.: Harvard University Press.

Traweek, Sharon. 1988. *Beamtimes and Lifetimes: The World of High Energy Physicists.* Cambridge, Mass.: Harvard University Press.

Tyack, David B., and Elisabeth Hansot. 1992. *Learning Together: A History of Gender in American Public Schools.* New York: Russell Sage Foundation.

Williams, Kathleen Broome. 2001. *Improbable Warriors: Women Scientists and the U.S. Navy in World War II.* Annapolis, Md.: Naval Institute Press.

Zuckerman, Harriet, and Jonathan R. Cole. 1991. *The Outer Circle: Women in the Scientific Community.* New York: Norton.

2 From "Engineeresses" to "Girl Engineers" to "Good Engineers": A History of Women's U.S. Engineering Education

Amy Sue Bix

Engineering education in the United States has had a gendered history, one that until relatively recently prevented women from finding a place in the predominantly male technical world. For decades, Americans treated the professional study of technology as men's territory (Bix 2000b; Ogilvie 1986; Rossiter 1982, 1995). Until World War II and beyond, many leading engineering schools, including Rensselaer Polytechnic Institute, the Georgia Institute of Technology, and the California Institute of Technology, remained closed to women. The few women admitted to the Massachusetts Institute of Technology (MIT) struggled against a hostile intellectual and social environment. Women studying or working in engineering were popularly perceived as oddities at best, outcasts at worst, defying traditional gender norms. As late as the 1960s, women still made up less than 1 percent of students studying engineering in the United States, and critics either dismissed or ridiculed women's interest in the profession. Throughout the last half of the twentieth century, activists fought to change that situation, to win acknowledgment of women's ability to become good engineers.

The reasons for the strongly masculine connotations of engineering work stem, to a significant extent, from its distinctive origins. Throughout much of the nineteenth century in the United States, it was relatively rare for practitioners to have earned a formal engineering degree. Instead, individuals acquired credentials through on-the-job experience in a machine shop, railroad yard, or surveying crew. Such work environments excluded most women. More than that, many engineering chores involved hard, even dangerous, physical exertion, something perceived as inappropriate for respectable women. Other trends reinforced the masculinity of engineering; in the twentieth century, makers of erector sets and model trains marketed these technological toys specifically as a way of turning boys into future engineers. Girls who expressed technical interests were often steered instead into the science side of home economics (Bix 2002; Oldenziel 1999, 2000; Purcell 1979; Wajcman 1991).

"Engineeresses"

In the late 1800s and early 1900s, a handful of women ventured into engineering studies, primarily at land grant institutions (Goff 1946; Ingels 1952; LeBold and LeBold 1998; Trescott 1990). For instance, Olive Dennis earned her civil engineering degree in 1920 from Cornell, then worked more than twenty years at the Baltimore and Ohio railroad. She served the B&O as a "draftsman" (as she described herself), designing some of the railroad's terminals and (more in line with stereotypes about feminine nature) designing china used in train restaurants (Dennis 1948; Handy 1940). Women such as Dennis attracted a certain attention, since they were a rarity, a curiosity. Commenting on that female presence, a 1920s newspaper headline read, "Three Coeds Invade Engineering Courses and Compete with Men at Cornell University: Stand Well in Their Studies" (*Cornell Daily Sun* 1937, 1). The term "invade" turns up repeatedly at a number of schools in popular references to enrollment of a few female engineering students during the 1920s and again in the 1940s. That word's common use following World War I and during World War II is not surprising, but does underline the extent to which women in engineering appeared as the *other*, entering a field that everyone assumed was and must be male territory (Bix 2000a).

The issue of women's venturing into strange space came to a head with World War II, when the United States suddenly faced a "manpower" crisis. As essential men were called up for service, industry desperately needed people at drawing boards and in engineering shops to keep planes, tanks, and other war materiel rolling off the assembly line. There simply were not enough male engineers available. Just as manufacturers turned to "Rosie the Riveter" on the shop floor, companies sought to hire female engineers. But of course, managers immediately encountered the obvious difficulty—they could not find enough women trained for technical work.

Companies such as General Electric began recruiting women who possessed at least basic math and science skills, then gave those women emergency crash courses to turn them into wartime engineering aides. In one of the largest and most elaborate of such plans, in 1942, the Curtiss-Wright airplane company announced its initiative for training what it called "Curtiss-Wright Cadettes." Seven colleges—Cornell, Iowa State, Minnesota, Penn State, Purdue, Rensselaer Polytechnic Institute (RPI), and the University of Texas—agreed to work with the firm and teach its specially prepared curriculum to more than six hundred women. Program representatives recruited sophomore, junior, and senior coeds through advertisements in college papers, calling especially for those with training in mathematics at least through algebra. Candidates underwent a ten-month immersion in classes on engineering mathematics, job terminology, aircraft drawing, engineering mechanics, airplane materials, theory of flight, and aircraft production. After that intensive exposure, Curtiss-Wright assigned Cadettes to plants to work in airplane design research, testing, and production (*War Training Programs* 1945).

Six of the seven campuses participating in the Curtiss-Wright program already had women enrolled. Granted, Cornell or Iowa State coeds usually majored in teaching or home economics rather than engineering, but at least students and faculty were accustomed to seeing women around campus. At these schools, announcement of the Cadette program elicited some joking about the notion of female engineering students. But Cadettes could claim to be doing their part for the war effort, and on that patriotic ground they were welcomed. By contrast, at all-male RPI, the arrival of "engineeresses" created a culture shock. Local newspapers carried giant headlines: "RPI Opens Doors to Women: Institute Breaks 116 Year Old Rule Due to War Need . . . Curtiss Wright Women . . . Invade RPI Campus" (*Rensselaer Polytechnic* 1943). Soon RPI discovered advantages to having "Katie Kaddettes" on campus. Cadettes threw themselves into school culture, joining the acting troupe and cheerleading squad. The Curtiss-Wright story represented a perfect wartime morale booster: Cadettes proved temptingly photogenic, and *Life* published a special feature on them. The "engineeresses" were a curiosity, but acceptable as a temporary war measure (Bix 2000a).

"Girl Engineers"

As World War II drew to a close, returning male veterans flooded into American engineering programs, and the wartime emergency rationale for encouraging women to develop their technical talents vanished. More than that, conservative gender modes of the postwar decades brought a prevailing expectation that women's career ambitions must give way to the goal of marrying and raising children. Young girls who did express technical interests were often deliberately discouraged by negative remarks from family or teachers (Bix 2000a).

At places where engineering's macho culture had become most ingrained, such as Georgia Tech, talk of women engineers seemed ridiculous. The college humor magazine ran entire issues poking fun at the very notion, with cartoons depicting shapely women getting their curls caught in equipment, using a hydraulic testing machine to crack nuts, and invading the campus pool where men enjoyed skinny-dipping. Nevertheless, the issue of coeducation came to the fore in 1948, with rumors that Atlanta women's groups were "raising funds with which to carry through the courts the question of compelling the regents to admit women to Georgia Tech" (Van Leer 1948b). Georgia Tech president Blake Van Leer observed, "The Attorney General tells me that if they do, they are certain to win" (1948b). A test case loomed, centered on a technically talented high school woman who wished to study engineering at Georgia Tech. Van Leer commented,

> My personal feelings sway me in one direction, whereas my official position influences me another way. I have been associated with coeducational institutions practically all of my life, and I have always felt it was wrong to discriminate against a student because she happened to be a woman. I feel that way about Miss Bonds.

She is obviously a Georgia citizen and a qualified and responsible engineering student; this makes it seem wrong . . . for her to be denied an engineering education in her native state simply because she is a woman. On the other hand, Georgia Tech is traditionally a man's school. The majority of students, faculty, and alumni are opposed [to coeducation]. (1948a)

Reportedly under the influence of his wife and daughter (who both had technical interests of their own), Van Leer followed his personal inclination and started advocating women's admission. Van Leer pointed out that Georgia Tech had plenty of physical room to house women, who had already been admitted to night school and extension classes. Meanwhile, the Women's Chamber of Commerce of Atlanta passed a resolution calling on the state to let female students enter Georgia Tech. Regents immediately objected: "Here is where the women get their noses under the tent. . . . We'll have home economics and dressmaking at Tech yet" (*Atlanta Journal* 1952). One later explained, "I didn't want to see . . . Tech [become] a campus full of 'debutantes' looking for a husband. I still feel very strongly that too many girls at Tech will all but destroy the seriousness of purpose in the lives of many young men at school" (Arnold 1961). In 1952, over such resistance, the board passed a measure admitting women to Georgia Tech under limited conditions (Bix 2000a).

The decision roused protests among students and alumni who felt passionately loyal to Georgia Tech's engineering-school traditions, which they felt defined the school as inherently and necessarily male-only. The first coeds to appear on campus caused a sensation; papers published photos showing women trying on traditional freshman "rat caps." The *Atlanta Constitution* ran a cartoon showing lingerie drying on a clothesline strung from the main campus tower. The girl engineers were a curiosity; as one article explained, "A petite blonde is the first woman to attempt invasion of the home of the 'Ramblin' Wreck' since that male stronghold became coed last week. She is golden-haired Mary Joan Coffee . . . who makes it plain that she is going to Tech strictly to study and not to look for boy friends" (McNatt 1952, 1).

Such comments made it evident that female engineers of the postwar years would need to fight to be taken seriously, and, in that battle, would need to band together as allies. In 1946, about twenty female engineering students at Iowa State organized a local group called the Society of Women Engineers to assist "in orienting new women students in the division" (*Iowa Engineer* 1946, 222). That same year, female students at Syracuse and Cornell vented their frustration at being excluded from several major engineering honor societies (or restricted to a "woman's badge" instead of full membership). The new honorary society they created, Pi Omicron, soon established chapters at colleges and universities around the nation, where members held orientations to welcome new female engineering majors. The society's mission was "to encourage and reward scholarship and accomplishment . . . among the women students of engineering . . . ; to promote the advancement and spread of education in . . . engineering among women" (*Cornell Engineer* 1946, 14).

In 1950, female engineers in New York, Boston, Philadelphia, and Washington, D.C., began gathering on a semiregular basis, officially incorporating in 1952 as the Society of Women Engineers (SWE), a professional, nonprofit educational service organization. The organization defined its objectives as "to inform the public of the availability of qualified women for engineering positions; to foster a favorable attitude in industry toward women engineers, and to contribute to their professional advancement; to encourage young women with suitable aptitudes and interest to enter the engineering profession, and to guide them in their educational programs" (SWE 1953). One of SWE's first steps was to establish a Professional Guidance and Education Committee, which poured enormous energy into reaching potential converts. Members personally wrote to dozens of high school girls, sending information about engineering and replying to questions. In 1958, Boston's chapter of SWE published a pamphlet containing biographical sketches of a few "typical" women engineers and explanations of how girls could prepare to enter engineering. Its authors concluded, "If this pamphlet shall have inspired one young woman to consider an engineering career . . . and one parent to 'encourage' the daughter's desire to enter the technical field, this pamphlet will then have been a worthwhile venture" (Miller 1964).

Such volunteer guidance reflected one of SWE's primary beliefs: that girls often shied away from technical pursuits because they simply did not realize that women could and did go into engineering. Irene Carswell Peden, associate professor of electrical engineering at the University of Washington, wrote,

> It is important to think of women engineers as real people doing real jobs which the student could do, too. . . . A girl is not likely to choose a career field disapproved by her parents, teachers, classmates, and friends. All of these people . . . seem to be responding in part to an erroneous but popular image of the woman engineer as a cold, . . . aggressive female who trudges through life in her flat-heeled shoes without a man in sight (away from the job). . . . Many women engineers are very attractive; most represent a perfectly normal cross section of femininity. The only way that this image can be brought into line with reality . . . is by . . . personal contact. (1965, 2)

In 1954 and 1955, members of Cleveland's SWE appeared on local television programs as living proof of women's small but persistent presence in the engineering profession. At a time when many Americans perceived female engineers as odd, manlike creatures, SWE representatives took pains to offer a presentable feminine image, emphasizing that many of them were married and had children. Advocates believed that women engineers could gain greater acceptance in society simply by making themselves more visible in a professional, positive way (Bix 2000b).

In the mid-1950s, SWE expanded its efforts to include more active outreach. Members volunteered to assist at "Junior Engineer and Scientist Summer Institute" (JESSI) programs, which brought high schoolers to college to explore science and receive educational guidance. At one JESSI session in Colorado, fifty-three girls listened to a five-woman panel discuss why they had chosen engineering careers. Female engineers led JESSI students on visits to industry and gave the girls (and boys) tours of their laboratories (Rutherford 1954).

By 1957, female engineering students at Drexel, Purdue, the University of Colorado, City College of New York, the University of Missouri, and in Boston had founded student sections of SWE, and the parent organization welcomed its junior counterparts. Established women engineers vividly remembered how intimidating it felt to be the sole woman in an engineering class. They knew, as Helen O'Bannon wrote, that "being one of a small group following a path that appears to violate society's norms is lonely" (1975). Mildred Dresselhaus argued that such young women deserved support from older mentors, who could provide the encouragement necessary to "keep going when the going gets rough or when [a girl] begins to ask, 'Is it worth it?'" Successful role models could give new students a boost in confidence, a chance "to see by example that women can 'make it' in engineering" (1975a). Older professionals especially sympathized with those young women just entering Georgia Tech, and in 1958, Atlanta's SWE chapter sent several members to participate in Georgia Tech's start-of-the-year camp for first-year women. "One must realize that there are this year approximately 1,300 freshmen at Georgia Tech and only 19 freshman coeds. There will be numerous problems and SWE Atlanta Section is proud to play an integral part in the quite difficult assimilation of female engineering students in an almost all-male school" (Dresselhaus 1975b, 30).

"Good Engineers"

More than simply pressing all-male schools to admit coeds, more than just encouraging young women to consider engineering studies, SWE and other advocates wanted to convince doubters that women could be good engineers. They strove to win respect, knowing that opposition remained quite visible. In 1955, Eric Walker, Penn State University's engineering dean, wrote a column titled "Women Are NOT for Engineering." Walker declared that most women did not have the "basic capabilities" needed for engineering. He concluded that teaching them didn't make sense; since "[t]he most evident ambition of many women is to get married and raise a family . . . few companies are willing to risk $10,000 on a beautiful blonde engineer, no matter how good she may be at math" (1955, 18).

The effort needed to counter such critics can be seen in the story of women at the Massachusetts Institute of Technology. The school had actually been coed since 1871, and between the 1920s and 1940s, MIT averaged fifty female students on campus each year, amidst approximately five thousand men. Coeds represented a curiosity. The student newspaper introduced a 1940 class member as a New York "glamour girl" who hoped to work in cancer research and won a $100 bet from fellow debutantes by gaining admission to MIT. Officially, coeds remained invisible. President Karl Compton told incoming students, "In choosing MIT, you've taken on a man-size job" (MIT 1941, 5). Campus traditions represented masculinity itself; as an official welcome, the institution held a "smoker" for freshmen and their fathers. Initiation took place at MIT camp, featuring water fights with the sophomores, baseball games with faculty, and plenty of male-bonding rituals. Including coeds in engineering programs seemed to present significant problems. Civil engineering students learned surveying and other field techniques at a rough camp, with accommodations judged unsuitable for females. Mechanical engineering class required round-the-clock observations of engine performance; generations of male students turned the "twenty-four-hour boiler tests" into beer parties. Women's staying overnight with men in the lab seemed inappropriate (Bix 2000b).

In the years following World War II, MIT kept admitting a few women per year, then proceeded to ignore their existence as a minor anomaly. In 1947, the dean of students defined MIT as intended "to prepare men for . . . engineering . . . [and] educate . . . men for responsible citizenship" (Baker 1947). Throughout the foreseeable future, coeds would "continue to be grossly outnumbered by men in classroom and lab," officials admitted. As Florence Stiles, adviser to women students, explained, the sense was that "women in general do not make acceptable engineers" (1946). One observer later wrote, "Before 1960, women entered MIT at their own risk. If they succeeded, fine; if they failed—well, no one had expected them to succeed" (Wick 1970). The few coeds around hesitated to rock the boat. One of MIT's female graduates from this era remembered,

> I was very conscious of having to represent women in each class. If I did anything wrong, . . . said anything stupid, it would be ammunition for all the men who didn't want us there in the first place. . . . Discriminatory events were so common that it didn't occur to us to object; . . . [besides], other engineering schools weren't accepting women . . . , so even though MIT was only accepting twenty a year . . . I felt MIT was doing us an enormous favor to have us there at all. (Jansen 1977)

Through the mid-1950s, many at MIT argued for ending coeducation, citing the high female drop-out rate. Margaret Alvort, the women's-house supervisor, wrote that her "doubt as to whether [coeds] belong . . . has grown into certainty that they do not." If MIT wanted to serve the nation by graduating as many top-notch engineers and scientists as possible, then "there is little in the records of the girls . . . to justify their continuance" (Alvort 1956). The school medical director agreed: "[W]hen there is such a shortage of engineers, one wonders if we are justified in taking positions away from male students for female." Coeds might bring "pleasure and ornamentation" to campus, but usually proved unable to hold their own against MIT men's competitiveness and "high-grade intellects." In short, he concluded, "except for the rare individual woman, [MIT] is an unsuitable place" (Keller 1981, 12).

Significantly, MIT President James Killian believed some women could succeed in scientific and technical fields and therefore argued for their continued access. He wrote, "I do not see how the Institute, having admitted women for so long, can now change," nor should it, considering that America's Cold War race with the Soviet Union called for development of *all* professional talent. Striving to "think more boldly . . . about recognizing [women's] presence," Killian sought support for coeducation (1956). In 1960, alumna Katharine Dexter McCormick pledged $1.5 million to build MIT's first on-campus women's dorm. McCormick knew that in her day, MIT had enrolled forty-four women, a figure that had barely risen five decades later (Bix 2000b).

Dedication of McCormick Hall in 1963 attracted national publicity. "Hardly anyone imagines girls attending mighty MIT," *Time* reported. "Yet last week Tech . . . dedicated its first women's dormitory to go with its first women's dean, an attractive blonde lured from nearby Radcliffe." MIT used McCormick Hall's opening to draw attention to its female students. Noting that "opportunities for women in science [and] engineering . . . are clearly increasing," 1963's catalog mentioned up front that MIT was coed. Women's applications jumped 50 percent in 1964. Backers of coeducation hailed McCormick Hall as a "vote of confidence, testimony . . . that women are to remain a permanent part of MIT" (*Time* 1963, 51).

Now that the university had finally created a physical place on campus for female students, women's dean Jacquelyn Mattfeld called on MIT to integrate coeds intellectually and socially. A "conservative . . . Wall Street attitude toward women still runs through MIT's veins," she declared. Many male professors and students regarded female undergrads as "incompetent, unnatural, and intruders" (1965).

Rather than waiting and hoping for such hostility to vanish, a new generation of MIT coeds began to band together to consider remedies. They began addressing issues such as employment discrimination, after facing corporate interviewers who openly questioned how long a woman engineer would remain on the job, doubted whether she could think about mechanical details "like a man," and offered distinctly lower salaries. To discuss such problems, the school's newly invigorated Association of Women Students (AWS) helped organize a "Symposium on American Women in Science and Engineering" at MIT in 1964. Planners hoped to attract widespread media coverage, teaching industry, the public, and young women themselves that women could be good engineers and scientists. The symposium attracted college faculty and administrators, high school students, and guidance counselors, plus more than 250 delegates from Smith, Radcliffe, Wellesley, the University of California, Georgia Tech, Northwestern, Purdue, and other institutions. The then-novel coming together of such a large group served an important purpose in itself; one mechanical engineer from Michigan State University found it "reassuring to see so many other women in the same situation" (*Michigan State News* 1964, 1). Speakers such as University of Chicago professor Alice Rossi called on society to encourage independence, curiosity, and reasoning power in girls, while Radcliffe President Mary Bunting called on employers to provide day care and flexible schedules to help women balance motherhood and work (Bix 2000b).

Corporations of that period did not prove particularly proactive in initiating accommodations for career women. Indeed, through most of the 1960s, company recruitment for engineers ignored the existence of women in the profession. A 1954 advertisement in MIT's engineering magazine for the Ramo-Wooldridge communications corporation featured a photograph of its senior staff, twenty-three "key men" clad almost identically in dark suits and white shirts. A 1968 ad centered on a photograph of a serious-looking middle-aged man in white shirt, dark tie, and thick glasses, under the headline, "This is the image of a Kodak mechanical engineer." Such advertisements reinforced the attribution of mechanical interests and aptitude to men, while relegating women to the sidelines. In one 1958 ad for the Avionics Division of ITT, a cartoon showed a little boy building a complicated machine, while a girl carrying a doll looked up admiringly and asked him, "Have you always been a genius?"

In attempting to catch the attention of engineering majors, companies pandered to the masculine identity and image of their profession. The University of Michigan's engineering magazine ran a 1966 advertisement for Douglas Aircraft that showed a drawing of a young man floating happily along a beach in an inner tube, while two young women in skimpy swimsuits gazed at him adoringly. Such appeals often crossed the line into sheer sexism. Another Douglas ad, under the headline "Intrigued by Exotic Designs?" featured a cartoon of a man sitting at a drawing board, carefully laying out an illustration of a naked woman outstretched like an airplane in flight, with measurements accenting the curves of her breasts, legs, and behind. One advertisement for Chance Vought Aircraft used a bizarre illustration of a young man having an intimate meal with a female robot, complete with flirtatious eyes and pointed breasts. The ad copy read, "Do mechanical brains intrigue you? Do those intellectual vamps arouse your engineering instincts? Then why go on ogling? Especially if you're an electrical or mechanical major! Plan to enjoy the company of the best mechanical computers. Create your own electronic brains for missile guidance."

Within such a recruiting climate, the equation of "engineer" with "male" amounted to a real obstacle for female professionals. Despite the fact that the 1960s were characterized by a national shortage of qualified engineers, employers often refused even to consider women candidates. According to *Industrial Relations News* in 1961, interviews with personnel administrators and corporate managers revealed that "81 percent wouldn't hire female engineers, and most of the remaining respondents would be dead set against permitting them to reach middle-management levels" (*SWE Newsletter* 1961, 1).

The fact that employers expected to hire male engineers did not escape female job candidates. Gerda Kohlheb, a chemical engineering student at City University of New York in the mid-1960s, commented, "Most women who do go through an engineering curriculum find a great deal of opposition during their college career. . . . [M]ale students tend to regard her as somewhat of an intruder. Upon entering a classroom, the looks that sometimes greet her are not unlike the ones she would get if she walked into a men's room" (1967, 13). Hoping to avoid such situations, Kohlheb went abroad to seek a summer job in the chemical industry, saying, "Most American companies were hesitant to accept a woman to work in the field, and I hoped I would be accepted as a woman more readily in Europe." Kohlheb, "as extra insurance . . . decided to commit the sin of omission"; on her job application, she used only a first initial, rather than her first name (1967, 13).

Given the persistence of employers' doubts about women engineers, many activists of the late 1960s and early 1970s designed plans to help female students break through the barriers. Working in MIT's engineering school, professors Mildred Dresselhaus and Sheila Widnall inaugurated a freshman seminar aimed at acclimatizing coeds to engineering. To make women comfortable with manual skills that boys traditionally picked up from hobbies or from fathers, the syllabus included lab projects in electronics, welding, and model building. Dresselhaus further helped organize meetings titled "Let's Talk about Your Career," where female students consulted faculty, staff, and guests for advice on graduate school, employment, and the perpetual issue of blending marriage with work. Arguing that male students' familiarity with the business world gave them a competitive advantage, MIT alumnae started an annual seminar, "Getting the Job You Want in Industry: A Woman's Guerrilla Guide to the Pin-Striped World." By advising coeds on résumé writing and interview techniques, alumnae hoped to level the playing field (Bix 2000b).

Similar efforts at promoting women's advancement took place at schools such as Purdue University, where an intensive recruiting and retention campaign had raised the enrollment of female engineering students from 46 in 1968, to 280 in 1974, to more than 1,000 in 1979, the largest such class in the nation. Purdue also had one of the country's most active student SWE chapters, which published its own newsletter and ran a "big sister" program pairing entering women with upperclass mentors. SWE offered help in locating summer jobs and produced an annual "résumé book," showcasing members' credentials, which it sold to potential employers (Bix 2000a).

Meanwhile, important changes were under way. The 1964 federal Civil Rights Act had included language barring employment discrimination on the basis of sex and created the Equal Employment Opportunity Commission. When enforcement lagged, the National Organization for Women mobilized to pressure the agency to secure women's workplace rights. Those legal and political developments had a direct impact on the official culture of engineering, transforming the tone of recruiting. Where formerly corporations had casually deployed sexist imagery and equated the concepts of "engineer" and "male," leading companies of the 1970s carefully stipulated that they were equal opportunity employers (Woloch 1999).

In many such advertisements, corporations highlighted the stories of women engineers already in their employ as a means of attracting the attention of female students entering the job market. This tactic also fostered favorable public relations for the firm, bolstering an impression of compliance with equal opportunity rules. Kodak, which just a few years before had defined its engineering image as the stereotypical nerd, ran a 1973 advertisement featuring a group photo of fifteen women engineers, each identified by degrees earned, specialty, and responsibilities. A University of Minnesota mechanical engineer "designs and troubleshoots hydraulic systems, bearings, and shaft seals. She is a specialist on friction, wear, and lubrication." An electrical engineer from the South Dakota School of Mines handled "machines [that] are three stories high, a football field long, and work to the tolerances of an expensive watch in depositing emulsion layers on color film." Tongue in cheek, the ad told readers, "This picture could be misleading. Engineering jobs at Kodak are not restricted to ladies."

One of the most significant questions, of course, was whether female engineers could enter training for management and receive promotion opportunities alongside their male colleagues. General Electric promised a readiness to bring women into the higher ranks of business through programs of rotating assignments and field experience. A 1974 ad showed a drawing of a young woman smiling broadly and leaning back confidently in her executive chair, while the headline read, "We're Looking for Engineers Who Were Born to Lead."

In playing up to women's aspirations, many companies deliberately integrated feminist imagery and language into their recruitment of female engineers. In 1975, one power systems corporation offered the slogan "Women Engineers: You've Come A Long Way, But You'll Be Surprised How Much Farther You Can Go With Gibbs & Hill." Another 1975 ad, for semiconductor manufacturer INMOS, used a drawing of a young boy in a baseball uniform, saying, " 'When I grow up, I want to be an engineer, like my Mom.' . . . A new wish for a different time." Not surprisingly, such feminist sentiments were especially prominent in the advertising that companies increasingly placed in the *SWE Newsletter* and SWE conference programs of the 1970s, appealing directly to the growing constituency of female professionals. Under the headline "We make products other than appliances—and hire people other than men," GE specifically invited female "engineering and manufacturing professionals . . . to work with us in creating, manufacturing, and marketing . . . advanced, high technology jet aircraft engines. . . . You'll have the same opportunities for professional advancement as your male counterparts, the same pay and the same status" (GE 1978).

Some young female engineering students just completing their studies in the early 1970s were impressed by this dramatic rhetorical shift, ready to believe that a cultural revolution in attitudes toward women engineers was under way. A sophomore majoring in chemical engineering at Penn State University declared in 1971 that "probably with the Women's Lib movement, previous discrimination will be lessened or non-existent by the time I apply for an engineering job" (*Centre Daily Times* 1971, 7). But other engineering graduates and older women were more skeptical about the likelihood of fundamental change in the workplace climate. Moreover, SWE cautioned that with a perception that companies were targeting female job candidates, "unfortunately, we're seeing a little backlash (. . . on the part of male students) which we don't like very much" (*SWE Newsletter* 1979, 1). As SWE observers were well aware, federal legislation and new recruiting language would not instantaneously convert engineering into a feminist paradise. Underneath equal opportunity rhetoric, many female engineers ended up underemployed, working in positions that failed to use their full capacity, and frustrated by deeply ingrained discrimination (Durkin 1975; Gitschier 1973; Gluch 1977; Mathis and Mathis 1972). Women engineers frequently complained that male colleagues and supervisors initially tended to doubt their ability. Thus a female engineer was forced to defend her right to employment and demonstrate her skill through hard work, while "her male counterpart . . . is basically accepted as able to do the work successfully, unless he himself proves otherwise" (Bugliarello 1971, 8).

Activists at MIT warned that the environment on campus remained unfriendly toward women. An ad hoc committee, co-chaired by Dresselhaus and engineering major Paula Stone, drew on fundamental feminist principles to declare that "a discriminatory attitude against women is so institutionalized in American universities as to be out of the awareness of many of those contributing to it" (MIT Ad Hoc Committee 1972, 3). The report noted that women at MIT faced both open opposition and silent prejudice, concluding,

> If many people (professors, staff, male students) . . . persist in feeling that women jeopardize the quality of MIT's education, that women do not belong in traditionally male engineering and management fields, that women cannot be expected to make serious commitments to scientific pursuits, that women lack academic motivation, that women can only serve as distractions in a classroom, . . . then MIT will never . . . be a coed institution with equal opportunities for all. (MIT Ad Hoc Committee 1972, 3)

The document represented a self-directed rallying cry, telling MIT women that gender discrimination would change only when female students, faculty, and staff organized to demand improvement. The early 1970s brought a burst of activism, as MIT women drew strength from the national feminist movement to assert their presence physically, intellectually, socially, and politically. Advocates carefully listed all the awards coeds received in order to document that women could indeed be good engineers, and to show remaining doubters that women could lead and succeed in the most difficult technical studies. MIT's Admissions Office had revised photographs and text in the catalog to highlight coeds and sent special recruiting material to all female National Merit and National Achievement semifinalists. AWS feared, however, that it would require more "high-powered" efforts to increase female enrollment, to overcome social forces pushing girls away from science and engineering, and "to demythify incorrect assumptions about women at MIT" (MIT Association of Women Students 1977). AWS produced its own pamphlets encouraging high school girls to apply and urged members to contact hometown seniors over Thanksgiving and Christmas vacation. "The women in particular may just need an encouraging word from you before taking the plunge" (MIT Association of Women Students 1977). MIT coeds also volunteered to sit in the Admissions Office during the peak interview period, ready to chat with interested girls (Bix 2000b).

To help MIT women maintain a positive sense of identity within a male-dominated atmosphere, campus women's groups initiated monthly colloquia addressing wide-ranging feminist subjects such as the nature of androgyny, sexism in popular culture, and the strengths and difficulties of two-career marriages. Dresselhaus and Wick created a new organization, the Women's Forum, which brought together undergraduates, graduate students, faculty, staff, and wives to develop "consciousness-raising skits" and express concerns about women's health, athletic opportunities, day care, and career planning. Women's Dean Jacquelyn Mattfeld and her successor, Professor Emily Wick, served as administrative advocates for MIT's coeds, ready "to assist women students as they make their way through this very male institution" (Wick 1971). Mattfeld and Wick stepped in to mediate when coeds encountered trouble dealing with advisers, professors, or teaching assistants. Similarly, many of MIT's few women faculty considered it their responsibility, as successful professionals, to lobby on behalf of other women on campus. Widnall complained, "Engineers may have a view of engineering which is twenty years out-of-date, and they communicate that to other people. Engineers have an image of engineering that is very masculine . . . [and] takes a long time to change" (1976, 12). She described women's activism as a "very exciting" force which could open opportunities for new generations of girls. "There's obviously a direct connection between militant feminism in the junior highs and the ultimate enrollment of women in engineering. . . . Everybody, mothers in particular, . . . are much more aware of the importance of encouraging their daughters to take life seriously" (13).

In 1973, MIT convened another workshop on women in science and engineering. Embracing feminist language, President Jerome Wiesner spoke about a need "to encourage women's participation in every aspect of our technological society. This is another front in the almost universal battle for equality of opportunity" (Ruina 1973, 3). Commemorating the hundredth anniversary of MIT's awarding degrees to women, advocates considered 1973 an occasion for celebration. Though they still saw much room for improvement, female enrollment had more than tripled over just two decades. By the late 1970s, women made up 17 percent of MIT undergraduates and 12 percent of engineering majors. The sheer increase in population mattered; as women became more of a presence on campus, activists gained a critical mass for organization and for visibility. Female graduate students formed their own society, as did women at MIT's Lincoln Laboratory research center. Such groups kept women's issues on the front burner, providing an identity and a cause for many (especially valuable to female faculty and graduate students based in departments with few other women) (Bix 2000b).

MIT activists engaged the national women's movement, thinking about how their immediate interest in supporting women's engineering and science education connected with broader feminist issues. In 1974, the Women's Forum invited Gloria Steinem to speak, and her speech to a packed auditorium defined feminism as the struggle for political, social, and educational autonomy for women (Brandeau 1975, 1). Activists found food for thought in Steinem's message, given that female students still frequently felt like second-class citizens. In subsequent years, MIT women would gain courage to start complaining about sexual harassment, about obscene mail sent over computers, and about male colleagues who refused to take women seriously, made them feel invisible, and undermined their self-confidence. The women claimed the right to demand change.

On a broader scale, this same force of energized activism found expression nationwide by the early 1970s, as female engineers followed specific strategies to promote women's place in the field. Established professionals offered support for juniors; for instance, the Los Angeles section of SWE provided speakers and counselors to student sections at USC, UCLA, Loyola Marymount, Harvey Mudd, Cal State Long Beach, Pomona, Fullerton, and Cal Poly San Luis Obispo. Such campus SWE groups provided vital intellectual, social, and psychological support for incoming female engineering majors. Karen Lafferty Instedt, an Ohio State student from 1968 to 1971, later wrote that SWE gave her "an opportunity to meet the other female engineers who, like me, were isolated in their respective fields and classrooms. The SWE section functioned as a refuge of sorts—where one could find an understanding ear from a peer or a kindhearted, encouraging professor or dean" (1978, 1). By the end of the 1970s, student sections had been chartered in more than 170 colleges, universities, and technical institutes. SWE held an annual national student conference featuring technical sessions and exhibits, professional workshops, industrial tours, even sessions on career planning, power dynamics, management, personal assertiveness training, and how to "dress for success." By decade's end, SWE's membership totaled more than ten thousand women and men. As SWE grew, its leaders were able to mobilize outside support, collecting money to help finance college education for young women pursuing technical studies. SWE administered annual scholarship competitions for female engineering majors, funded by RCA, Westinghouse, other major companies, and women engineers themselves. By publicizing such awards and showcasing the winners, SWE sent a message that women had won a permanent place in American engineering, and were fully deserving of social and financial support (Bix 2000a).

Activists of the 1970s organized dozens of conferences, open houses, and other public events in many states to celebrate and advance women's achievements in engineering. Some meetings were organized to bring female students together with each other and with older mentors. For instance, the University of Washington (with almost 450 women engineering students in 1977) hosted an annual conference where those coeds met with working professionals such as Rockwell ceramics engineer Bonnie Dunbar. The SWE section at the University of North Dakota brought in corporate representatives to talk about how to project a professional image, how to have a successful interview, and how to balance work and marriage (Bix 2000b).

Other conferences were organized by college engineering women themselves for younger girls as a way of encouraging them to pursue technical interests. SWE's philosophy generally assumed that girls and boys possessed essentially similar abilities to excel in math and science. They blamed girls' relative lack of interest in engineering on socialization that handed dolls to girls and toy tools to boys, that put girls in home economics and boys in shop class. SWE further attributed girls' underrepresentation in engineering to failures of the school system, finding fault with guidance counselors who failed to take girls' ambitions seriously or let them drop math and science. To counter such problems, a 1973 University of Illinois conference, titled "Women in Engineering: It's Your Turn Now," gave high school junior and senior girls a chance to participate in "rap sessions," informal conversations with college SWE members. A 1974 symposium sponsored by SWE sections at the Universities of Florida and South Florida featured a tour of the Kennedy Space Center, plus discussions of student financial aid, co-op programs, and career openings. Promotional material read,

> As an engineering student you'll gain something most women don't get in college, a professional skill which can be used immediately upon graduation . . . [with] the highest starting salary bracket of the major professional job categories for women holding a bachelor's degree. . . . You owe it to yourself to look into the possibilities and opportunities offered by engineering. (*SWE Newsletter* 1979, 1)

Other SWE chapters went directly into the high schools as self-described "missionaries," seeking to spread the message that women could be good engineers. Starting in 1976, Berkeley's SWE section sent teams to visit local junior high classes. Presenters described how they became interested in engineering and sought "to dispel myths about women in engineering" (SWE Berkeley 1980). Other SWE activists sought to influence even younger girls, to encourage curiosity and technical enthusiasm in elementary-age children. Boston's SWE published a coloring book titled *Terry's Trip*, the story of a girl visiting her aunt, a mechanical engineer. After talking to engineers, male and female, who worked at her aunt's toy factory, Terry concludes, "Maybe some day I'll be an engineer like Aunt Jennifer." In a similar project, North Carolina SWE produced a booklet titled *Betsy and Robbie*, the story of a girl who visits a university engineering fair and becomes fascinated with the robot designed by a female student. Such material emphasized that women were fully qualified for engineering, a discipline requiring creativity and logic more than physical strength. Photos documented the daily activities of women engineers working in industry, government, and academia, providing role models to win young women's interest and create respect for female engineers as good engineers (Sloan 1979).

Amidst this climate of activism, it is worth emphasizing that while many female engineers, such as Widnall, embraced the philosophy of feminism, others actively rejected the label (Mack 2001). Engineering tends to be a conservative field, and many women shied away from anything that might be judged as too radical. They worried about popular perceptions of "women's lib" and feared that being active in SWE would get them branded as "troublemakers." Nevertheless, such women benefited from the efforts of activists who did identify with the feminist movement and undertook conscious, passionate campaigns to break down institutional barriers.

At the end of the twentieth century, however, women were still nowhere close to proportional representation in the profession. In 1979, women made up

12.1 percent of undergraduates enrolled in engineering across the United States; by 1998, that percentage had gradually risen to 19.7 percent. In 1996, 11,316 women earned bachelor's degrees in engineering, 17.9 percent of the nationwide total. By occupation, women constituted 9 percent of all engineers in 1998 (NSF 2000).

Evidence confirms suspicions that while equal opportunity sentiment and federal legislation might have helped open some doors for some women in the 1970s, the changing language and imagery of employment ads to denote hiring diversity did not solve more fundamental problems. In 1993, SWE reported that "in some respects, women in engineering begin careers at parity with men or better, but as one looks at more experienced people, this picture changes" (SWE 1993b). A survey of more than 1,700 professionals showed that although the average annual base salary of female engineers under age thirty was actually higher than men's by a couple of thousand dollars, average male salaries surpassed women's in the thirty to thirty-nine age group and kept rising at a faster pace. The pay differential among senior professionals favored male engineers by almost $15,000. SWE analysts concluded, "Women engineers appear to fare poorly. . . . [C]omparison of salaries suggests that nothing is being done about long-recognized inequities" (SWE Report on Employment). In a finding clearly related to pay, SWE research documented that at every age level, a greater percentage of male than female engineers had moved into management (SWE 1993a). Data from the National Science Foundation (NSF) suggested that for female engineers less than ten years after graduation, median annual salaries were $2,000 to $3,000 below those of men in the same cohort. In older generations, the salary gap widened to $5,000 or more; for example, among engineers twenty to twenty-four years out of school, men earned a median salary of $68,600 versus women's $60,000 (NSF 2000).

Despite the optimistic predictions of female undergraduates in the 1970s that gender bias in the engineering workplace would soon disappear, the issue remained very much alive at the start of the twenty-first century. The SWE survey reported a distinct gender gap in perceptions of job discrimination. More than half of all male engineers surveyed, 55 percent, said that men and women were always treated equally in engineering, while less than one-third of women, 26 percent, agreed. Fifteen percent of female engineers reported that they saw consistent inequities, a statement supported by just 5 percent of men. Another 2 percent of men surveyed (and zero women) indicated that they saw "reverse discrimination" in current conditions (SWE 1993a).

Ironically, it would be in engineering education itself, where advocates had pushed so hard for undergraduate women to gain access, that employment disparities remained particularly stubborn. Observers had long noted the relative absence of women in engineering departments, especially at levels above assistant professor. The National Science Foundation 2000 report confirmed such perceptions and concluded that although part of the gender gap in rank and tenure might be attributable to the relative scarcity of women earning engineering doctorates, that factor alone could not account for the difference. On occasions when a woman was promoted to dean or even department chair in engineering, the event still appeared sufficiently unusual in 2003 to make headline news (Jerousek 2003).

The reasons for female underrepresentation in academic engineering posts are numerous and complex, yet one persistent factor remains the discipline's chilly climate. In 1994, three tenured women at MIT began comparing notes on disturbing experiences. Though the professors worried about "putting a life-time of hard work and good behavior at risk . . . [and] feared being seen as radical trouble makers" (Goldberg 1999, A1), they proceeded to collect data (an approach suiting MIT's scientific mind-set). As MIT's official self-examination ultimately acknowledged, female faculty had received smaller financial resources, less work space, and fewer rewards than male colleagues. Engineering professor Sallie Chisholm described the phenomenon as "microdiscrimination": small but cumulative assaults on women's careers, "unrecognized assumptions and attitudes that work systematically against women faculty" (Koerner 1999, 56). The report declared,

> Marginalization increases as women progress through their careers at MIT . . . [and] this pattern repeats itself in successive generations. . . . Each generation of young women, including those who are currently senior faculty, began by believing that gender discrimination was "solved" in the previous generation and would not touch them. Gradually, however, their eyes were opened to the realization that the playing field is not level after all. (MIT 1999)

In working to document systematic inequities and demand change, the MIT women of the 1990s were following in the footsteps of others who had organized to improve conditions for women trying to enter engineering. In the twenty-first century, it will take ongoing attention to neutralize the more subtle and hence more stubborn problems of "microdiscrimination." In that fashion, the debate goes on concerning women's place in engineering, one of the most traditionally male professions.

References

Alvort, Margaret. 1956. Letter to L. F. Hamilton, June 21. AC220, box 2, folder 2, MIT Archives.

Arnold, Robert D. 1961. Letter to Fuller E. Callaway, Jr., June 27. Box 9, folder "coed," 85-11-01, Georgia Tech Archives.

Atlanta Journal. 1952. "Regents Vote Tech as Coed." April 9: 1.

Baker, Everett. 1947. Memo, January 26. AC4, box 26, folder 12, MIT Archives.

Betsy and Robbie. n.d. [ca. 1983]. Box 119, file "Betsy and Robbie," SWE collection, Wayne State University Archives.

Bix, Amy Sue. 2000a. " 'Engineeresses' Invade Campus: Four Decades of Debate over Technical Coeducation." *IEEE Technology and Society Magazine* 19(1): 20–26.

———. 2000b. "Feminism Where Men Predominate: The History of Women's Science and Engineering Education at MIT." *Women's Studies Quarterly* 28(1–2): 24–45.

———. 2002. "Equipped for Life: Gendered Technical Training and Consumerism in Home Economics, 1920–1980." *Technology and Culture* 43(4): 728–54.

Brandeau, Margaret. 1975. "Steinem: Castes Trap Women." *The Tech* (MIT), January 15: 1.

Bugliarello, George. 1971. "Women, Technology, and Society." In *Women in Engineering: Bridging the Gap between Society and Technology,* ed. George Bugliarello, Vivian Cardwell, Olive Salembier, and Winifred White, 1–18. Chicago: University of Illinois at Chicago Circle.

Centre Daily Times. 1971. "More Coeds Preparing to Become Engineers." July 28: 7.

Cornell Daily Sun. 1937. "Three Coeds Invade Engineering Courses and Compete with Men at Cornell University." November 12: 1.

Cornell Engineer. 1946. "Pi Omicron." April: 14.

Dennis, Olive. 1948. "Modernization of Railroad Passenger Facilities." *Cornell Engineer* 14(2): 7–9, 34, 36.

Dresselhaus, Mildred S. 1975a. "A Constructive Approach to the Education of Women Engineers." Box 128, file "Women in Engineering—Beyond Recruitment Conference Proceedings, June 22–25, 1975," SWE collection, Wayne State University Archives.

———. 1975b. "Some Personal Views on Engineering Education for Women." *IEEE Transactions on Education* 18(1): 30–34.

Durkin, Glen C. 1975. "Engineering—A 'Weird' Career for Women?" *Penn State Engineer,* January: 19.

GE (General Electric). 1978. Advertisement in *SWE Newsletter,* May: 5.

Gitschier, Jane. 1973. "Sex: Female, Major: Engineering." *Penn State Engineer,* November: 31.

Gluch, Bonnie. 1977. "Women in Engineering: A Personal Perspective." *Penn State Engineer,* October: 26–27.

Goff, Alice. 1946. *Women Can Be Engineers.* Youngstown, Ohio: n.p.

Goldberg, Carey. 1999. "MIT Acknowledges Bias against Female Professors." *New York Times,* March 23: A1, A16.

Handy, Adelaide. 1940. "Woman Designer of Bridges Has Enhanced Rail Travel." *New York Times,* December 22: 40.

Ingels, Margaret. 1952. "Petticoats and Slide Rules." *Midwest Engineer,* August: 2–16.

Instedt, Karen Lafferty. 1978. "How Should SWE Serve Undergraduates?" *SWE Newsletter,* June–July: 1.

Iowa Engineer. 1946. "New Society Organizes." May: 222.

Jansen, Christina. 1977. Interview by Shirlee Shirkow. MC86, box 9, MIT Archives.

Jerousek, Madelaine. 2003. "Engineering College at ISU Gets First Female Department Head." *Des Moines Register,* May 30: B1.

Keller, Evelyn Fox. 1981. "New Faces in Science and Technology: A Study of Women Students at MIT." August. Folder "women students M3," MIT Archives.

Killian, J. R., Jr. 1956. Letter to J. A. Stratton, October 22. AC220, box 212, MIT Archives.

Koerner, Brendan I. 1999. "The Boys' Club Persists." *U.S. News & World Report,* April 5: 56.

Kohlheb, Gerda. 1967. "Some Thoughts from a 'Lady Engineer.'" *Chemical Engineering,* September 11: 13–15.

LeBold, William K., and Dona J. LeBold. 1998. "Women Engineers: A Historical Per-
   spective." *ASEE Prism,* March: 30–32.
Mack, Pamela. 2001. "What Difference Has Feminism Made to Engineering in the
   Twentieth Century?" In *Feminism in Twentieth-Century Science, Technology,
   and Medicine,* ed. Angela N. H. Creager, Elizabeth Lunbeck, and Londa
   Schiebinger, 149–68. Chicago: University of Chicago Press.
Mathis, Betty Ann, and Harold F. Mathis. 1972. "Women Enrolled in Engineering Cur-
   ricula." File "women," Pennsylvania State University Archives.
Mattfeld, Jacquelyn. 1965. Notes. AC134, box 1, folder "Academic council 6/64–6/65,"
   MIT Archives.
McNatt, Bob. 1952. "Petite Blonde Is First Tech Coed Candidate." *Atlanta Journal and
   Constitution,* April 13: 1.
Michigan State News. 1964. "Female Scientist Image Blasted." November 4: 1.
Miller, Joy. 1964. "Women Engineers: They're Feminine and So Bright." *Perth Amboy
   N.J. News,* July 30: 30.
MIT (Massachusetts Institute of Technology). 1941. *Massachusetts Institute of Tech-
   nology Handbook.*
———. 1999. "A Study on the Status of Women Faculty in Science at MIT." Special is-
   sue of the *MIT Faculty Newsletter* 11(4).
   http://web.mit.edu/fnl/women/women.html. Accessed February 16, 2005.
MIT Ad Hoc Committee. 1972. "Report of MIT Ad Hoc Committee on the Role of
   Women at MIT." MC485, box 13, file "Ad Hoc Comm. on Women's Admis-
   sion," MIT Archives.
MIT Association of Women Students. 1977. Notes. File "AWS," MIT Archives.
NSF (National Science Foundation). 2000. *Women, Minorities, and Persons with Dis-
   abilities in Science and Engineering: 2000.* Arlington, Va.: National Science
   Foundation (NSF 00-327). http://www.nsf.gov/sbe/srs/nsf00327. Accessed
   February 16, 2005.
O'Bannon, Helen. 1975. "The Social Scene: Isolation and Frustration." Box 128, file
   "Women in Engineering—Beyond Recruitment Conference Proceedings, June
   22–25," SWE collection, Wayne State University Archives.
Ogilvie, Marilyn Bailey. 1986. *Women in Science.* Cambridge, Mass.: MIT Press.
Oldenziel, Ruth. 1999. *Making Technology Masculine: Men, Women, and Modern Ma-
   chines in America, 1870–1945.* Amsterdam: Amsterdam University Press.
———. 2000. "Multiple Entry Visas: Gender and Engineering in the U.S., 1879–1945."
   In *Crossing Boundaries, Building Bridges: Comparing the History of Women Engineers, 1870s–1990s,* ed. Annie Canel, Ruth Oldenziel, and Karin Zachmann,
   11–50. Harwood Academic.
Peden, Irene Carswell. 1965. "Women in Engineering Careers." Booklet, SWE collection, Wayne State University Archives.
Purcell, Carroll. 1979. "Toys, Technology, and Sex Roles in America, 1920–1940." In
   Dynamos and Virgins Revisited: Women and Technological Change in History,
   ed. Martha Moore Trescott, 163–82. Metuchen, N.J.: Scarecrow.
Rensselaer Polytechnic. 1943. "Curtiss Wright Women Enter Rensselaer to Begin Ten
   Month Aeronautics Course." February 22: 1.
Rossiter, Margaret W. 1982. *Women Scientists in America: Struggles and Strategies to
   1940.* Baltimore, Md.: Johns Hopkins University Press.
———. 1995. *Women Scientists in America: Before Affirmative Action, 1940–1972.* Baltimore, Md.: Johns Hopkins University Press.

Ruina, Edith. 1973. *Women in Science and Technology: A Report on the Workshop on Women in Science and Technology, held May 21, 22, and 23, 1973, at the Massachusetts Institute of Technology, Cambridge, Massachusetts.* Cambridge, Mass.: MIT Press.

Rutherford, Alta. 1954. "Women Engineers in Redlands Spotlight." *Detroit News,* April 19: 27.

Sloan, Sarah. 1979. "Terry's Trip." *SWE Newsletter,* November–December: 2.

Stiles, Florence. 1946. Letter to Carroll Webber, Jr., March 28. AC220, box 2, folder 2, MIT Archives.

SWE (Society of Women Engineers). 1953. Brochure, ca. 1953. SWE collection, Wayne State University Archives.

———. 1993a. Society of Women Engineers Report, 1993. http://www.swe.org/SWE/ProgDev/stat/stathome.html. Accessed July 10, 2003.

———. 1993b. *A National Study of Women and Men Engineers: A Study of the Members of 22 Engineering Societies.* Executive summary. http://www.swe.org/Publications/NationalSurvey.doc. Accessed April 4, 2005.

SWE Berkeley (SWE Student Section, University of California, Berkeley). 1980. *Junior High School Outreach: A Practical Guide.* Box 118, file "Junior High School Outreach 1980," SWE collection, Wayne State University Archives.

SWE Newsletter. 1961. "81% of Male Bosses Won't Hire Gal Engineers: Remainder Take Dim View of Middle Management Spots." March: 1.

———. 1979. "Entry and Turnover in Employment." March–April: 1.

Terry's Trip. n.d. [ca. 1979]. Box 131, file "Terry's Trip," SWE collection, Wayne State University Archives.

Time. 1963. "Where the Brains Are." October 18: 51.

Trescott, Martha Moore. 1990. "Women in the Intellectual Development of Engineering: A Study in Persistence and Systems Thought." In *Women of Science: Righting the Record,* ed. G. Kass-Simon and Patricia Farnes, 147–87. Bloomington: Indiana University Press.

Van Leer, Blake R. 1948a. Letter to Sandy Beaver, Nov. 8. Box 2, folder "9," 86-01-08, Georgia Tech Archives.

———. 1948b. Letter to Harmon Caldwell, Dec. 28. Box 2, folder "9," 86-01-08, Georgia Tech Archives.

Wajcman, Judy. 1991. *Feminism Confronts Technology.* University Park: Pennsylvania State University Press.

Walker, Eric. 1955. "Women Are NOT for Engineering." *Penn State Engineer,* May: 18.

War Training Programs. 1945. *War Training Programs—World War II: Curtiss-Wright Engineering Cadette Training Program.* Report, Program Series A, Iowa State College, v. A I. April 1.

Wick, Emily L. 1970. "Proposal for a New Policy for Admission of Women Undergraduate Students at MIT." March 9. AC220, box 2, file 2, MIT Archives.

———. 1971. Letter to Paul Gray, November 16. MC485, box 13, file "MIT," MIT Archives.

Widnall, Sheila. 1976. Interview by Shirlee Shirkow. MC86, box 8, MIT Archives.

Woloch, Nancy. 1999. *Women and the American Experience.* New York: McGraw Hill.

Part Two.

Institutional and Cultural Barriers for Women in STEM

3 Using POWRE to ADVANCE: Institutional Barriers Identified by Women Scientists and Engineers

Sue V. Rosser

On January 29, 2001, a number of presidents, chancellors, and provosts of the most prestigious research universities (the California Institute of Technology, the Massachusetts Institute of Technology, the University of Michigan, Princeton University, Stanford University, Yale University, the University of California at Berkeley, Harvard University, and the University of Pennsylvania), together with twenty-five women scientists from those institutions, held a special meeting at MIT. At the close of the meeting, they issued a statement saying, "Institutions of higher education have an obligation, both for themselves and for the nation, to fully develop and utilize all the creative talent available. . . . We recognize that barriers still exist [to women]." They agreed

- To analyze the salaries and the proportion of other university resources provided to women faculty
- To work toward a faculty that reflects the diversity of the student body
- To "recognize that this challenge will require significant review of, and potentially significant change in, the procedures within each university, and within the scientific and engineering establishments as a whole." (Campbell 2001)

For the first time, in public and in print, the leaders of the nation's most prestigious universities suggested that institutional barriers have prevented women scientists and engineers from having a level playing field and that science and engineering might need to change to accommodate women.

Almost simultaneously, the National Science Foundation (NSF) initiated ADVANCE, a new awards program, at a funding level of $17 million for 2001, with two categories including institutional, rather than individual, solutions to empower women to participate fully in science and technology. NSF encouraged institutional, rather than individual, solutions because of "increasing recognition that the lack of women's full participation at the senior level of academe is often a systemic consequence of academic culture" (NSF 2001a, 2). Under ADVANCE, Institutional Transformation Awards ranging up to $750,000 per year for up to five years promote the increased participation and advancement of women; Leadership Awards recognize the work of outstanding organizations and individuals, and enable them to initiate, sustain, and intensify new activity (NSF 2001a).

Several decades of federal funding have supported NSF programs at the K–16 level to attract girls and women to science, mathematics, engineering, and technology and retain them in those fields, primarily through the Program for Gender Equity (NSF 2001a). At the graduate level, NSF directed fellowships toward women: initiatives such as Faculty Awards for Women (FAW), Visiting Professorships for Women (VPW), Career Advancement Awards (CAA), and Professional Opportunities for Women in Research and Education (POWRE) support the research of individual women scientists at critical junctures during their careers. Although the proportion of female majors in scientific and technological fields has increased since the 1960s to 49 percent in 1998, the percentage of women in computing, the physical sciences, and engineering remains significantly lower than that in other disciplines (NSF 2003).

The most prestigious and mainstream universities and foundations stated that the parts of institutions and the particular professions remaining most closed to women must become more woman-centered, and admitted that this will require significant institutional changes to empower women scientists and engineers. Creating a woman-friendly university is a daunting task because of the historical, statistical, and cultural traditions that have built the university to fit male needs, developmental stages, and interests. Given this historical legacy, it is not surprising that many women have wondered whether they or the institution are the cause of their failure to fit in and succeed within the university.

For many years, most of the research approaches, including that of the national funding agencies, suggested that the problems faced by women in science resulted from individual choices and decisions. Research on gender differences in time to tenure, publication productivity, and receipt of prestigious awards, coupled with assertiveness training and reentry programs for women, led to the model of woman as deficient. Not surprisingly, many women internalized this model and wondered whether the obstacles to success in their academic scientific careers meant that something was wrong with them as individuals. At best, they lost confidence in themselves and wondered where they had gone wrong. At worst, they dropped out of academia and science.

As a dean at a Research I institution, and as a scholar who has worked for a quarter of a century on theoretical and applied problems of attracting and retaining women in science and engineering, I have heard such doubts and dilemmas expressed in a variety of forms by diverse women scientists and engineers in all types of institutions. Virtually all of the women are united in their love for science and desire to sustain their interest in the physical, natural world that attracted them to the study of science initially. Most would like nothing better than to pursue that love through their research and teaching in academia. But as the women themselves know, and as the statistics about gender and science document, more women than men are lost from science at every level of the pipeline. The women scientists question whether their individual choices, decisions, and will power, or institutional obstacles and barriers, prevent them from fulfilling their research potential and attaining their career goals.

In what follows, I first review the status of women in science and engineering. I subsequently discuss the barriers to full participation in these fields identified by POWRE awardees. Failure to significantly change the percentage of women by applying individual solutions suggests the need for systemic institutional changes to facilitate the careers of individual women scientists and engineers. I discuss possible changes in the last section.

Statistical Profile of Women in Science and Engineering

Currently, 55.9 percent of undergraduates and 54.3 percent of graduate students are women (NSF 2000). In these days of attention to statistics and interest in meeting the needs of the student as consumer, universities have made some accommodations for the statistical majority, who are women students. However, resistance to both curricular and extracurricular changes for equity can be seen, in part, in the struggle of women's studies (Carroll 2001; de Groot and Maynard 1993; Pryse et al. 1999; Zimmerman 2000) and women's sports (Heckman 1997; Sandler 1997) to obtain equity and legitimacy.

In some disciplines and whole colleges within the university, women students remain in the minority. Although the proportion of women majoring in scientific and technological fields has increased since the 1960s to reach 49 percent in 1998 (NSF 2003), the percentage of women in computing, the physical sciences, and engineering remains small. In 1998 women received 74.4 percent of the bachelor's degrees in psychology, 52.5 percent in the social sciences, 52.7 percent in the biological and agricultural sciences, 39 percent in the physical sciences, 37 percent in the geosciences, and 18.6 percent of the degrees in engineering (NSF 2003). The percentage of computer science degrees awarded to women actually dropped from 37 percent in 1984 to 26.9 percent in 1998 (NSF 2003).

The percentage of graduate degrees in these fields earned by women remained lower. While women earned 55.5 percent of the M.S. degrees in all fields, they obtained only 39.3 percent of the degrees in science and engineering fields (NSF 2000). And women earned 40.6 percent of the Ph.D. degrees in all fields but only 32.8 percent of the Ph.D.s in science and engineering (NSF 2000).

Women faculty still represent a statistical minority; data reported in 2000, based upon 1995 figures, by the Commission on Professionals in Science and Technology show that women constituted 34.7 percent of faculty overall, with the vast majority holding positions at the lowest ranks in the less prestigious institutions. Women faculty constituted the following percentages of full-time instructional faculty: 50.4 percent of the instructors, 54.3 percent of the lecturers, 43.6 percent of the assistant professors, 31.8 percent of the associate professors, and 17.8 percent of the full professors at all institutions. In the academic year 1996–97, 71.8 percent of men faculty and 51.6 percent of women faculty had tenure. At four-year institutions, 70.9 percent of men and 46.9 percent of women had tenure, while at universities, 74.6 percent of men and 47.4 percent of women faculty had tenure (CPST 2000, tables 5–10). With their male colleagues continuing as the majority and in the more powerful positions, women faculty have met considerable resistance to requests for policies to stop the tenure clock during child-bearing, on-site day care, and dual-career hiring to make the university more female-friendly.

The small number of women receiving degrees in the sciences and engineering results in an even smaller percentage of women faculty in these fields; for example, only 19.5 percent of science and engineering faculty at four-year colleges and universities are women; 10.4 percent of the full professors, 21.9 percent of the associate professors, and 32.9 percent of the assistant professors in science and engineering at these institutions are women (NSF 2000). Although many have read these statistics as suggesting that women will reach parity with men in these fields as they advance through the ranks, other information indicates that more substantial changes must occur to make the climate more female-friendly to retain senior women in science and engineering.

Perhaps it is not surprising that the male dominance in these fields is reflected not only in men's statistical majority, but also in a continued tradition of male-centered approaches in labs, practices, and departmental cultures. The extent to which these approaches, practices, and cultures present institutional barriers for women scientists and engineers has been underlined by the MIT report released in 1999 and recent anecdotal reports that some women scientists actively choose to avoid research universities because of the hostile climate (Schneider 2000). Data document that women make up 40 percent of tenure-track science faculty in undergraduate institutions (Curry 2001). Although the bulk of science and technology research occurs at institutions formerly classified as Research I, women at most of these institutions face barriers such as decreased lab space, lower salaries, and fewer prestigious opportunities. A dawning recognition that these barriers can best be addressed by institutional, rather than individual, changes is evident in the statement released after the MIT meeting on January 29, 2001, and in the focus of NSF's ADVANCE initiative.

Barriers Identified by POWRE Awardees

To be most effective, institutional changes should address the institutional barriers identified as most problematic by women scientists and engineers. Data from almost four hundred respondents to an e-mail survey of POWRE awardees for the fiscal years 1997, 1998, 1999, and 2000 reveal what these barriers are. Since POWRE was the NSF initiative that ADVANCE replaced in 2001, the quantitative and qualitative data from the entire POWRE awardee cohort are particularly relevant in exposing the barriers that institutions should change to empower and enable women scientists and engineers.

Established in 1997, the POWRE program had two main objectives, intended to address the need to fully develop and use the nation's human resources for science and engineering:

To provide opportunities for further career advancement, professional growth, and increased prominence of women in engineering and in the disciplines of science supported by NSF; and to encourage more women to pursue careers in science and engineering by providing greater visibility for women scientists and engineers in academic institutions and in industry. (NSF 1997, 1)

Women scientists and engineers who were U.S. citizens at any rank in tenured, tenure-track, or non-tenure-track positions at any four-year college, comprehensive school, or research university were eligible to apply to POWRE. Although a few tenured full professors, faculty at four-year institutions, and women not on the tenure track received awards, the vast majority of POWRE awardees were untenured assistant professors in tenure-track positions at research universities.

While it is not known to what extent the issues facing POWRE awardees can be generalized to other women scientists and engineers, POWRE awardees represent a relatively successful group of women who hold positions with high potential at good institutions and have received special awards and peer-reviewed funding from a premier governmental scientific foundation. Issues that POWRE awardees identify as significant barriers for them therefore may be assumed to be equally or more problematic for women scientists and engineers who are relatively less successful or prominent.

Methods of Study

All 598 recipients of new POWRE grants for fiscal years 1997, 1998, 1999, and 2000 were sent a questionnaire via e-mail. The questionnaire included the following two open-ended questions:

1. What are the most significant issues/challenges/opportunities facing women scientists today as they plan their careers?
2. How does the laboratory climate (or its equivalent in your subdiscipline) impact upon the careers of women scientists?

Over those four years, 389 of the 598 awardees responded to the questionnaire: 71.6 percent of the 1997 awardees, 76.6 percent of the 1998 awardees, 65.5 percent of the 1999 awardees, and 63.5 percent of the 2000 awardees. The women receiving POWRE awards in all four years represented all seven directorates of NSF disciplines (biological sciences [BIO], computer and information science and engineering [CISE], education and human resources [EHR], engineering [ENG], geosciences [GEO], mathematical and physical sciences [MPS], and social, behavioral, and economic sciences [SBE]). The success rate (percentage of applicants receiving funding) ranged from 14 percent to 47 percent among the seven directorates, and the overall success rate increased from 20 percent in 1997 to 26 percent in 1998, 27 percent in 1999, and 33 percent in 2000.

Sixty-seven of the 96 POWRE awardees for FY '97, 119 of the 173 awardees for FY '98, 98 of the 159 FY '99 awardees, and 105 of the 170 FY '00 awardees to whom the e-mail survey was sent responded. The non-response rate ranged between 23 percent and 37 percent over the four-year period; some failures to respond were the result of invalid e-mail addresses. In addition to failures to respond, life circumstances prevented acceptance of the award in some cases. For example, in FY 2000, one awardee was killed in an accident, and one responded that personal circumstances surrounding a divorce required her to postpone her acceptance of the award. The sample responding to the e-mail questionnaire in all four years appeared to be representative of the population of awardees with regard to discipline, and the non-respondents did not appear to cluster in a particular discipline. The limited data available from the e-mail responses revealed no other respondent or non-respondent bias (Rosser 2001).

Results

Question 1: What Are the Most Significant Issues/Challenges/ Opportunities Facing Women Scientists Today as They Plan Their Careers?

The details of the procedure used to develop the sixteen basic categories of responses to Question 1 have been previously published for FY '97 awardees (Rosser and Zieseniss 2000); the same codes and categories were applied to the responses from FY '98, '99, and '00 awardees. Although most respondents replied with more than one answer, in some years at least one awardee gave no answer to the question. While the survey data are categorical and therefore not appropriate for means testing, differences in responses across award years and across directorates clearly emerge when response frequencies are examined.

As table 3.1 documents, overwhelming numbers of respondents across all four years found "balancing work with family" (response 1) to be the most significant challenge facing women scientists and engineers. In all four years, large percentages of respondents ranked "time management issues" (response 2), "isolation and lack of camaraderie and mentoring due to small numbers" (response 3), "gaining credibility and respectability from peers" (response 4), and "two-career placements" (response 5) as major challenges. After 1997 "time management" (response 2) appeared to be less of a problem, whereas "affirmative action backlash/discrimination" (response 9) was cited more often by 1998, 1999, and 2000 awardees. FY 2000 awardees reported "low numbers of women" (response 3) and "positive responses" (response 10) more frequently than awardees in previous years. Table 3.2 shows the grouping of the responses to Question 1 into four categories:

CATEGORY A

Pressures women face in balancing career and family. Adding restrictions because of spousal situations (responses 5 and 7) to "balancing work with family responsibilities" (response 1) suggests that category A—pressures women face in balancing career and family—is the most significant barrier identified by women

74 *Sue V. Rosser*

Table 3.1. Total Responses to Question 1

Question 1: What are the most significant issues/challenges/opportunities facing women scientists today as they plan their careers?

	Categories	1997 % of responses	1998 % of responses	1999 % of responses	2000 % of responses
1	Balancing work with family responsibilities (children, elderly relatives, etc.)	62.7 (42/67)	72.3 (86/119)	77.6 (76/98)	71.4 (75/105)
2	Time management/balancing committee responsibilities with research and teaching	22.4 (15/67)	10.1 (12/119)	13.3 (13/98)	13.3 (14/105)
3	Low numbers of women, isolation and lack of camaraderie/mentoring	23.9 (16/67)	18.5 (22/119)	18.4 (18/98)	30.5 (33/105)
4	Gaining credibility/respectability from peers and administrators	22.4 (15/67)	17.6 (21/119)	19.4 (19/98)	21.9 (23/105)
5	"Two career" problem (balance with spouse's career)	23.9 (16/67)	10.9 (13/119)	20.4 (20/98)	20.0 (21/105)
6	Lack of funding/inability to get funding	7.5 (5/67)	4.2 (5/119)	10.2 (10/98)	8.6 (9/105)
7	Job restrictions (location, salaries, etc.)	9.0 (6/67)	9.2 (11/119)	7.1 (7/98)	5.7 (6/105)
8	Networking	6.0 (4/67)	<1 (1/119)	0 (0/98)	4.8 (5/105)
9	Affirmative action backlash/discrimination	6.0 (4/67)	15.1 (18/119)	14.3 (14/98)	12.4 (13/105)
10	Positive: active recruitment of women/more opportunities	6.0 (4/67)	10.1 (12/119)	9.2 (9/98)	14.3 (15/105)
11	Establishing independence	3.0 (2/67)	0 (0/119)	6.1 (6/98)	2.9 (3/105)
12	Negative social images	3.0 (2/67)	3.4 (4/119)	2.0 (2/98)	<1 (1/105)
13	Trouble gaining access to nonacademic positions	1.5 (1/67)	1.7 (2/119)	1.0 (1/98)	1.9 (2/105)
14	Sexual harassment	1.5 (1/67)	<1 (1/119)	2.0 (2/98)	1.9 (2/105)
15	No answer	0 (0/67)	<1 (1/119)	1.0 (1/98)	1.9 (2/105)
16	Cut-throat competition	—	—	1.0 (1/98)	1.9 (2/105)

scientists and engineers regardless of directorate or year of award. The follow-ing quotations exemplify this category:

- At the risk of stereotyping, I think that women generally struggle more with the daily pull of raising a family or caring for elderly parents, and this obviously puts additional demands on their time. This is true for younger women, who may struggle over the timing of having and raising children, particularly in light of a ticking tenure clock, but also for more senior women, who may be called upon to help aging parents (their own or in-laws). Invariably they manage, but not with-out guilt. (2000, respondent 63)
- In contrast to other issues related to women choosing careers in science, the two-body problem has received far too little public as well as governmental attention. Universities are basically tackling the problem individually; some act progres-sively, others don't. The fates of these capable women depend too much on the individual deans or department chairs involved. (1998, respondent 45)
- Managing dual career families (particularly dual academic careers). Often women take the lesser position in such a situation. Ph.D. women are often married to Ph.D. men. Most Ph.D. men are not married to Ph.D. women. (2000, respondent 16)

CATEGORY B

Problems faced by women because of their low numbers and stereotypes held by others regarding gender. A second grouping (responses 3, 4, 8, 10, and 12) appears to result from the low numbers of women scientists and engineers and conse-quent stereotypes surrounding expectations of their performance. Isolation and lack of mentoring, as well as difficulty gaining credibility and respectability from peers and administrators, typify category B:

- Although possibly less now than before, women scientists still comprise a small proportion of professors in tenure-track positions. Thus, there are few "models" to emulate and few to get advice/mentoring from. Although men could also men-tor, there are unique experiences for women that perhaps can only be felt and shared by other women faculty, particularly in other Ph.D.-granting institutions. Some examples of this: a different (i.e., more challenging) treatment by under-graduate and graduate students of women faculty than they would of male fac-ulty; difficulties in dealing with agencies outside of the university who are used to dealing with male professors; difficulties related to managing demands of scholarship and grantsmanship with maternity demands. More women in a department would possibly allow a better environment for new women faculty members to thrive in such a department through advice/mentoring and more awareness of issues facing women faculty members. (2000, respondent 26)
- There remains a disconnect between women faculty and the upper administra-tion of Universities, which is male dominated. The natural tendency to pass on information in casual networks can lead to exclusion of women from the inner circles of information, not necessarily maliciously, but just due to human nature. (2000, respondent 51)
- The biggest challenge that women face in planning a career in science is not be-ing taken seriously. Often women have to go farther, work harder and accomplish more in order to be recognized. (2000, respondent 21)

Table 3.2. Categorization of Question 1 across Year of Award

Question 1: What are the most significant issues/challenges/opportunities facing women scientists today as they plan their careers?

Categories	Response numbers[b]	Means of responses			
		1997	1998	1999	2000
A Pressures women face in balancing career and family	1, 5, 7	31.9%	30.8%	35.0%	32.4%
B[a] Problems faced by women because of their low numbers and stereotypes held by others regarding gender	3, 4, 8, 10, 12	12.3%	10.1%	9.8%	14.5%
C[a] Issues faced by both men and women scientists and engineers in the current environment of tight resources, which may pose particular difficulties for women	2, 6, 16	10.0%	4.8%	8.2%	7.9%
D More overt discrimination and harassment	9, 11, 13, 14	3.0%	4.4%	5.8%	4.8%

[a]The alphabetic designation for categories B and C have been exchanged, compared with earlier papers (Rosser and Zieseniss 2000), to present descending response percentages.

[b]Given the responses from all four years, after receiving faculty comments at various presentations of this research, and after working with the data, we exchanged four questions between categories B and D to better reflect the response groupings. Responses 10 and 12 (considered in category D in Rosser and Zieseniss 2000) were moved to category B, and responses 11 and 13 (included in category B in Rosser and Zieseniss 2000) were placed in category D.

- In my field, [. . .] women are so poorly represented that being female certainly creates more notice for you and your work, particularly when presenting at conferences. This can be beneficial, as recognition of your research by your peers is important for gaining tenure; it can also add to the already large amount of pressure on new faculty. (2000, respondent 70)

CATEGORY C

Issues faced by both men and women scientists and engineers in the current environment of tight resources, which may pose particular difficulties for women. Category C (responses 2, 6, and 16) includes issues faced by both men and women, but which may pose particular difficulties for women, because of either their low numbers or their need to balance career and family. For example, time management and balancing committee responsibilities with research and teaching (response 2) can be difficult for both male and female faculty. However, because of their low numbers in science and engineering, women faculty are often asked to serve on more committees to meet gender diversity needs, even while they are still junior, and to advise more students, either formally or informally (NSF 1997). Cut-throat competition makes it difficult for both men and women to succeed and obtain funding. Gender stereotypes that reinforce women's socialization to be less overtly competitive may make it more difficult for a woman scientist or engineer to succeed in a very competitive environment.

- I have noticed some problems in particular institutions I have visited (or worked at) where women were scarce. As a single woman, I have sometimes been viewed as "available," rather than as a professional co-worker. That can be really, really irritating. I assume that single men working in a location where male workers are scarce can face similar problems. In physics and astronomy, usually the women are more scarce. (1997, respondent 26)
- I still find the strong perception that women should be doing more teaching and service because of the expectation that women are more nurturing. Although research as a priority for women is given a lot of lip service, I've not seen a lot of support for it. (2000, respondent 1)

CATEGORY D

More overt discrimination and harassment. Category D (responses 9, 11, 13, and 14) identifies barriers of overt harassment and discrimination faced by women scientists and engineers. Sometimes even a positive response, such as active recruitment of women or making more opportunities available to them (response 10) leads to backlash and difficulties gaining credibility from peers who assume the woman obtained her position because of affirmative action.

- There are almost no women in my field, no senior women, and open harassment and discrimination are very well accepted and have never been discouraged in any instance I am aware of. (1998, respondent 53)
- I have often buffered the bad behavior of my colleagues—and over the years I have handled a number of sexual harassment or "hostile supervision" cases

where a more senior person (all of them male) was behaving inappropriately toward a lower social status woman (or in rarer cases a gay man). (1999, respondent 59)

- The discrimination they continue to face in the workplace. We seem to be making virtually no gains in terms of rates at which women are granted tenure or promotion to full professor. The older I get, the more depressing these statistics become. Women's research is often marginalized. Women's approaches are not recognized. Men scientists want to judge women by "their" standard (i.e., the white male way of doing things!). Most men have no appreciation for the power and privilege of their whiteness and maleness. (1999, respondent 70)

COMPARISONS OF RESPONSES AMONG WOMEN FROM DIFFERENT
DISCIPLINES AND FIELDS

Table 3.3 shows the responses to Question 1 when the data from all four years are pooled and the responses are categorized by the NSF directorate of the awardee; this categorization assumes that the NSF directorate granting the POWRE award serves as an indicator of the discipline or field of the awardee. (Note that for data interpretation, Education and Human Resources [EHR] is removed, since the numbers are smaller and all awardees come from disciplinary backgrounds included in other NSF directorates.) Perhaps the most striking finding is the overall similarity among the directorates. Balancing work with family responsibilities stands out overwhelmingly as the major issue for women from all directorates, just as it did for awardees for all years.

The top six responses were fairly consistent across all directorates, with few exceptions. For MPS, response 3, "low numbers of women, isolation and lack of camaraderie/mentoring," was lower (11.9%) and response 10, "positive: active recruitment of women/more opportunities," was higher (15.5%) than for other directorates. These frequencies are curious, given that MPS includes physics, where low numbers of women have been a problem. However, MPS also includes chemistry and mathematics, fields where women have increased substantially and where job opportunities are plentiful; this may account for the positive response. Both ENG (13.0%) and GEO (15.8%) also have relatively high frequencies for response 10. Again, this may reflect the positive job opportunities in these fields at the time of the survey, although both of these directorates gave high responses (15.9% and 23.7%, respectively) to 9, "affirmative action/backlash/discrimination." CISE and BIO awardees also gave a more frequent response to 9 (20.0% and 11.8%), although a less frequent response to 10 (8.6% and 3.5%, respectively).

Contrary to expectations, the higher frequency of response 9, "affirmative action/backlash discrimination" and response 10, "positive: active recruitment of women/more opportunities" was not always accompanied by a higher frequency of response 3, "low numbers of women, isolation and lack of camaraderie/ mentoring," within a particular directorate group. This suggests that perceptions of both negative discrimination and positive opportunities may not necessarily be correlated with low numbers in a field. This finding contradicts an

Table 3.3. Responses to Question 1 according to Directorate

Question 1: What are the most significant issues/challenges/opportunities facing women scientists today as they plan their careers?

Categories		SBE % of responses		MPS % of responses	
1	Balancing work with family responsibilities (children, elderly relatives, etc.)	60.3	(38/63)	77.4	(65/84)
2	Time management/balancing committee responsibilities with research and teaching	15.7	(10/63)	13.1	(11/84)
3	Low numbers of women, isolation and lack of camaraderie/mentoring	23.8	(15/63)	11.9	(10/84)
4	Gaining credibility/respectability from peers and administrators	17.5	(11/63)	20.2	(17/84)
5	"Two career" problem (balance with spouse's career)	14.3	(9/63)	28.6	(24/84)
6	Lack of funding/inability to get funding	4.8	(3/63)	7.1	(6/84)
7	Job restrictions (location, salaries, etc.)	3.3	(2/63)	7.1	(6/84)
8	Networking	1.6	(1/63)	1.2	(1/84)
9	Affirmative action backlash/ discrimination	7.9	(5/63)	6.0	(5/84)
10	Positive: active recruitment of women/ more opportunities	7.9	(5/63)	15.5	(13/84)
11	Establishing independence	3.3	(2/63)	4.8	(4/84)
12	Negative social images	1.6	(1/63)	2.4	(2/84)
13	Trouble gaining access to nonacademic positions	1.6	(1/63)	2.4	(2/84)
14	Sexual harassment	3.3	(2/63)	1.2	(1/84)
15	No answer	4.8	(3/63)	0	(0/84)
16	Cut-throat competition	0	(0/63)	0	(0/84)

[a]Because of the low numbers of awardees, the EHR directorate should be carefully interpreted here. Many of the women representing this directorate have other disciplinary training and could be classified in other directorates. For this reason, we have chosen not to interpret the EHR responses.

ENG % of responses		EHR[a] % of responses		CISE % of responses		BIO % of responses		GEO % of responses	
65.2	(45/69)	91.7	(11/12)	60.0	(21/35)	82.4	(70/85)	73.7	(28/38)
11.6	(8/69)	0	(0/12)	17.1	(6/35)	12.9	(11/85)	21.1	(8/38)
21.7	(15/69)	33.3	(4/12)	31.4	(11/35)	20.0	(17/85)	39.5	(15/38)
24.6	(17/69)	25.0	(3/12)	31.4	(11/35)	16.5	(14/85)	13.2	(5/38)
13.0	(9/69)	16.7	(2/12)	22.9	(8/35)	11.8	(10/85)	21.1	(8/38)
8.7	(6/69)	0	(0/12)	5.7	(2/35)	8.2	(7/85)	10.5	(4/38)
5.8	(4/69)	8.3	(1/12)	5.7	(2/35)	11.8	(10/85)	10.5	(4/38)
0	(0/69)	8.3	(1/12)	5.7	(2/35)	2.4	(2/85)	5.3	(2/38)
15.9	(11/69)	8.3	(1/12)	20.0	(7/35)	11.8	(10/85)	23.7	(9/38)
13.0	(9/69)	8.3	(1/12)	8.6	(3/35)	3.5	(3/85)	15.8	(6/38)
1.4	(1/69)	0	(0/12)	2.9	(1/35)	3.5	(3/85)	0	(0/38)
2.9	(2/69)	0	(0/12)	5.7	(2/35)	2.4	(2/85)	0	(0/38)
0	(0/69)	0	(0/12)	5.7	(2/35)	1.2	(1/85)	0	(0/38)
0	(0/69)	0	(0/12)	5.7	(2/35)	0	(0/85)	2.6	(1/38)
1.4	(1/69)	0	(0/12)	0	(0/35)	0	(0/85)	0	(0/38)
0	(0/69)	0	(0/12)	2.9	(1/35)	1.2	(1/85)	2.6	(1/38)

earlier paper in which data from only the 1997 awardees were used to compare engineers with scientists (see Rosser and Zieseniss 2000).

Question 2: How Does the Laboratory Climate Impact upon the Careers of Women Scientists?

Question 2 of the e-mail survey attempted to explore women's perceptions of their work environment. As with Question 1, data from Question 2 are not amenable to standard tests of means for award years and directorates. While statistical differences between years or directorates cannot be inferred, notable trends do emerge when the frequencies of responses are analyzed by award year and directorate. Across all award years, the answer "balancing career and family/time away from home" (the same response as for Question 1) was given by more respondents than any other. As table 3.4 documents, in contrast to Question 1, the responses given to this question reflect less consensus. Awardees from all years, but particularly 1997 awardees, had some difficulty understanding the question. Although many women did not mention problems in either their laboratory or work environment related to gender issues (responses 3, 4, and 9), the largest number of responses did suggest that to some degree their gender led to their being perceived as a problem, an anomaly, or deviant in the laboratory or work environment.

Awardees for 1998 and 1999 ranked "hostile or intimidating environment" (response 7) higher than 1997 and 2000 awardees. Awardees for 1999 ranked the "boys' club atmosphere" (response 6), "lack of numbers/networking" (response 11), and "lack of funding" (response 16) as more problematic than did 1997, 1998, or 2000 awardees. In contrast, 1998 awardees ranked "have not experienced problems" (response 3) and "positive impact" (response 10) higher than either 1997 or 1999 awardees. 2000 awardees ranked "positive impact" (response 10) and "lack of camaraderie/communications and isolation" (response 5) higher than any of the previous three years of awardees. Awardees for 1999 and 2000 also mentioned new issues not articulated by 1997 or 1998 awardees, such as "space" (response 21), "cultural/national stereotypes for women" (response 20), and "department doesn't understand basic issues" (response 19).

Table 3.5 shows the responses to Question 2 when the data from all four years are pooled and categorized by the NSF directorate of the awardee. As with Question 1, the most striking finding is the similarity of responses among the awardees from different directorates. However, some differences in responses emerge, which might be predicted, based upon the disciplines. (Note that for this analysis EHR is removed, since the numbers are small and the awardees come from different disciplinary backgrounds.) For example, large numbers of awardees from SBE and CISE indicate that they are not in a lab atmosphere or cannot answer the question (response 4), or simply give no answer (response 9). Some responses seem peculiar, or even contradictory. For example, awardees from ENG give the highest response rates both to "have not experienced problems" (response 3) and to "hostile environment/intimidating/lack of authority"

Table 3.4. Total Responses to Question 2

Question 2: How does the laboratory climate (or its equivalent in your subdiscipline) impact upon the careers of women scientists?

Categories		1997		1998		1999		2000	
		% of responses		% of responses		% of responses		% of responses	
1	Don't know/question unclear	16.4	(11/67)	4.2	(5/119)	7.1	(7/98)	5.7	(6/105)
2	Balancing career and family/time away from home	13.4	(9/67)	19.3	(23/119)	16.3	(16/98)	13.3	(14/105)
3	Have not experienced problems	11.9	(8/67)	16.8	(20/119)	10.2	(10/98)	9.5	(10/105)
4	Not in lab atmosphere/can't answer	11.9	(8/67)	5.9	(7/119)	1.0	(1/98)	8.6	(9/105)
5	Lack of camaraderie/communications and isolation	9.0	(6/67)	11.8	(14/119)	9.2	(9/98)	14.3	(15/105)
6	"Boys' club" atmosphere	9.0	(6/67)	9.2	(11/119)	18.4	(18/98)	9.5	(10/105)
7	Hostile environment/intimidating/lack of authority	9.0	(6/67)	14.3	(17/119)	15.3	(15/98)	8.6	(9/105)
8	Establishing respectability/credibility	9.0	(6/67)	10.9	(13/119)	10.2	(10/98)	3.8	(4/105)
9	No answer	7.5	(5/67)	6.7	(8/119)	5.1	(5/98)	<1	(1/105)
10	Positive impact	6.0	(4/67)	10.1	(12/119)	6.1	(6/98)	11.4	(12/105)
11	Lack of numbers/networking	4.5	(3/67)	6.7	(8/119)	12.2	(12/98)	4.8	(5/105)
12	General problem with time management	4.5	(3/67)	1.7	(2/119)	5.1	(5/98)	3.8	(4/105)
13	Safety concerns/presence of toxic substances (health concerns)	3.0	(2/67)	0	(0/119)	4.1	(4/98)	1.9	(2/105)
14	Benefit by working with peers	3.0	(2/67)	2.5	(3/119)	3.1	(3/98)	5.7	(6/105)
15	Problem of wanting research independence	3.0	(2/67)	0	(0/119)	1.0	(1/98)	<1	(1/105)
16	Lack of funding	1.5	(1/67)	<1	(1/119)	5.1	(5/98)	<1	(1/105)
17	Benefit from time flexibility/determine own lab hours	3.0	(2/67)	1.7	(2/119)	3.1	(3/98)	1.9	(2/105)
18	Did not answer	0	(0/67)	0	(0/119)	3.1	(3/98)	0	(0/105)
19	Department doesn't understand basic issues	—		—		—		<1	(1/105)
20	Cultural/national stereotypes for women	—		—		—		6.7	(7/105)
21	Space	—		—		1.0	(1/98)	0	(0/105)
22	Better bathroom facilities	—		—		—		<1	(1/105)

Table 3.5. Responses to Question 2 according to Directorate

Question 2: How does the laboratory climate (or its equivalent in your subdiscipline) impact upon the careers of women scientists?

Categories		SBE % of responses		MPS % of responses	
1	Don't know/question unclear	9.5	(6/63)	6.0	(5/84)
2	Balancing career and family/time away from home	11.1	(7/63)	14.3	(12/84)
3	Have not experienced problems	14.3	(9/63)	7.1	(6/84)
4	Not in lab atmosphere/can't answer	19.0	(12/63)	4.8	(4/84)
5	Lack of camaraderie/ communications and isolation	4.8	(3/63)	20.2	(17/84)
6	"Boys' club" atmosphere	7.9	(5/63)	14.3	(12/84)
7	Hostile environment/intimidating/ lack of authority	4.8	(3/63)	16.7	(14/84)
8	Establishing respectability/ credibility	9.5	(6/63)	9.5	(8/84)
9	No answer	11.1	(7/63)	3.6	(3/84)
10	Positive impact	3.2	(2/63)	6.0	(5/84)
11	Lack of numbers/networking	7.9	(5/63)	11.9	(10/84)
12	General problem with time management	1.6	(1/63)	3.6	(3/84)
13	Safety concerns/presence of toxic substances (health concerns)	1.6	(1/63)	3.6	(3/84)
14	Benefit by working with peers	4.8	(3/63)	1.2	(1/84)
15	Problem of wanting research independence	1.6	(1/63)	0	(0/84)
16	Lack of funding	0	(0/63)	1.2	(1/84)
17	Benefit from time flexibility/ determine own lab hours	3.2	(2/63)	1.2	(1/84)
18	Did not answer	0	(0/63)	2.4	(2/84)
19	Department doesn't understand basic issues	0	(0/63)	1.2	(1/84)
20	Cultural/national stereotypes for women	1.6	(1/63)	2.4	(2/84)
21	Space	1.6	(1/63)	0	(0/84)
22	Better bathroom facilities	0	(0/63)	1.2	(1/84)

[a]Because of the low numbers of awardees, the EHR directorate should be carefully interpreted here. Many of the women representing this directorate have other disciplinary training and could be classified in other directorates. For this reason, we have chosen not to interpret the EHR responses.

ENG % of responses		EHR[a] % of responses		CISE % of responses		BIO % of responses		GEO % of responses	
10.1	(7/69)	8.3	(1/12)	5.7	(2/35)	7.1	(6/85)	5.3	(2/38)
10.1	(7/69)	16.7	(2/12)	8.6	(3/35)	28.2	(24/85)	18.4	(7/38)
18.8	(13/69)	8.3	(1/12)	11.4	(4/35)	8.2	(7/85)	18.4	(7/38)
1.4	(1/69)	8.3	(1/12)	17.1	(6/35)	0	(0/85)	2.6	(1/38)
11.6	(8/69)	33.3	(4/12)	5.7	(2/35)	9.4	(8/85)	5.3	(2/38)
10.1	(7/69)	25.0	(3/12)	8.6	(3/35)	12.9	(11/85)	10.5	(4/38)
18.8	(13/69)	0	(0/12)	14.3	(5/35)	11.8	(10/85)	5.3	(2/38)
7.2	(5/69)	8.3	(1/12)	8.6	(3/35)	8.2	(7/85)	5.3	(2/38)
4.3	(3/69)	8.3	(1/12)	11.4	(4/35)	1.2	(1/85)	0	(0/38)
7.2	(5/69)	16.7	(2/12)	2.9	(1/35)	12.9	(11/85)	21.1	(8/38)
5.8	(4/69)	8.3	(1/12)	2.9	(1/35)	7.1	(6/85)	0	(0/38)
2.9	(2/69)	0	(0/12)	5.7	(2/35)	5.9	(5/85)	2.6	(1/38)
2.9	(2/69)	0	(0/12)	0	(0/35)	2.4	(2/85)	0	(0/38)
2.9	(2/69)	0	(0/12)	2.9	(1/35)	4.7	(4/85)	7.9	(3/38)
1.4	(1/69)	0	(0/12)	0	(0/35)	1.2	(1/85)	2.6	(1/38)
0	(0/69)	0	(0/12)	0	(0/35)	3.5	(3/85)	10.5	(4/38)
1.4	(1/69)	0	(0/12)	0	(0/35)	3.5	(3/85)	5.3	(2/38)
1.4	(1/69)	0	(0/12)	0	(0/35)	0	(0/85)	0	(0/38)
0	(0/69)	0	(0/12)	0	(0/35)	0	(0/85)	0	(0/38)
1.4	(1/69)	0	(0/12)	5.7	(2/35)	1.2	(1/85)	0	(0/38)
0	(0/69)	0	(0/12)	0	(0/35)	0	(0/85)	0	(0/38)
0	(0/69)	0	(0/12)	0	(0/35)	0	(0/85)	0	(0/38)

(response 7). MPS awardees give high response rates to "lack of camaraderie/ communications and isolation" (response 5) and to "lack of numbers/networking" (response 11). Although this response to Question 2 is internally consistent, it contradicts the response of MPS awardees to Question 1, where they gave a relatively low response rate to "low numbers of women, isolation and lack of camaraderie/mentoring" (see table 3.3, response 3).

Again, quotations from the responses of the women scientists and engineers explain the context and provide specific illustrations of the problems and difficulties that affect their careers:

- There is little recognition of the contradiction that researchers are expected to spend personal time in the lab doing research, when especially women are expected to spend their personal time for family obligations. (2000, respondent 1)
- The laboratory climate in my field negatively impacts the careers of women scientists. Many of my colleagues are foreign males who do not take females seriously and do not collaborate with them. (2000, respondent 62)
- We do a lot of work with agencies outside of the university that are predominantly dominated by men (police, courts, correctional agencies, legislators). These agencies have also been used to dealing with male professors. Thus, it is difficult for women to establish links and work with these agencies—I am still working on establishing ties with agencies around my area, working closely with other established women and men faculty members. (2000, respondent 26)

In contrast to the quotations above, some women scientists and engineers, as suggested by responses 10, 14, and 17, find the laboratory environment extremely positive and productive.

Many women have life experiences that differ from those of their male colleagues, and these experiences may lead women scientists and engineers to different approaches to, interests in, and questions about their research than those traditionally taken up by men (Keller 1983; Rosser 1990, 1997). As they do in identifying the difficulties, the words of the respondents themselves provide the most convincing evidence for the potential of new ideas and approaches women can contribute to science and engineering:

- The most significant challenge I face is favoring "hacker" experience. In the computer science discipline in which I work, respect is conferred upon those who possess knowledge obtained primarily through countless hours investigating the nuances of hardware and operating systems. To many in my peer group, this is a relaxing hobby and way of life. Though I learn these nuances as I need them for my research outside of my work, I read literature, am deeply interested in social issues and am committed to being involved in my child's life. I see this alternate experience base as an asset to my field. As Rob Pike of C language fame recently said, "Narrowness of experience leads to narrowness of imagination." But for now, the perception is still tilted against me. (1999, respondent 68)
- I've built a project and a lab with a group of female scientists. It was a mere coincidence (or was it?) to form an interdisciplinary research visualization group in applied medicine (e.g., virtual surgical training, teaching anatomy via 3D visualization, at [my university's] medical school. Because our group consists of com-

puter scientists, computational linguists, cognitive psychologists, anatomists, we had to establish communication between these disciplines . . . somehow we managed to develop an amazing climate to collaborate and also attract female graduate students to do research with us. (1998, respondent 50)

• I find the laboratory climate more liberal than, say, the "office climate." I also feel autonomous, powerful and free in this environment (maybe it's because I get to use power tools?). In the laboratory climate, I am able to create and build. I am also able to ask for help and delegate responsibility. Sometimes my colleagues ask me for help. There is a hierarchical structure at the laboratory in which I work, but it is more fluid, roles switch as projects come through. Sometimes I will take the lead and other times I will follow. In terms of my career, working in a laboratory offers a fantastic opportunity to work alone, work with a large group and manage a project, offer support to a colleague, and to build a small community. (1997, respondent 27)

Policy Considerations for ADVANCE

Data from the almost four hundred awardees from FY '97, '98, '99, and '00 who responded to the e-mail questionnaire provide insights into barriers that institutions must seek to remove or at least lower in order to better retain women scientists and engineers and to attract more women to the disciplines. In FY 2001, the National Science Foundation made nine ADVANCE Institutional Transformation Awards (NSF 2001b); an additional eight awards had been given for FY 2002.

The experiences of the POWRE awardees and the findings of this research suggest various policy considerations that might help ADVANCE participants remove the institutional barriers that prevent women from being full participants in science and technology disciplines. After grouping together similar responses across all four years of the study, at least four distinct policy issues and accompanying recommendations emerge.

Balancing Career and Family

The most pressing, immediate concern that institutions must alleviate is the difficulty women face in balancing family and career. Though this issue affects many women (and increasingly men also), it is particularly challenging for women in competitive fields such as science and engineering (Wasserman 2000).The balancing act is required of more than just the women juggling children and their jobs; it also affects women's decision on when (or whether) to have children. For many women, the decision affects the likelihood of their getting tenure (Cook 2001). Family-friendly policies that stop the tenure clock, provide on-site day care, and facilitate dual-career hires should help both men and women faculty (Wenniger 2001; Wilson 2001). AAUP policy is to stop the tenure clock for parental leave (available to both women and men), postponing the time when faculty come up for tenure. The University of California system-wide policy offers faculty an option called "active service–modified duties" that permits a parent, spouse, or partner with substantial responsibilities caring for a young child to request a quarter or semester of active service–modified duties around the time of birth or adoption (Cook 2001). Since balancing the tenure clock with the biological clock challenges women scientists and engineers who want to become biological mothers in ways never faced by men, such policies will benefit women more.

In a similar fashion, since most female scientists and engineers (62%) are married to male scientists or engineers, who are also often in the same field, such women experience more problems with the two-career issue (Williams 2001) than their male colleagues, most of whom are married, but not to women scientists and engineers (Sonnert and Holton 1995). Although "balancing career with family" and "dual-career" relationships appear at first blush to be the result of the individual choices made by women alone or in conjunction with their spouses or partners, the predominance of these responses among awardees from all four years in response to an open-ended question suggests that addressing the problem at the level of the individual is inadequate. Institutional responses are needed to resolve these family-centered issues identified by overwhelming numbers of POWRE awardees each year.

A few institutions have begun to formalize policies to facilitate partner hires (Wilson 2001). The University of Arizona, for example, has taken a proactive stance by negotiating a set of guidelines for partnership hires, as a means to attract and retain couples; these guidelines include provisions such as the provost's or vice president's paying up to one-third of a partner's salary for three years, when funds are available (Riley 2001).

Low Numbers of Women and Stereotyping

Problems resulting from low numbers of women in science and engineering can lead to stereotypes surrounding their performance, isolation, lack of mentoring, and difficulty gaining credibility among their peers and administrators. Such problems become increasingly difficult to address at the institutional level because of the considerable variations among fields. As the numbers of women have increased quite markedly in some disciplines (psychology, sociology) and begun to approach parity in others (life sciences), while remaining relatively small in others (engineering and computer science), it may be important to focus on the different challenges women face in different disciplines. Small numbers make women very visible; visibility draws attention to successful performance, but it also spotlights errors. The variance in numbers from field to field suggests that institutions may need to establish different priorities and policies for women in different disciplines in sciences and engineering. For example, a one-size-fits-all policy may not work as well for women in engineering as it does for their counterparts in biology. The MIT Report and the statement issued at the end of the January 29, 2001, meeting have prompted not just the nine institutions represented at the meeting but several others as well, such as the University of Arizona and Georgia Tech (Riley 2001), to undertake studies of salaries, space, and other resources provided to women scientists and engineers on their campuses.

Continuing low numbers offer particular challenges and some opportunities. Because the unwritten rules of academia often go unlearned by women in academe until professional disaster strikes (Aisenberg and Harrington 1988), increasing the number of women in science and engineering becomes even more critical to ensuring that such rules are learned. Low numbers mean that these women are often the first woman, or one of only a few, in their department or college. They may have no senior women colleagues to act as role models and to serve as mentors, to provide them access to networks of necessary professional information. These low numbers also lead to their being asked to serve on more committees (even at the junior level) and to advise more students. Although these service activities provide opportunities for women to be visible and to experience leadership and administration at an early stage in their career, they may not be valued by the institution for promotion and tenure and may make time management more difficult. Thus, it is crucial for the Research I institutions to ensure either that junior women faculty are not given extra teaching and service responsibilities, or that the tenure and promotion committees recognize and validate such work to compensate for lost research time and focus.

Overt Discrimination and Harassment

The low numbers that result in active recruitment of women into many areas may have both positive and negative consequences. Demand may give women engineers starting salaries that are equal to or higher than those of their male counterparts (Vetter 1996). On the other hand, the negative perception of affirmative action policies and active recruitment of women can lead to various forms of backlash, ranging from overt discrimination to difficulties gaining credibility with peers and administrators who assume a woman was given the position to fill a quota.

The situations that women encounter of overt and subtler harassment must be dealt with at the institutional level. Institutions and professional societies need to establish policies against sexual harassment and gender discrimination, including policies protecting pregnant faculty in hiring, promotion, and tenure, if such policies do not exist (Elliott 2001). Not only may flexibility and acceptance of differences between men and women be crucial for maintaining and increasing the numbers of women and promoting the careers of individual women in sciences and engineering, but such tolerance may also serve as the key for new approaches to collaboration and creative generativity.

Institutional policies against sexual harassment and gender discrimination must be implemented and enforced. Senior administrators play critical roles in allocating human, financial, physical, and time rewards for those who enforce such policies. For example, it is not appropriate to remove a documented harasser from a problem situation by giving him an outstanding research award or providing him with a research sabbatical. Upon rare occasions, when the senior administrator is the harasser, the institution must be particularly responsible to insure that action is taken. Indiana University South Bend demoted Daniel Cohen from his position as chancellor after he lost a sexual harassment suit. When the faculty voted not to censure him, current Chancellor Kenneth Perrin banned Cohen from campus (Wenniger 2001). In many fields, sexual harassment and gender discrimination workshops should include substantial focus on cultural and national differences in gender roles and on appropriate professional behavior in U.S. universities, including collaboration with women colleagues.

Decreased Funding Issues

The recent trend toward tightening the federal budget for research and the resulting competitive environment affects both men and women scientists and engineers. However, women may face a disproportionate disadvantage in this area due to issues related to their low numbers and family-balancing act. Women also tend to work in teams more than men do. Though collaborative research is becoming more common, it is critical to establish oneself as an independent researcher in order to secure grants and funding; thus women may actually be less successful when they tend to collaborate. Women are also socialized to be less overtly competitive—a trait often associated with success—which may make it more difficult for them to succeed in a highly competitive environment. Thus, the lack of social and professional connections available to most women in academic science and engineering departments, overt and covert gender bias, and differences in socialization create special and unique problems for women (Fox 2001). To enhance funding opportunities, academic departments can develop grant-writing seminars for new and even existing faculty or encourage faculty to attend existing seminars offered through campus offices of sponsored research. Although all faculty should be encouraged to collaborate where research topics make it appropriate, institutions must also foster women's independent research. Retention of the Fellows Awards category within ADVANCE continues the opportunity provided by POWRE for women to receive support for their independent research initiatives after their careers have been interrupted (NSF 2001b).

Recognition of these policy issues is only a first step in overcoming the institutional barriers that keep women from fully participating in science and technology. The POWRE data provide important information for policymakers at the institutional level to use in identifying and implementing appropriate interventions; they suggest that the talent of women scientists and engineers is fostered and unleashed by institutional policies and guidelines such as those offered above. A tremendous love for science and technology and extreme dedication to their research and profession strongly characterize the responses of the overwhelming majority of POWRE awardees in all four years. Most seek to have the barriers removed so that they can be productive researchers who take creative approaches to the physical, natural world. The ADVANCE projects provide institutions with incentives to facilitate the removal of these barriers so that women scientists and engineers can demonstrate their power.

Note

This chapter re-presents a keynote address given at the NSF-funded "Conference on Retaining Women in Early Academic STEM Careers." It combines parts of previously published papers: Sue V. Rosser and Eliesh O'Neil Lane, "Key Barriers for Academic Institutions Seeking to Retain Women Scientists and Engineers," *Journal of Women and Minorities in Science and Engineering* 8(2): 2002; Sue V. Rosser and Eliesh O'Neil Lane, "Funding for Women's Programs at NSF: Using Individual POWRE Approaches for Institutions to ADVANCE," *Journal of Women and Minorities in Science and Engineering* 8(3–4): 2003; and Sue V. Rosser, "Institutional Barriers for Women Scientists and Engineers: What Four Years of Survey Data of NSF POWRE Awardees Reveal," in *Women in Higher Education: Empowering Change*, ed. Melanie McCoy and JoAnn Di Georgio-Lutz (Westport, Conn.: Greenwood, 2002). Part of this research was supported by NSF grant HRD-9807594.

References

Aisenberg, Nadya, and Mona Harrington. 1988. *Women of Academe: Outsiders in the Sacred Grove.* Washington, D.C.: Joseph Henry.

Campbell, Kenneth. 2001. "Leaders of 9 Universities and 25 Women Faculty Meet at MIT, Agree to Equity Reviews." MIT News Office. http://web.mit.edu/ newsoffice/nr/2001/gender.html. Accessed January 31, 2001.

Carroll, Berenice. 2001. "Reflections on '2000 Subversions: Women's Studies and the 21st Century.'" *NWSA Journal* 13(1): 139–49.

Cook, Sarah G. 2001. "Negotiating Family Accommodation Practices on Your Campus." *Women in Higher Education* 10(4): 25–26.

CPST (Commission on Professionals in Science & Technology). 2000. *Professional Women & Minorities: A Total Resources Data Compendium.* Washington, D.C.: Commission on Professionals in Science & Technology.

Curry, D. 2001. "Prime Numbers." *Chronicle of Higher Education,* July 6: A9.

de Groot, Joanna, and Mary Maynard, eds. 1993. *Women's Studies in the 1990s: Doing Things Differently?* New York: St. Martin's.

Elliott, Saranna Thornton. 2001. "Does Your School Discriminate against Pregnant Faculty?" *Women in Higher Education* 10(7): 23–24.

Fox, Mary Frank. 2001. "Women, Men, and Engineering." In *Gender Mosaics,* ed. Dana Vannoy, 249–57. Los Angeles, Calif.: Roxbury.

Heckman, D. 1997. "On the Eve of Title IX's 25th Anniversary: Sex Discrimination in the Gym and Classroom." *Nova Law Review* 21 (Winter): 545–661.

Keller, Evelyn Fox. 1983. *A Feeling for the Organism: The Life and Work of Barbara McClintock.* New York: W. H. Freeman.

NSF (National Science Foundation). 1997. *Professional Opportunities for Women in Research and Education.* Program announcement. Washington, D.C.: National Science Foundation.

———. 2000. *Women, Minorities, and Persons with Disabilities in Science and Engineering: 2000.* Arlington, Va.: National Science Foundation (NSF 00-327). http://www.nsf.gov/sbe/srs/nsf00327. Accessed February 16, 2005.

———. 2001a. *ADVANCE. Program Solicitation.* Arlington, Va.: National Science Foundation.

———. 2001b. *ADVANCE Institutional Transformation Awards.* http://www.nsf.gov/advance. Accessed October 1, 2001.

———. 2003. *Women, Minorities, and Persons with Disabilities in Science and Engineering: 2002.* Arlington, Va.: National Science Foundation (NSF 03-312). http://www.nsf.gov/sbe/srs/nsf03312. Accessed February 16, 2005.

Pryse, Marjorie, and members of the Task Force on Faculty Roles and Rewards of the National Women's Studies Association. 1999. *Defining Women's Studies Scholarship.* A Statement of the National Women's Studies Association Task Force on Faculty Roles and Rewards. College Park, Md.: National Women's Studies Association.

Riley, Marla D. 2001. "University of Arizona's Millennium Project to Assess Campus Equity." *Women in Higher Education* 10(4): 1–2.

Rosser, Sue V. 1990. *Female-Friendly Science: Applying Women's Studies Methods and Theories to Attract Students.* Elmsford, N.Y.: Pergamon.

———. 1997. *Re-engineering Female Friendly Science.* Elmsford, N.Y.: Pergamon.

———. 2001. "Balancing: Survey of Fiscal Year 1997, 1998, and 1999 POWRE Awardees." *Journal of Women and Minorities in Science and Engineering* 7(1): 1–11.

Rosser, Sue V., and M. Zieseniss. 2000. "Career Issues and Laboratory Climates: Different Challenges and Opportunities for Women Engineers and Scientists (Survey of Fiscal Year 1997 POWRE Awardees)." *Journal of Women and Minorities in Science and Engineering* 6(2): 1–20.

Sandler, Bernice. 1997. "Too Strong for a Woman: The Five Words that Created Title IX." *About Women on Campus* 5(2): 5.

Schneider, A. 2000. "Female Scientists Turn Their Backs on Jobs at Research Universities." *Chronicle of Higher Education,* August 18: A12–A14.

Sonnert, Gerhard, and Gerald Holton. 1995. *Who Succeeds in Science? The Gender Dimension.* New Brunswick, N.J.: Rutgers University Press.

Vetter, Betty. 1996. "Myths and Realities of Women's Progress in the Sciences, Mathematics, and Engineering." In *The Equity Equation: Fostering the Advancement of Women in the Sciences, Mathematics, and Engineering,* ed. Cinda-Sue Davis et al., 29–56. San Francisco: Jossey-Bass.

Wasserman, Elga R. 2000. *The Door in the Dream: Conversations with Eminent Women in Science.* Washington, D.C.: Joseph Henry.

Wenniger, M. D. 2001. "Partner Hires: A Fact of Life on Most Campuses." *Women in Higher Education* 10(4): 5.

Williams, Joan. 2001. "What Stymies Women's Academic Careers? It's Hiring Couples." *Chronicle of Higher Education,* December 15: B10.

Wilson, Robin. 2001. "The Backlash against Hiring Couples." *Chronicle of Higher Education,* April 13: A16.

Zimmerman, Bonnie. 2000. "Building NWSA." *NWSA Journal* 12(1): 165–68.

4 Telling Stories about Engineering: Group Dynamics and Resistance to Diversity

Cynthia Burack and Suzanne E. Franks

Naming Diversity

Professions, businesses, and educational institutions increasingly promote the cause of diversity and commit resources to enhancing the success of members of different social groups. This is certainly true of the engineering profession in general and of colleges, departments, and programs throughout the United States that train and educate engineers. At the same time, there remains a great deal of misunderstanding within engineering about "diversity"— what it is, why institutions should be concerned with it, and how to achieve it. In this paper, we apply insights drawn from psychodynamic group theory to engineering and engineering education to investigate some common problems concerning the institutionalization of diversity. In particular, we argue that understanding resistance to diversity is enhanced by a group-psychodynamic perspective and that leaders of in-groups play a key role in conducting group responses—consciously and unconsciously, positively and negatively.[1]

Although concern with diversity has deep historical and philosophical roots (Mill 1994), diversity in the United States is usually identified with contemporary feminism and, thus, for many in male-dominated professions, carries a host of negative associations of radicalism and misandry. Critics of diversity also may assume that it is anti-individualist, respecting group membership more than individuality and individual achievement. However, for feminist advocates of diversity, there is no inconsistency between respecting individual achievement and carefully nurturing the conditions for wide participation by members of underrepresented groups.

There is some support for this feminist perspective within the engineering profession. Speaking on diversity in the engineering workforce, the president of the National Academy of Engineering, William A. Wulf, explicitly relates the concern with diversity of thought to the diverse contexts and experiences that are legacies of group identity in most cultures. Wulf notes that "[a]t a fundamental level, men, women, ethnic minorities, racial minorities, and people with handicaps experience the world differently. Those differences in experience are the 'gene pool' from which creativity springs" (1999a, 10). If engineering lacks diversity, he argues, then "[s]ince the products and processes we create are limited by the life experiences of the workforce, the best solution—the *elegant* solution—may never be considered because of that lack!" (1999b). As Wulf suggests, organizations diversified by race, ethnicity, religion, class, and gender are the best hope for problem solving and creativity.

However, a quick look at current demographics shows that engineering—as a profession and as an educational enterprise—continues to be relatively homogeneous. As Beatriz Clewell and Patricia Campbell note in a recent review of the data on women's progress in science and engineering, "the more things change, the more they stay the same" (2002, 257). In the early 1970s, women constituted a mere 1 percent of engineering undergraduate enrollment. Following the enactment of Title IX, the enrollment of women grew, and presently women represent approximately 19 to 20 percent of engineering undergraduates. During the past ten years, however, the rate of increase has slowed, and indeed the percentage of women undergraduates has been essentially static for the last five years. The percentage of engineering African American and Native American undergraduates has remained virtually unchanged over the past ten years, at 6 to 7 percent and less than 1 percent, respectively. The percentage of undergraduates identified as Hispanic has increased slightly over the same period, from about 6 to 8 percent (WEPAN 2002).

In computer science, the situation is even more distressing, as the percentage of women has actually declined since the mid-1980s, when women earned nearly 40 percent of the B.S. degrees, compared to approximately 28 percent currently. The percentage of B.S. degrees in computer science awarded to African Americans and Hispanics has increased over the last ten years by about 1 percentage point, to 10 and 5 percent respectively, while that of Native Americans has remained unchanged at less than 1 percent.[2]

This astonishing lack of improvement and, in some cases, actual deterioration in the participation of women and minorities in engineering has occurred during the same time period when women have flocked to medicine, law, business, and veterinary medicine; when corporate America has touted the benefits of diversity and articulated its need for a diverse workforce; and when programs for women and minorities in engineering have been established and maintained at universities across the country. Two questions inescapably come to mind: What is it about engineering that renders it so resistant to diversity in its students, educators, and practitioners? How can feminists and others who work to make engineering more diverse understand and respond to resistance to diversity?

A Group-Psychoanalytic Approach to Diversity

There are many fruitful perspectives from which to study diversity and the obstacles to diversification in science and engineering. Feminists have used a variety of methods and conceptual tools to do this kind of work, including social, political, philosophical, and educational theories (see, e.g., Hacker 1989; Harding 1986; Longino 1990; McIlwee and Robinson 1992; Rosser 1995; Spanier 1995; Tuana 1989). Feminists have also used the insights of psychoanalytic theory to theorize about the internalization of gender roles and the implications of gender relations and attributions for identity construction and gendered subjectivity (Keller 1985). However, although feminist thought has employed psychoanalytic perspectives to theorize about the roots of gender identity and other important topics, feminists have not drawn consistently upon the literature in psychodynamic theories of groups (Burack 1997). Like work on gender development, much of this literature also proceeds from the relational tradition and is influenced by such theorists as Wilfred Bion (1989) and A. K. Rice (1965). Rather than being fundamentally individual and developmental, group psychoanalysis uses clinical and social observation to analyze the ways in which shared defenses and unconscious assumptions influence processes and outcomes in groups, organizations, and larger social collectivities.[3]

Observers often take for granted that organizational efforts toward greater diversity trigger resistance in those who are already present—and whose racial, ethnic, gender, and other identities are already represented—in these diversifying organizations. However, our understanding of the causes and meanings of this resistance is often sparse and vague, and our responses to resistance either hortatory or punitive. For those who value diversity, it is difficult to concede that resistances to it might be predictable, even understandable, and that not all resistance is merely a function of deliberate bias against out-groups (Rustin 1991).

Group life has both conscious and unconscious dimensions. The conscious public agenda of most groups is not difficult to specify. For engineers, this agenda might be described as analogous to medicine: engineers serve as "doctors" to the needs of society. Engineers are problem solvers, who diagnose situations and apply their training, experience, and common sense to design solutions and produce creative innovations in the hope of improving life for all people. Groups also have unconscious emotional processes and agendas, and although this emotional agenda may differ from group to group, many patterns emerge that hold across groups.

Certain shared emotional operations are commonly observed in groups. As many observers have noted, these operations become more entrenched and more difficult to ameliorate in circumstances in which a shared sense of threat bonds group members to each other and to their group identity. Orientation toward us-vs.-them thinking, defense of group boundaries, and the tendency to idealize the group characterize many social groups, although these processes may emerge in a wide variety of ways, both in intra-group relations and in relations between a group and outsiders (Volkan 1988). There are many circumstances in which these tendencies of groups are functional for certain purposes, as when identifications with sports teams reinforce "team spirit" and boost team-related sales. It is also common in some venues for leaders to deliberately exaggerate these emotional operations and create, for example, the group dynamics of military life (Segal 1995). However, it is also common for the ordinary group dynamics of professional, religious, and social groups to become socially dysfunctional and to create patterns of thought and behavior that individual group members can neither explain nor justify.

It is important that, for psychoanalytic group theorists, group psychology is not only a function of particular gatherings of individuals who can be understood as belonging to a group. Rather, the physical gathering of collectivities *displays* group processes, rather than creating them. As Bion explains, group processes are always present in the life worlds of those who understand themselves as members of a group:

> [I]t is important that the group should come together so that the characteristics of the group and the individual in it should be demonstrable. I attach no intrinsic importance to the coming together of the group. . . . Now this congregation of the group in a particular place at a particular time is obviously very important . . . but it has no significance whatsoever in the production of group phenomena. (1989, 132)

Bion's attention to the ways in which group phenomena construct the subjectivity of those who identify with groups is at the core of psychodynamic theories of groups. Although they usually describe the collective behaviors associated with various group positions and fantasies, these theories are not merely behavioral. Rather, they attend closely to the fears, fantasies, anxieties, and defenses of those whose group identifications produce the observable phenomena of group psychology.

Besides documenting and analyzing the psychodynamics of groups as a whole, psychoanalytic theories of groups give central attention to leadership and to the often-unconscious dynamics between leaders and other group members. But who are the leaders in classrooms, workplaces, or professions? If we are seeking to understand the effects of leadership on group members, psychoanalysts of groups would instruct us to employ a broad definition of leadership that, in education, can include administrators, professors, and even students. Such a broad definition makes sense because leadership does not need to be institutionalized to affect the processes and outcomes of groups. Hence, leadership need not be limited to those who teach the overt curriculum. Leadership embraces all those who answer the question "who are we?"—all those who tell the story of the group—in ways that resonate with group members and that influence subsequent group practices.

Leaders of groups "create a story about the world," helping group members to shape and pass down particular versions of the group's ideals and ethos (Alford 1994). This role makes the leader not merely a teacher of cognitive concepts, but a kind of interpreter and historian. Together, leaders and group members engage in the mutual work of crafting and verbalizing a meaningful group memoir, with the leader as the "conductor" who sets the intellectual and emotional agenda (Foulkes 1990).

Leadership and Language

It is important to pay attention to both content and language when leaders relate the group's story, because language carries with it unconscious meanings that influence the thinking of group members and, in turn, convey their group dynamics. The language employed by group members both influences and is influenced by perceptions of group attributes. Calling typical uses of language into question interrupts the mutually constituting relationship between language and perceptions, and creates conceptual space for rethinking language that underpins in-group attitudes and values. Interrogating language use also gives leaders opportunities to interpret back to group members the multiplicity of unconscious meanings that may be contained within familiar terms and expressions. Two examples of language use in engineering are especially meaningful in the context of the predominance of men in the discipline.

It is common for engineers to counter narrow perceptions of what they do, and the kinds of aptitudes that engineering requires, by pointing out that engineers use two quite different kinds of skills: "hard" and "soft." Hard skills are technical, mathematical, and scientific; soft skills are interpersonal and communicative. The use of the terms "hard" and "soft" as modifiers for skills or areas of expertise in engineering and technology is so widespread that their meanings are understood and shared, though left unspoken, by the community.[4]

On the surface, "hard" refers to that which has mathematical content or involves the use of hands-on skill with technological equipment. "Soft" refers to what is devoid of mathematical content and does not involve technological equipment. A closer inspection of such usage, however, reveals that there is a hierarchy, with hard skills ranked more highly, despite the recent trend to describe soft skills as just as important and more difficult to learn than hard skills.[5]

These uses of the modifiers "hard" and "soft" have no obvious connection to the skills they denote in engineering. There is no strong intuitive connection between mathematics and "hardness" that those outside the science and engineering professions would make and that would affirm the usage as reflecting a commonsense parallel. However, connections between masculinity, virility, male sexuality, and hardness are culturally engrained, have unconscious emotional resonance, and are widely and immediately understood. Likewise, the connection of softness with femininity is a cultural signifier with both conscious and unconscious meaning. Nor are "hard" and "soft" understood as equal terms. Perhaps because of their status as already gendered, hardness and softness are hierarchically ordered, with what is hard commanding greater respect and recognition than the soft. It is no accident of language that enemy groups frequently express ridicule by describing each other as soft or that groups express anxiety about their own softness or permeability in the face of threat. The unspoken charge is of effeminacy—the desexing and degrading of men through metaphorical impotence.[6]

Do engineers intend to express these gendered meanings when they talk of hard and soft skills? Perhaps not consciously, but the reality of our widely shared and unconscious understandings of these terms works to inflect skills language with gendered meanings. Because of this inflection, it would not be surprising to find male engineers ranking these skills (hard = high, soft = low), resisting proficiency in soft skills, expressing anxiety about the acquisition of soft skills, and—implicitly or explicitly—identifying women with lesser ability to achieve proficiency in the hard skills that denote masculine competence. If we acknowledge these unconscious group identifications, it makes sense that leaders in engineering choose to avoid the embedded gendered connotations of these words and select terms for skills that do not reinforce the gendered divisions that already exist in the profession.

A second example is widespread in ordinary usage even though it is the object of sustained criticism in academia, publishing, and other segments of social life: the use of putatively "gender-neutral" language. Certainly there are those who continue to defend the gender-neutrality of "man" and masculine pronouns in the face of criticisms of these usages. However, most use of gender-neutral language is imitative and undeliberated—the effect of overwhelmingly consistent linguistic practices rather than of thoughtful linguistic choices. When leaders in engineering speak of "the engineer" in putatively gender-neutral terms as "he," they reinforce widely held stereotypes of men's and women's competence and foster an immediate and unconscious image of the typical, and perhaps ideal, engineer as male.

Is this image of the ideal engineer an obstacle to diversity? Suzanne Franks was invited recently to conduct a class for first-year and sophomore engineering students on the general topic of gender issues in engineering. Students were separated into four groups. Each group was given a list of forty adjectives, taken from the Bem (1993) Sex Roles Inventory. Groups were given written instructions to sort the words into two categories. Two of the groups were asked to sort the words into the categories "masculine" and "feminine," while the other two were asked to sort into "engineer" and "non-engineer." Across groups, the sorting into masculine and engineer resulted in virtually the same lists, as did the sorting into feminine and non-engineer, suggesting a common understanding both of gender and of the ideal attributes of engineers. The students, all of whom considered themselves relatively unprejudiced and bias-free, were astonished to find that their groups' understanding of what it means to be an engineer and what it means to be feminine were mutually exclusive. It should be noted that all of the students in the class were members of underrepresented minority groups, and approximately 25 percent were female, unusual for engineering classrooms. Thus, it is not just in-group members who understand and accept the unspoken definition of who "belongs" in the in-group and who does not.

Language that bolsters unconscious images of the group member is particularly pernicious if it is true, as William Wulf (1999b) suggests, that in some cases members of different social groups may actually conceptualize problems, or seek to solve them, in different ways. If the unconscious image of "the engineer" is white and masculine, in-group members are likely to have great difficulty conceptualizing the diverse forms of problems and solutions brought to the group by women and minority men as intellectually legitimate. It is very difficult for an in-group to draw its boundaries, reinforce them unconsciously through language, images, and stereotypes, and then be able to transcend them through rational evaluation. Taking into account the unconscious emotional valuations that often are expressed in common forms of language use, we suggest the following alterations in the ways that leaders tell the story of their profession and of their own engineering programs:

"Engineering meets the needs of society" instead of "Engineering just doesn't appeal to women."

"Women (and/or minority men) succeed in our program" instead of "Some of our best students are women (and/or minority men)."

"Successful engineers have strong communication and technical skills" instead of "To succeed, an engineer needs soft skills as well as hard skills."

Such locutions subtly subvert the unconscious group equation of engineering with whiteness and masculinity at the same time that they embrace an inclusive definition of the benefits of engineering to all.

One of the concerns about criticisms of language use in the context of diversity is that such criticisms are wielded by "thought police" who seek to undermine personal freedom in the cause of a political crusade. Indeed, because speech is often idiosyncratic and bodily at the same time that it is shared and profoundly social, criticisms of language use are particularly likely to be construed as attacks on personal identity. However, such reactions are misleading. They personalize critiques of collective phenomena and deflect attention away from the consequences of language use for marginalized or underrepresented group members. What are always at issue in feminist critiques of in-group language are uses of language that, far from being idiosyncratic, are strikingly consistent and widely shared by in-group members at the same time that they are defended as merely expressions of personal belief or individual thought. It is this consistent, shared—indeed, deeply unconscious—quality of group language use that so stubbornly resists examination and reform.

The Values of Diversity

When we encourage diversity we want, at a minimum, to increase the number of people of different groups who choose a particular profession or participate in activities in which those like themselves are not well represented. But inviting in members of underrepresented groups is only one of the values of diversity education and openness. Equally important is the goal of addressing the preconceptions and unconscious images held by those members of a profession whose social identities are well represented. It is easier to understand why this second goal is as important as the first if we take group psychodynamics seriously.

Only present engineers, or those currently in training to become engineers, can welcome newcomers and out-group members into the group. Distant administrators, such as university presidents, provosts, recruitment officers, and diversity specialists, cannot. Engineers, including students, make up the everyday psychodynamic environment of the group. Their repertoire of responses shapes the "climate" of the group.

Diversity initiatives must respond to the kinds of questions that are constantly being posed within insular groups: "Why is it necessary to permit entry to these 'others'? If they belonged here with us wouldn't they be here already?" These questions suggest that group members perceive diversity not only as an opportunity, but also—and perhaps in many circumstances primarily—as a threat. In terms of group psychology, the mandate to diversify a group may easily be experienced as a threat to the integrity of the group. This is particularly true because the diversity mandate is, tacitly or explicitly, a kind of criticism undertaken by outsiders and imposed upon an in-group. Regardless of the actual makeup of the in-group, such a criticism is likely to evoke an emotional reaction that is akin to "circling the wagons." When this reaction occurs, what is enacted is an amalgam of us-vs.-them thinking, defense of group boundaries, and the tendency to idealize the group *as it is,* not as it might be at some point in the future.

In situations of threat or perceived threat, leaders are torn between representing the "best" of the group's aspirations and reinforcing the "worst" of the group's fears and anxieties. If leaders do not understand the pressures the group is under, they may easily end up doing both: sending group members the message that although they must give lip service to the goals of diversity, leaders and group members tacitly are in solidarity against these goals. One way in which this dynamic occurs in engineering education, often without the deliberate collusion of leaders, is through the emphasis on the image of engineering students and engineers as intellectually superior to those in other fields of endeavor.

It is common for professors and other leaders in engineering education to tell the story of engineering by relating to the next generation of engineers the legacy of intellectual superiority in which they are now implicated. Nor is this narrative about the superiority of engineers exclusive to chauvinistic accounts of the discipline. In a recent article on strategies to recruit and retain more women in science and engineering, Cathy Trower argues that "[s]cientific discovery must not be limited to a select few with homogeneous pedigrees and demographic characteristics" (2002). She notes a National Academy of Sciences (NAS 2000) report which states that science has predominantly been carried out by white males, and that to continue in this way risks alienating citizens from science. However, she quotes from the same report to make a point that underscores the myth of superiority: "Recognizing that 'science must be an elitist enterprise' because 'it needs the very best minds' does not mean that we must discourage or turn away potential scientists because they don't fit a mold to which we have become accustomed" (2002, 1).

The NAS calls for diversity in a way that explicitly reinforces the notion that scientists and engineers are superior and that implicitly conveys the notion that those who do not fit the familiar mold should be welcomed in cautiously to ensure the elitist state of the enterprise. What conversations such as this communicate is that the "very best minds" are already doing science and engineering, and that what the outsiders might have to offer is obtained at great risk to the integrity of the enterprise as a whole. However, the notion that scientists and engineers are an elite, superior group is not supported by data that show that students, both men and women, who leave science and engineering do not differ in GPA or standardized test scores from those who stay (Seymour and Hewitt 1997). Yet the assumption that engineering students are "the cream of the crop" is embedded even in critical accounts of engineering education, where it can pass unnoticed as merely an empirical statement about the relative abilities of engineers and non-engineers (Tonso 1996, 223).

The superiority myth can be both helpful and harmful to women and to men from underrepresented groups struggling to survive and succeed in engineering. While a graduate student, Suzanne Franks noted of her graduate school education,

> I had convinced myself . . . that I was special because of what I was doing . . . that other women who *weren't* . . . engineers or "hard" scientists were just wimps who weren't trying and weren't as good as me. . . . [But] your status as exceptional is distorted . . . you're not equal [to male classmates]. . . . You really begin to believe that you are different and superior, at the same time that you feel different and inferior. This effectively blocks you from uniting with other women, having any sense of solidarity, and from doing anything to change society. (O'Barr and Wyer 1992, 67–68)

This fractured self-conception is not unusual among engineers. During a question-and-answer session at a seminar Cynthia Burack presented to engineering faculty and students, a female student reported that the idea that engineers were smarter than other students made her feel good and helped her keep going when faced with a long and difficult homework set. This same student, however, reported privately to Suzanne Franks that her belief that engineering students are smarter than other students was often accompanied by the belief that she was not as smart as all the other engineering students and that, as a result, she was not cut out to be an engineer.

Such assertions of superiority as are commonly found in engineering culture sidestep contentious debates in psychology about intelligence(s) and the measurements that seek to quantify them (Gould 1996). More important, attempts to diversify engineering education confront an additional problem that speaks to the importance of minimizing in-group defenses against outsiders. The message of superiority reinforces in-group idealization—"we are smarter than those outsiders"; "the kinds of people smart enough to be members of this group are already represented here"—at the same time that diversity seeks to introduce outsiders into the fold. Therefore, we suggest this additional alteration in the way that leaders tell the story of engineering:

> "Engineers work hard and solve problems" instead of "Engineers are the cream of the crop."

From a group-psychodynamics perspective, the mixed message presented by the call for diversity in a culture that holds the superiority myth reinforces durable unconscious convictions about the unfitness of underrepresented group members rather than challenging them. With idealizing convictions reinforced in this manner, hortatory messages about the goods to be derived from diversity are unconvincing. Diversity is likely to be dismissed by those who have received the "real" underlying message: group members must be on the alert to defend the boundaries of their empire of worth against the unworthy interlopers. Once mixed messages of superiority and the mandate of diversity are broadcast, in-group members can hardly be blamed for believing that their leaders are not serious about diversity. Trusted leaders will not betray and damage the group by insisting on admitting the unqualified, or so the group will believe. And we should not be surprised when group members act in accordance with this belief in a variety of ways that include ridicule, sabotage, scapegoating, and racial or sexual harassment. These strategies are common tools in the arsenal of group conflict. These particular tools have a long history in American social, political, and economic life as ways of marginalizing and controlling disempowered groups such as men and women of color and white women. But the tools may be wielded effectively whenever there are in-groups and out-groups, regardless of the particular identities of those involved. It is actual or perceived power, insider status, or a sense of threat, and not whiteness, maleness, or any other particular identity formation, that situates groups to employ such strategies effectively.

Diversity: Benefit and Threat

Related to the belief in and expression of intellectual superiority is the threat presented by self-segregating practices such as the establishment of institutionalized programs for women and minority students as well as lectures, classes, and other events directed at out-group students. In her whimsical definition of "womanism," Alice Walker indicates that the womanist is "not a separatist, except periodically, for health" (1984, xi). Walker's concern with the availability of separatism to people of color "for health" can be generalized to the situation of out-group members in engineering. The unconscious group culture of engineering and its social residues can be difficult for women and minority males to negotiate without the ability to appeal to segregated spaces and the interpretive possibilities they hold. Unfortunately, when minority group members get together, even in informal ways, they are likely to be seen by in-group members as "self-segregating" and to arouse suspicion as to their motives. This happens in a myriad of ways, both small and large, creating a form of "death by a thousand paper cuts" (Knight Higher Education Collaborative 2001, 2). One need not seek far for examples:

Women are standing in public spaces (hallways, campus sidewalks) engaging in conversation with other women faculty or administrators. Men who know the women comment as they pass by: "Say, no self-segregating here!" "What are you three up to?" "What are you two plotting now?" Groups of two or three men engineers gathered in public spaces for conversation are unlikely to be assailed by similar comments from either men or women.

An announcement for a pizza party for women engineering students is vandalized, and male students complain to their female classmates about being excluded.

In a set of office moves on campus, the space for a multicultural student study lounge is eliminated. The story circulates that other space needs had priority and that multicultural students should be encouraged to mingle with white students, not to self-segregate. The need of those students for a place within their very white educational world that they could call their own is not recognized.

In-group members often complain: "we don't have a special lounge (or program, or scholarship) for whites (or men), so why should *they* get one?" These group members are unable to see how all of the institutional structures and spaces they occupy are already their own—this is as invisible and unremarkable to them as the air they breathe. They are unable to see how those structures and spaces exclude and disadvantage out-group members, because they themselves are so very comfortable and welcomed within them. In addition, neither the recipients of these kinds of comments nor leaders of engineering culture consistently address such comments as signs of in-group anxiety. Instead, denigrating discourse about self-segregation or the privileges of out-group members is most often either ignored or read exclusively as hostility. Left uninterpreted, dimensions of anxiety and defensiveness remain unavailable to the group and cannot be integrated into members' understanding of the group and of themselves.

How can leaders in engineering acknowledge the need of minority out-group members for "safe" space while achieving the goals of integration, teamwork, and mutual respect? First, leaders must be persuaded of the benefits to the program, department, or team as a whole of voluntary and temporary self-segregation. Self-segregation may work to interrupt group processes that marginalize or stereotype those who represent minority social groups. Those who would offer safe spaces and special programs to women and underrepresented minority men must defend many fronts at once. They must articulate the need for such programs to those who control access to resources, while simultaneously addressing stereotypes about the unsuitability of these groups for engineering. And all leaders—whether of the group as a whole or of those who constitute out-groups—should speak openly and receptively with group members about the anxieties and defenses latent in the explicit goals and institutional practices of engineering.

When someone asks, "why special programs for women?" the answer must be that women and underrepresented minority men do not need special help to be good engineers, but they do benefit from assistance in dealing with engineering culture. Thus, the special programs are not for *them* so much as they are for *engineering*. It is engineering education, as it is currently constituted, that cannot effectively welcome out-group members, and that needs an array of programs to help counteract the conscious and unconscious biases of science and engineering culture, practice, and pedagogy that squelch the interests of those who are different, and ultimately drive them away.

How can leaders acknowledge the discomforts of increasing diversity without betraying the goals of increasing diversification? Peter Skerry argues that "diversity is typically associated with dissent and conflict" and then asks, "Why is this downside of diversity so consistently overlooked?" Under a heading that bluntly announces, "Diversity Brings Conflict," Skerry notes that ethnic and racial diversity is associated with "emotional conflict among co-workers" (2002, 20–22). For psychoanalytic theorists of groups, the kind of resistance to diversity noted by Skerry is ordinary rather than extraordinary, although the form and intensity of resistance in all groups is shaped and conditioned by particular social and historical circumstances.

Some observers of higher education and the professions in the United States note that many formerly homogeneous white and male disciplines have been more successful in diversifying than has engineering. Such a social and historical reality is likely to be determined by many factors, and students of the engineering profession continue to analyze the ways in which engineering identity is constructed out of histories of class, race, and gender participation and foreclosure (Eglash 2002; Florman 1994; Hacker 1989; Oldenziel 1999). Analyzing the particular context of engineering and engineering identity also requires empirical and theoretical examinations of the multiple versions of masculinity—"masculinities"—that different disciplines construct, perpetuate, and defend (Eglash 2002; Faulkner 2000).

In order to construct a full account of resistance to diversity in engineering, it is not sufficient to employ a psychodynamic group analysis that ignores the particular social and relational context of the group under study. However, without a psychodynamic analysis, it is difficult for those who teach the benefits of diversity to acknowledge and understand the resistance of in-group members, including leaders, to diversification. This is particularly true when feminists, diversity experts, and other interested parties accurately point out that the interests of engineering are better served by diversity than by continued group homogeneity and resistance. Like individuals, groups are both rational and irrational, but even the irrationality of groups may be predictable and transparent to analysis.

Notes

The authors acknowledge research support from the Coca Cola Fund for Women's Studies Scholars.

1. We use the terms "in-group" and "out-group" to refer, respectively, to those who are in a position to defend their group identity against outsiders and to those who, although they may be a small percentage of the group's membership, may be perceived as outsiders, and whose identities are underrepresented in the in-group.
2. Data available from sources such as the National Science Foundation (NSF) and the Commission on Professionals in Science and Technology (CPST), summarized and made available on the WEPAN Web site (2002), have not been disaggregated by both race and gender in the past. Therefore, it is difficult to discuss the situation of women of color separate from that of white women or of men of color. However, it appears from data cited in Clewell and Campbell's review (2002) and an NSF report, *Women, Minorities, and Persons with Disabilities: 2002,* that data disaggregated by both gender and race are becoming available.
3. The literature in group psychoanalysis is large, but see Anzieu 1984; Bion 1989; Hinshelwood 1987; Kreeger 1975; Liff 1975; Rice 1965; and Rutan and Stone 2001.
4. Another variation on this theme is the dichotomy of "hard mastery" and "soft mastery." See Mangan (2002) and Turkle (1984). For another perspective on the content of the dichotomy see Faulkner (2000), who presents the distinction between hard and soft technology as one of scale and power. In all versions, the gendered associations of these terms remain intact.
5. See, for example, Rossheim 2002: "'The core competencies now are the soft skills. These qualifications are accented by hard skills.' Most people can learn the hard skills, but communications-intensive soft skills are hard to teach."
6. The suggestion that the term "hard skills" actually refers to the difficulty associated with engineering curricula and practice is a common strategy of resistance to the implications of the connection between "hardness" and masculinity. If "hard" did actually refer to difficulty, then the analogous term would be "easy," not "soft." Because the connection between "hard" and "difficult" does exist, however, it functions to enhance the hierarchical ordering of "hard" and "soft" skills. Mastery of that which is more difficult is more highly valued in engineering culture.

References

Alford, C. Fred. 1994. *Group Psychology as Political Theory.* New Haven, Conn.: Yale University Press.

Anzieu, Didier. 1984. *The Group and the Unconscious.* London: Routledge and Kegan Paul.

Bem, Sandra Lipsitz. 1993. *The Lenses of Gender: Transforming the Debate on Sexual In-equality.* New Haven, Conn.: Yale University Press.

Bion, Wilfred R. 1989. *Experiences in Groups and Other Papers.* London: Routledge. (Orig. pub. 1961.)

Burack, Cynthia. 1997. "Crossing Boundaries: Black Feminism and Group Leadership." *Gender and Psychoanalysis* 2(3): 343–67.

Clewell, Beatriz Chu, and Patricia B. Campbell. 2002. "Taking Stock: Where We've Been, Where We Are, Where We're Going." *Journal of Women and Minorities in Science and Engineering* 8(3–4): 255–84.

Eglash, Ron. 2002. "Race, Sex, and Nerds: From Black Geeks to Asian American Hipsters." *Social Text* 20(2): 49–64.

Faulkner, Wendy. 2000. "The Power and the Pleasure? A Research Agenda for 'Making Gender Stick' to Engineers." *Science, Technology, and Human Values* 25: 87–119.

Florman, Samuel C. 1994. *The Existential Pleasures of Engineering.* 2nd ed. New York: St. Martin's.

Foulkes, S. H. 1990. *Selected Papers of S. H. Foulkes: Psychoanalysis and Group Analysis.* Ed. Elizabeth Foulkes. London: Karnac.

Gould, Stephen J. 1996. *The Mismeasure of Man.* New York: Norton.

Hacker, Sally. 1989. *Pleasure, Power, and Technology: Some Tales of Gender, Engineering, and the Cooperative Workplace.* Boston, Mass.: Unwin Hyman.

Harding, Sandra. 1986. *The Science Question in Feminism.* Ithaca, N.Y.: Cornell University Press.

Hinshelwood, R. D. 1987. *What Happens in Groups: Psychoanalysis, the Individual, and the Community.* London: Free Association Books.

Keller, Evelyn Fox. 1985. *Reflections on Gender and Science.* New Haven, Conn.: Yale University Press.

Knight Higher Education Collaborative. 2001. "Gender Intelligence." *Policy Perspectives* 10(2): 1–9.

Kreeger, Lionel, ed. 1975. *The Large Group: Dynamics and Therapy.* Ithaca, N.Y.: F. E. Peacock.

Liff, Zanvel A., ed. 1975. *The Leader in the Group.* New York: Jason Aronson.

Longino, Helen E. 1990. *Science as Social Knowledge: Values and Objectivity in Scientific Inquiry.* Princeton, N.J.: Princeton University Press.

Mangan, K. S. 2002. "Horse Sense or Nonsense?" *Chronicle of Higher Education* 48(43): A8–A9.

McIlwee, Judith Samson, and J. Gregg Robinson. 1992. *Women in Engineering: Gender, Power, and Workplace Culture.* Albany: State University of New York Press.

Mill, John S. 1994. *On Liberty and Other Writings.* Cambridge: Cambridge University Press. (Orig. pub. 1859.)

NAS (National Academy of Sciences). 2000. *Enhancing the Postdoctoral Experience for Scientists and Engineers.* Washington, D.C.: National Academy Press.

NSF (National Science Foundation). Division of Science Resources Statistics. 2003. *Women, Minorities, and Persons with Disabilities in Science and Engineering: 2002.* Arlington, Va.: National Science Foundation (NSF 03-312). http://www.nsf.gov/sbe/srs/nsf03312. Accessed February 16, 2005.

O'Barr, Jean, and Mary Wyer, ed. 1992. *Engaging Feminism: Students Speak Up and Speak Out.* Charlottesville: University Press of Virginia.

Oldenziel, Ruth. 1999. *Making Technology Masculine: Men, Women, and Modern Machines in America, 1870–1945.* Amsterdam: Amsterdam University Press.

Rice, A. K. 1965. *Learning for Leadership: Interpersonal and Intergroup Relations*. London: Tavistock.

Rosser, Sue V., ed. 1995. *Teaching the Majority: Breaking the Gender Barrier in Science, Mathematics, and Engineering*. New York: Teachers College Press.

Rossheim, John. 2002. "What It Takes to Temp." http://ct.monster.com/articles/ whatittakes/. Accessed January 31, 2005.

Rustin, Michael. 1991. *The Good Society and the Inner World: Psychoanalysis, Politics, and Culture*. New York: Verso.

Rutan, J. Scott, and Walter N. Stone. 2001. *Psychodynamic Group Psychotherapy*. New York: Guilford.

Segal, Hanna. 1995. "From Hiroshima to the Gulf War and After: A Psychoanalytic Perspective." In *Psychoanalysis in Contexts: Paths between Theory and Modern Culture*, ed. Anthony Elliott and Stephen Frosh, 191–204. London: Routledge.

Seymour, Elaine, and Nancy M. Hewitt. 1997. *Talking about Leaving: Why Undergraduates Leave the Sciences*. Boulder, Colo.: Westview.

Skerry, Peter. 2002. "Beyond Sushiology: Does Diversity Work?" *Brookings Review* 20(1): 20–23.

Spanier, Bonnie B. 1995. *Im/Partial Science: Gender Ideology in Molecular Biology*. Bloomington: Indiana University Press.

Tonso, Karen L. 1996. "The Impact of Cultural Norms on Women." *Journal of Engineering Education*, July: 217–25.

Trower, Cathy Ann. 2002. "Women without Tenure, Part 4: Why It Matters, What to Do." http://nextwave.sciencemag.org/cgi/content/full/2002/04/11/2. Accessed January 31, 2005.

Tuana, Nancy, ed. 1989. *Feminism and Science*. Bloomington: Indiana University Press.

Turkle, Sherry. 1984. *The Second Self: Computers and the Human Spirit*. New York: Simon and Schuster.

Volkan, Vamik D. 1988. *The Need to Have Enemies and Allies: From Clinical Practice to International Relationships*. Northvale, N.J.: Jason Aronson.

Walker, Alice. 1984. *In Search of Our Mothers' Gardens: Womanist Prose*. New York: Harvest/Harcourt Brace Jovanovich.

WEPAN (Women in Engineering and Program Advocates Network). 2002. CPST, data derived from Engineering Workforce Commission *Engineering and Technology Enrollments, Fall 1990 through 2000*, from National Science Foundation *ISRS, SESTAT*, 1997, and from National Science Foundation, *Survey of Doctoral Recipients*. http://www.wepan.org/documents/protected_data.html. Accessed July 29, 2002.

Wulf, William A. 1999a. "Diversity in Engineering." In *Moving beyond Individual Programs to Systemic Change, WEPAN Annual Conference Proceedings*, 9–16. West Lafayette, Ind.: WEPAN Member Services.

———. 1999b. "Testimony to the Commission on the Advancement of Women and Minorities in Science, Engineering, and Technology Development." http:// www.nae.edu/NAE/naehome.nsf/weblinks/NAEW-4NHMH2?OpenDocument. Accessed February 16, 2005.

5 The Gender Gap in Information Technology

Mo-Yin S. Tam and Gilbert W. Bassett, Jr.

Concern about the technology gender gap has shifted from a focus on computer access and Internet connectivity to educational achievement and employment opportunities in information technology. While the gender gap in computer ownership and connectivity has narrowed in recent years (U.S. Department of Commerce 1995, 1998, 1999, 2000; Hoffman and Novak 1999), the gender divide in information technology (IT) education and employment remains (Angrist and Lavy 1999; Warschauer 2003).

The technology sector has rebounded since March 2003. The value of technology shares has surged, with the NASDAQ up roughly 40 percent between March and December of 2003. Also, individuals with technological skills and knowledge continue to have access to economic opportunities. The Bureau of Labor Statistics predicts this will continue into the twenty-first century. Weinman and Cain (1999) estimated that more than half of all new jobs will require technological skills. A recent poll of chief information officers (CIOs) and human resource (HR) managers of companies with sales of $20 million or more revealed that the demand for IT workers remained strong and the number of unfilled IT positions has tripled since 1999 (Pastore 2001).

This is particularly the case for what Cooke (2000) referred to as the "high-IT core" workforce, where there have been the greatest increases in employment and wages. The "high-IT core" workforce includes computer programmers, computer scientists, computer engineers, and system analysts whose jobs are directly involved in the "study, design, development, implementation, support or management of computer-based information systems, particularly software applications and computer hardware" (Carver 2000).

While employment in all IT occupations increased by 23 percent from 1992 to 1998 (double the increase in total U.S. nonfarm employment), employment in the "high-IT core" increased by 79 percent (Cooke 2000; Digital Divide Network 2000). With the increased demand for IT workers, there has also been a disproportionate increase in wages, with the highest increase occurring in the "high-IT core" sector. According to Cooke (2000), from 1992 to 1998, wages increased at 5.8 percent per year for all workers in IT industries (compared to a 3.6 percent annual rate for other private industry workers), while those in software and computer service industries, including computer programming services and software development, had increases of 6.7 percent. In 1998, the average wage of all IT workers was $58,000 (85 percent higher than the $31,400 average wage for all private workers), while for "high-IT core" workers it was $65,300. This represents a widening of the wage gap between the IT-producing workers and all private workers by more than $10,000 between 1992 and 1998. It is the gender gap in this sector of the IT workforce that has increasingly received attention from policymakers, researchers, and advocacy groups, including the Association of American University Women.

Women constitute a small portion of the increased employment in technology fields. In 1997 women made up 46 percent of the workforce, but only 25 percent of the professional IT workforce (ITAA 1998). The gender gap in female participation in the IT workforce persists.

Moreover, college enrollment data suggest the gap will widen, since the number of women IT majors has decreased in recent years. National Science Foundation (NSF) statistics show that the proportion of bachelor's degrees in computer science received by women dropped from 37 percent in 1984 to 28 percent in 1996 (Martin and Wardle 1999). It dropped again to 27 percent in 1998 (NSF 2003). Also, only 22 percent of computer science majors and 13 percent of computer and electrical engineering majors in 1998 were women (ITAA 1998). Moreover, the proportion of women with a master's degree in computer science was similarly small in 1998, with few women pursuing doctoral degrees (Huang et al. 1998). NSF data show that in 1998, only 29 and 18 percent of those receiving master's and doctoral degrees, respectively, in computer science were women (NSF 2003).

This chapter examines the IT gender gap and explores factors behind the divide. Lack of data has precluded previous quantitative study of this issue. To overcome this deficiency, we obtained multiyear (1995–2000) data from a large public university. Using these data, we investigated the gender gap in IT majors in order to understand the gender gap in the "high-IT core" workforce, as many graduates with IT majors join the "high-IT core" workforce. We found, consistent with our earlier results (Tam and Bassett 2000), a significant gender gap in IT majors after controlling for students' racial/ethnic backgrounds. One surprising result is that the gap is most prominent among white students. We also found that while the gender gap in Internet usage has been closing (U.S. Department of Commerce 2000), there has been no corresponding decrease in the IT-major gender gap.

We also tested the hypothesis that math performance affects the choice of being an IT major. Students with higher math performance when entering college might believe that they would be more likely to succeed in IT. Advisers might encourage students with higher math scores to choose an IT major. We used math ACT scores as a measure of math performance. Our regression results show that all students—men and women—with higher math ACT scores are more likely to become IT majors. However, while math performance affects IT participation, we found that its effect is far weaker for women. Women students with the same math ACT scores as men are much less likely to become IT majors. This result is consistent with observations in earlier studies (Holland and Eisenhardt 1990). Our analysis shows that closing the gender gap in math performance would only eliminate 14 percent of the overall gap in majors. Factors other than math performance, therefore, need to be considered in order to understand and to close the technology gender gap.

Understanding the Nature and the Causes of the Technology Gender Gap

The gender gap in educational achievement and employment opportunities puts women at a distinct disadvantage in a society that values computer technology. Weinman and Cain (1999, 1) warned, "if nothing is done, girls and women will be bystanders in the 21st century." This issue has sparked increasing concern among researchers as well as women's interest groups (Teachers Clearinghouse for Science and Society Education 1999). In addition, the lack of women in IT implies untapped resources for the advancement of technology. It is, therefore, important to understand the nature and causes of the gender gap. While summary statistics on women's IT participation are available (NSF 2000, 2003), absence of data on other variables, such as student math performance, has precluded analysis of the factors influencing IT participation.

To begin to remedy this situation we obtained student data from a large public university and used them to analyze the nature and causes of the technology gender gap. We studied the existence and time trend of the gender gap for IT majors, since they supply labor to the "high-IT core" workforce. We subsequently examined math performance as a potential cause of the gap.

The Gender Gap in IT Majors

Data

The University of Illinois at Chicago (UIC) is a comprehensive four-year public university. Its undergraduate enrollment was almost constant from 1995 to 2000 at about sixteen thousand students. Slightly more than half of the undergraduate students were female throughout this period. This is consistent with the national data. While total enrollment remained constant, the proportion of IT majors almost doubled, from 7 percent in 1995 to 13 percent by 2000. IT majors at UIC include those in computer sciences, computer engineering, electrical engineering, and management information systems, as well as mathematics and computer science.

Findings

In the fall semester of 2000, the percentage of women students in the IT area was only 27, which is much smaller than the 55 percent of women who constituted the entire undergraduate enrollment. Hence, women in the IT field were very much underrepresented.

To more precisely indicate the IT gap, we used an index representing the percentage of women in the IT group relative to the percentage of women in the entire student body (for details, see appendix A). When the value of the index is 1, this means that women are represented in the same proportion in the IT field as they are in the entire student body. A value smaller than 1 implies over-representation of women relative to the entire group, while a value greater than 1 indicates underrepresentation. More specifically, the index represents the factor by which the number of women in IT would have to increase to make their representation there proportional to that in the overall student body. For example, at UIC in 2000 the value of the index was 3.3. This means that with the number of men held constant, the number of women in IT would have to be multiplied by 3.3 for women and men to be equally represented in the IT area.

The gender distribution was different for various racial/ethnic groups of undergraduate students at UIC in the fall semester of 2000. While women constituted about 55 percent of the entire undergraduate body, they were 73 percent of Blacks, 58 percent of Hispanics, 52 percent of Whites, and 51 percent of Asians. Blacks and Hispanics were underrepresented in the IT field. The percentages of IT students who were Black and Hispanic were, respectively, 7 and 12, while Blacks and Hispanics constituted 10 percent and 17 percent of the entire student body. This raises the question of whether the female underrepresentation in IT fields simply reflected the underrepresentation of Hispanics and Blacks.

To address the interrelationship between gender and ethnicity in the IT gender gap, we examined the percentages of female IT students of various racial/ethnic groups in the fall semester of 2000. As table 5.1 shows, women students in IT were less than 50 percent of majors in all racial/ethnic groups. Hence, a gender gap existed within *all* racial/ethnic groups, and women's underrepresentation in IT reflected more than just minority underrepresentation.

There were, however, considerable variations among the racial/ethnic groups. In 2000, the percentage of women students in IT was smaller for Hispanics (29 percent) than for Asians (30 percent) and Blacks (49 percent). A surprising finding was that Whites had the smallest percentage of women students in IT (only 19 percent). Again, as indicated in appendix A, these differences have to be compared to the entire student body to capture more precisely the extent of the gender gap. Table 5.2 indicates that the gender gap is largest for Whites. In 2000, the value of the women's participation index for Whites was 4.8. The corresponding index values for Asians, Blacks, and Hispanics were 2.2, 2.9, and 3.3, respectively. Hence, gaps were large for all groups, but largest for White students. While the number of women students would have to triple to eliminate the gender gap for Asians, Blacks, and Hispanics, it would have to almost quintuple to eliminate the gap for Whites.

It should be noted that women were also underrepresented in engineering and the physical sciences, though the gaps were largest in IT. In engineering, the

Table 5.1. Percentage of Female Undergraduate Students in IT[a] by Racial/Ethnic Groups
Fall semesters, 1995–2000, UIC

Fall of	Asian	Black	Hispanic	White	Total IT	Total U-grad	IT majors as % of Entire Student Body
1995	22%	36%	24%	16%	21%	53%	7%
1996	23%	42%	23%	15%	21%	54%	8%
1997	26%	42%	22%	16%	23%	57%	9%
1998	30%	47%	30%	15%	26%	55%	12%
1999	29%	54%	35%	16%	27%	54%	13%
2000	30%	49%	29%	19%	27%	55%	13%

a. Information technology (IT) includes majors in computer engineering, computer science, electrical engineering, management information systems, and mathematics and computer science.

Table 5.2. Gender Gap Indices in IT Areas by Racial/Ethnic Groups
Fall semesters, 1995–2000, UIC

$I_{IT}^{j(W)}$	Asian	Black	Hispanic	White	Total	IT majors as % of Entire Student Body
1995	3.4	3.7	4.1	5.4	4.2	7%
1996	3.2	3.4	4.4	6.0	4.3	8%
1997	2.8	3.4	3.8	5.7	3.8	9%
1998	2.3	3.1	3.2	6.3	3.4	12%
1999	2.5	2.9	2.8	5.7	3.4	13%
2000	2.5	2.9	3.3	4.8	3.3	13%

percentages of Asian, Black, Hispanic, and White women students were 34, 50, 24, and 19, respectively. In the physical sciences, the corresponding percentages were 44, 63, 43, and 48.

Time Trends of the IT-Major Gender Gap

Previous studies have linked out-of-school activities to career choice (Johnson 1987; Sjoberg and Imsen 1988). The 2000 edition of the Department of Commerce's *Falling through the Net* showed that the gender gap in Internet usage had disappeared by August 2000, with 44 percent of both men and women using the Internet (U.S. Department of Commerce 2000). A 2001 UCLA survey polling over 400,000 students in 717 colleges and universities nationwide also showed that among first-year students, 77.8 percent of women and 79.5 percent of men reported frequent computer use (UCLA 2001). Is the information technology gender gap narrowing as the gender gap in out-of-school computer activities disappears?

The UIC data indicate that the IT major gap has narrowed, though the decline is small and the gap remains large. Table 5.1 shows that between 1995 and 2000, the percentage of IT majors at UIC almost doubled, changing from 7 percent to 13 percent. While the percentage of women in the undergraduate student body as a whole remained rather constant, at around 55 percent, the percentage of women students in IT majors increased slightly, from 21 percent in 1995 to 27 percent in 2000. It was, however, still significantly below 55 percent. Table 5.2 presents the corresponding values of the gender index from 1995 to 2000. For the entire student body, it was 4.2 in 1995 and 3.3 in 2000. Hence, as the Internet usage gender gap closed, the IT-major gender gap narrowed, though it was still significant in 2000.

Similar general trends are evident for all racial/ethnic groups. The data presented in table 5.2 show that the relative sizes of the gender gap among them remained constant during the period 1995 to 2000, with the largest gap remaining for Whites.

Technology Gender Gap and Math Performance

In this section, we consider the relationship between IT participation and math performance. We show that math performance is a significant factor for IT participation. However, differences between math performance for women and men account for only a small part of the IT-major gender gap.

The Impact of Math Performance on the Odds of Being an IT Major

We used math ACT scores as an indicator of students' math performance.[1] Using UIC's fall 2000 data, we considered the percentage of IT majors for each ACT score. We then estimated the impact of math performance on the odds of being an IT major. The logistic model we used is described in appendix B and the regression results for all students in fall 2000 are listed in table 5.3. The results show that ACT score affects IT participation. It explains 94 percent of the variation in the log odds of becoming an IT major. Figure 5.1 illustrates this relationship between ACT score and IT participation.

The curve shown in figure 5.1 closely follows the data except for the few points at very high ACT scores.[2] It indicates that math performance affects IT participation. To understand the relationship between math performance and the IT-major gender gap, we examined first whether math performance affects IT participation differently for men students and women students and whether men's math performance is higher than that of women.

Table 5.3. Regression Results for Equations B-1 and B-2 All Students
Fall 2000, UIC

	Equation B-1		Equation B-2	
	Estimate of Coef.	t stat[a]	Estimate of Coef.	t stat
Constant	−4.899	−712.1	−3.93	−256.1
ACT	0.126	425.1	0.106	169.2
Gender[b]			−0.93	−45.7
ACT*Gender			−0.011	−12.1

a. The large value of the t statistics is due to the large sample size.
b: Gender = 1 for female students.

Figure 5.1. UIC Fall 2000: Probability of Being an IT Major.

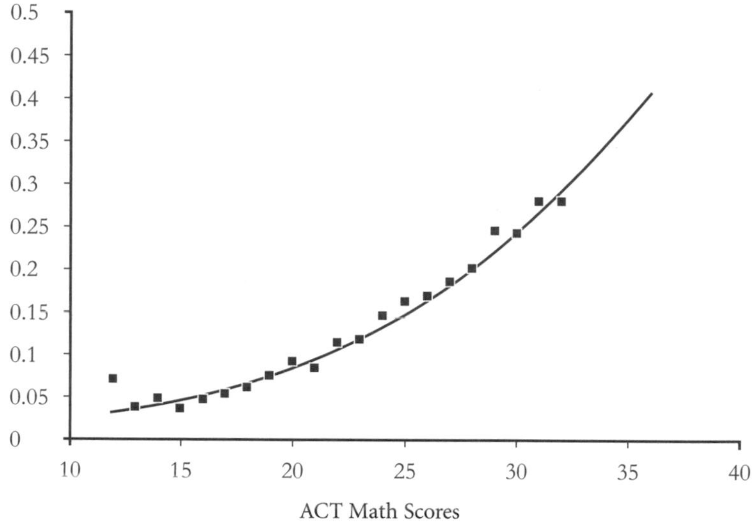

ACT Math Scores

Gender Gap in the Impact of Math Performance on the Odds of Being an IT Major

Does math performance affect women's IT participation differently from men's? To address this question, we considered an additional model (using fall 2000 UIC student data) that included the gender variable; see equation B-2 in appendix B. The regression results are presented in table 5.3.

The results show that ACT score is a positive determinant for being an IT major for both female and male students. Students with higher ACT scores, irrespective of gender, are more likely to become IT majors. However, the influence of a high math ACT score is far weaker on female students. This means

Figure 5.2. UIC Fall 2000: Probability of Being an IT Major, by Gender.

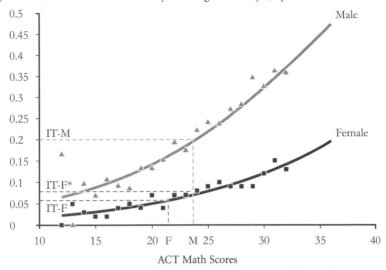

that a female student is less likely to become an IT major than a male student with the same ACT score.

This gender gap is illustrated in figure 5.2. The average math ACT scores for females (F) and males (M) are indicated on the horizontal axis of the figure. The probability of majoring in IT at various ACT scores is indicated on the vertical axis. The female curve [Female] and the male curve [Male] are regression lines for equation B-2 in appendix B. Both of the curves follow the data, indicating that math ACT score is a significant factor in becoming an IT major. However, the female curve is much flatter than the male curve, indicating that math performance is a far weaker factor for female students in becoming an IT major.

Math Performance Gender Gap

Recent data show that gender differences in math performance are narrowing, but still exist. The difference between mean SAT math scores for boys and girls, for example, was reduced from 48 points in 1990 to 36 in 1997. The 2000 NCES (National Center for Education Statistics) data show that while gender differences in math performance in grade 4 have become insignificant, female students still have lower scores in grades 8 and 12. An AAUW Education Foundation (1998) study indicates that while the gap in the number of math courses taken has narrowed, boys take more advanced courses. The NSF (2003) reports that while women have accounted for an increasing percentage of bachelor's degrees in sciences and engineering, the percentage of bachelor's degrees in math earned by women declined slightly, from 47.2 in 1995 to 46.8 in 1998.

The Gender Gap in Information Technology 115

Consistent with the national statistics, female students' mean ACT score at UIC in 2000 was 21.8 and male students' was 23.9.

IT-Major Gender Gap and Math Performance Gender Gap

The above results enabled us to determine the magnitude of the IT gap due to differences in math scores. Consider the regression curves for female and male students in figure 5.2. The probability that a female student with an ACT score of 21.8 (the mean ACT score for women students) will become an IT major is about 5 percent, as denoted by IT-F. On the other hand, the corresponding probability for a male student with a math ACT score of 23.9 (the mean ACT score for men) is 20 percent, as denoted by IT-M. The difference then is 20 – 5 = 15 percent.

How much of this gap is due to lower female math performance? An answer is given by supposing the math ACT score for female and male students to be equal. From the female curve in figure 5.2 we find that the probability that a female student with a math ACT score of 23.9 will become an IT major is 7 percent (as compared with a corresponding male probability of 20 percent). Hence, with the same math scores for men students and women students the technology gender gap would be reduced only slightly. The gap would be 13 percent instead of 15 percent, an absolute reduction of only 2 percent. That is, closing the gender gap in math performance eliminates only 14 percent of the technology gender gap. Evidently the magnitude of the technology gender gap reflects much more than math differences.

Time Trend for Effects of Gender Gap in Math Performance on Gap in IT Majors

Has the gender gap in the impact of math performance changed in the last few years? Figure 5.3 illustrates the trend from 1995 to 2000. It shows that women students' participation in IT has increased significantly since 1995, with a slightly larger increase among those with high math ACT scores. Women students with higher scores are increasingly likely to become IT majors. A further analysis of the data, however, indicated that the gender gap has not narrowed. The difference between the two solid lines in figure 5.4 represents the gender gap in math performance in 2000, while that between the two dashed lines represents the gender gap in 1995. The effect of math performance on IT participation has increased similarly for men and women, leaving a persistent gender gap.

Discussion

Using data from UIC we have documented the existence and persistence of the gender gap among IT majors. Since IT graduates are the primary source

Figure 5.3. UIC 1995–2000: Probability of Being an IT Major, Female.

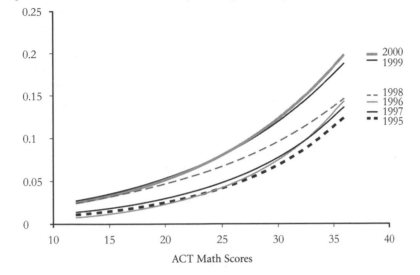

Figure 5.4. UIC 1995 and 2000: Probability of Being an IT Major, by Gender.

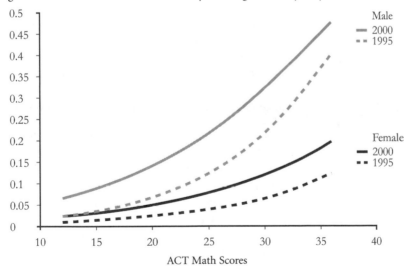

of the "high-IT core" workforce, the gender gap in IT majors will eventually translate into a gender gap in that sector of the workforce. Furthermore, we show that while the gap has diminished, it remains significant. Female IT majors would have to more than triple in order to eliminate the gender gap.

Our results also demonstrate that while math performance is a significant predictor for becoming an IT major, it is less important for female students. Moreover, the technology gender gap is not due to differences in math performance. This is consistent with the NSF (2003) data, which show that while the math performance gender gap is shrinking, the IT gender gap persists and is widening. The gap therefore must be due to other factors that influence the choice of major and occupation.

A number of studies show that IT gender differences begin in high school. Suzanne Silverman and Alice Pritchard (1996) reported from their two-year study of girls' participation in technology education in Connecticut that even in middle schools girls enjoyed technology education. However, they did not continue to participate when they reached high school (see also Margolis and Fisher 2002).

The drop of girls' interest in technology education is attributed by some researchers to their negative experience in computer science classes (Schofield 1995). Kimberly Weiner's experience in sixth grade provides a vivid illustration: "At free time, I would dash for the computer, only to find that I was the sole female jostling for a place in line. The boys were annoyed that a girl wanted in on 'their territory.' The other girls didn't understand why a girl would want to use a computer" (Weiner 1998). Similar incidents are documented by Jane Margolis and Allan Fisher (2002) and Dell'Angela (2001).

The interviews of Carnegie Melon University students conducted by Margolis and Fisher (2002) also revealed that women were much less likely than men to have learned from their fathers the use of computer technology. Hence, many more men than women had hands-on experience with technology before entering college, giving them a potentially intimidating advantage over girls. This, the researchers found, led women to doubt their technological ability and perpetuated a reluctance to participate in technology education.

Women's doubt of their math ability is also evidenced in the survey conducted by UCLA in 2001. It showed that women students were only half as likely as men students to rate their computer skills as "high" (CIRP 2001). This is consistent with Dorothy Holland and Margaret Eisenhardt's (1990) research, which suggests that women and men students evaluate their own math performance differently. In addition, other UCLA data (CIRP 1998) indicate that among all science and engineering majors, a significantly higher percentage of female first-year students than of men reported for remedial work in math (27.5 percent versus 16.4 percent).

Girls' perception that computers are "boy stuff" is another factor contributing to the IT gender gap (Kelly 1981, 1988). Women imagine that computer professionals are loners and antisocial, living in a solitary world. Also, as Anna Martinson argues in chapter 14 of this volume, violence and sexist representations in computer games help to cause women's relative lack of interest in computing. Finally, the lack of role models and the small proportion of female IT teachers lead to a perpetuation of the gap.

Hence, a change in societal and parental expectations and a concerted effort to encourage women's participation in technology education will be needed to diminish the technology gender gap. One example comes from Carnegie Melon University (CMU), which recently modified its admissions policy to encourage women's greater participation in the IT area. In 1995, the Computer Science Department at CMU began to deemphasize prior experience in computer programming in its admissions policy. As a result, female computer science students increased from 8 percent in the fall of 1995 to 42 percent in the fall of 2002. The CMU experiment also indicated that women, once enrolled, do well (Margolis and Fisher 2002).

Our study also compared the gender gap between various racial/ethnic groups. As indicated above, an unexpected and very interesting finding was that the gender gap in information technology differed greatly between racial/ethnic groups. Consistent with the ETS data (Coley 2001), differences in math scores for women and men students at UIC were similar among all ethnic groups. However, the IT gender gap for Whites was about twice as large as for other groups. One plausible explanation is that White women have better access to other professional fields, including law and medicine. For example, the UIC data show that White women's representation in biological sciences is relatively higher than other racial/ethnic groups'. However, more detailed data and further research are needed to examine the reasons for this result.

Appendix A

The index used to measure the IT major gap compares the percentage of women in an IT group with the percentage of women in the entire student body. Specifically, the index is given by

(A-1) $$I_{IT}^{(W)} = r^{(W)}/r_{IT}^{(W)}$$

where $r^{(W)}$ is the ratio of the number of women to the number of men in the entire student body and $r_{IT}^{(W)}$ is the ratio of the number of women to the number of men in the IT field. The index has the advantage of being similar to concentration indices used in other areas of study.

For comparisons over time or across different racial/ethnic groups the index is represented as follows:

(A-2) $$I_{IT}^{j(W)} = r^{j(W)}/r_{IT}^{j(W)}$$

where $r^{j(W)}$ is the ratio of the number of women to the number of men in the entire group j, and $r_{IT}^{j(W)}$ is the ratio of the number of women to the number of men in the IT field in group j.

For example, in the fall of 2000 at UIC, close to half of the Blacks in the IT field (49%) were women while only 30% of Asians in IT were women. However, there was also a larger percentage of females in the overall Black undergraduate student body (73%) than in the Asian (51%). The gender gap index enables us to figure out which group had a larger gender gap. Using equation A-2, the gender indices for Blacks and Asians can be calculated. They were respectively 2.9 and 2.5, indicating a bigger gender gap for Blacks.

Appendix B

The logistic model we estimate is

(B-1) $$\ln[p/(1-p)] = \alpha + \beta \, ACT + u$$

where p is the probability of being an IT major and ACT is the math ACT score. The ratio p/(1–p) represents the odds of being an IT major.

For the gender gap, we add a gender variable to the model:

(B-2) $$\ln[p/(1-p)] = \alpha + \beta \, ACT + \delta \, Gender + \gamma \, ACT^*Gender + u$$

where, as indicated above, p/(1–p) represents the odds of being an IT major, and Gender is 1 for female students and 0 for male students. The regression results are listed in table 5.3 and illustrated in figures 5.1 to 5.4.

Notes

The authors would like to thank the reviewers and the volume's editors for their comments and suggestions. This research is supported by a grant from the National Science Foundation 0074604.

1.   Notwithstanding the drawbacks of standardized tests, students' math ACT scores are the best quantifiable indicator of student math performance, given the data available.
2.   UIC has a small and very selective GGPA (Guaranteed Graduate Professional Admissions) program that guarantees undergraduate students with very high ACT scores admission to professional schools, including medical schools. These few observations at the very high ACT scores are omitted in figures 5.1 and 5.2 to avoid confusion.

   Each dot in Figure 5.1 represents the observed probability of being an IT major at a given ACT score. Note that the distribution of student ACT scores is bell-shaped, so that there are more points in the middle of the ACT axis than at the high and low ends.

References

AAUW (American Association of University Women) Education Foundation. 1998. *Gender Gaps: Where Schools Still Fail Our Children*. Executive summary. Washington, D.C.: American Association of University Women Educational Foundation. http://www.aauw.org/research/GGES.pdf. Accessed February 1, 2005.

Angrist, Josh, and Victor Lavy. 1999. "New Evidence on Classroom Computers and Pupil Learning." Working paper w7424, National Bureau of Economic Research.

Carver, Doris L. 2000. "Research Foundations for Improving the Representation of Women in the Information Technology Workforce." Report on a virtual workshop held September 27–November 5, 1999. Directorate for Computer and Information Science and Engineering, National Science Foundation Report. http://www.nsf.gov/cise/cns/cwardle/it_women/itwomen_final_report.pdf. Accessed April 4, 2005.

CIRP (Cooperative Institutional Research Program), UCLA. 1998. *The American Freshman: National Norms for Fall 1998*. Los Angeles: American Council on Education.

———. 2001. *The American Freshman: National Norms for Fall 2001*. Los Angeles: American Council on Education.

Coley, Richard J. 2001. "Differences in the Gender Gap: Comparisons across Racial/Ethnic Groups in Education and Work." Princeton, N.J.: Educational Testing Services, Policy Information Center. http://www.ets.org/research/pic/gender.pdf. Accessed February 1, 2005.

Cooke, Sandra. 2000. "The Information Technology Workforce." Appendix to chapter 5 of *Digital Economy 2000*. U.S. Department of Commerce, June. https://www.esa.doc.gov/Reports/de2Kapp.PDF. Accessed February 1, 2005.

Dell'Angela, Tracy. 2001. "A Puzzling Gender Gap Emerges in Test Scores." *Chicago Tribune*, November 16: A1.

Digital Divide Network. 2000. "Digital Divide Beat." July 12. http://www.digitaldividenetwork.org.

Hoffman, Donna L., and Thomas P. Novak. 1999. "The Evolution of the Digital Divide: Examining the Relationship of Race to Internet Access and Usage over Time." http://elab.vanderbilt.edu/research/papers/pdf/manuscripts/EvolutionDigitalDivide-pdf.pdf. Accessed February 16, 2005.

Holland, Dorothy C., and Margaret A. Eisenhart. 1990. *Educated in Romance: Women, Achievement, and College Culture*. Chicago: University of Chicago Press.

Huang, Amy, Ashland Ring, Shelly Toich, and Teresa Torres. 1998. "Gender Inequalities in Education." In "The Effect of Computers on the Gender Gap in Education," a class project in the form of a "special issue of *GREAT: Gender Relations in Educational Applications of Technology* 1(1), March 16." http://www-cse.stanford.edu/classes/cs201/Projects/gender-gap-in-education/index.htm.

ITAA (International Technology Association of America). 1998. *Building the 21st Century Information Technology Workforce*.

Johnson, Sandra. 1987. "Gender Differences in Science: Parallels in Interest, Experience, and Performance." *International Journal of Science Education* 9: 467–81.

Kelly, Alison. 1981. "Sex Differences in Science Achievement." In *The Missing Half: Girls and Science Education*, ed. Alison Kelly, 22–41. Manchester: Manchester University Press.

———. 1988. "Sex Stereotypes and School Science: A Three Year Follow-Up." *Education Studies* 14: 151–63.

Margolis, Jane, and Allan Fisher. 2002. *Unlocking the Clubhouse: Women in Computing*. Cambridge, Mass.: MIT Press.

Martin, C. Dianne, and Caroline Wardle. 1999. "Paradigms, Pitfalls, and the Pipeline: Gender Issues in the Information Technology Workforce." In *Women and Technology: Historical, Societal, and Professional Perspectives. Proceedings,* 343–46. IEEE International Symposium on Technology and Society. New York: IEEE.

NSF (National Science Foundation). 2000. *Women, Minorities, and Persons with Disabilities in Science and Engineering: 2000.* Arlington, Va.: National Science Foundation (NSF 00-327). http://www.nsf.gov/sbe/srs/nsf00327. Accessed February 16, 2005.

———. Division of Science Resources Statistics. 2003. *Women, Minorities, and Persons with Disabilities in Science and Engineering: 2002.* Arlington, Va.: National Science Foundation (NSF 03-312). http://www.nsf.gov/sbe/srs/nsf03312. Accessed February 16, 2005.

Pastore, Michael. 2001. "IT Worker Shortage Continues." October 3. http://cyberatlas. internet.com/markets/professional/article/0;s5;s9;s71—896731,00.html. Accessed February 16, 2005.

Robb, Drew. 2003. "IT Gender Gap Widening." January 6. http://www.itmanagement. earthweb.com/career/article.php/1564501. Accessed February 16, 2005.

Schofield, Janet Ward. 1995. *Computers and Classroom Culture.* New York: Cambridge University Press.

Silverman, Suzanne, and Alice Pritchard. 1996. "Building Their Future: Girls and Technology Education in Connecticut." *Journal of Technology Education* 7(2): 41–54. http://scholar.lib.vt.edu/ejournals/JTE/v7n2/pdf/. Accessed February 16, 2005.

Sjoberg, Avein, and Gunn Imsen. 1998. "Gender and Science Education." In *Development and Dilemmas in Science Education,* ed. Peter J. Fensham, 218–48. London: Falmer.

Tam, Mo-Yin S., and Gilbert W. Bassett, Jr. 2000. "The Gender Digital Divide in Information Technology: A Quantile Approach." Mimeo.

Teachers Clearinghouse for Science and Society Education. 1999. "AAUW Highlights Technology Gender Gap." http://freeinfo.org/tch/spring99/articles/gengap. html. Accessed February 1, 2005.

U.S. Department of Commerce. 1995. *Falling through the Net: A Survey of the "Have Nots" in Rural and Urban America.* Washington, D.C.: National Telecommunication & Information Administration.

———. 1998. *Falling through the Net II: New Data on the Digital Divide.* Washington, D.C.: National Telecommunication & Information Administration.

———. 1999. *Falling through the Net: Defining the Digital Divide.* Washington, D.C.: National Telecommunication & Information Administration.

———. 2000. *Falling through the Net: Toward Digital Inclusion.* Washington, D.C.: National Telecommunication & Information Administration.

UCLA Cooperative Institutional Research Program (CIRP) Freshman Survey. 2001. Los Angeles: University of California, Los Angeles.

Warschauer, Mark. 2003. "Demystifying the Digital Divide." *Scientific American* 42 (August): 42–47.

Weiner, Kimberly. 1998. "The Technology Gender Gap: Yes, Virginia . . . We Can E-mail Santa." http://www.womenofcolorado.com/Articles/tt110298.asp. Accessed February 16, 2005.

Weinman, Janice, and Lisa Cain. 1999. "Technology—the New Gender Gap." *Technos* 8(1): 9–12.

6 African American Women in Science: Experiences from High School through the Post-secondary Years and Beyond

Sandra L. Hanson

In the United States, young women's level of participation in many aspects of science education continues to be lower than that of young men. As a result, women remain underrepresented in science and science-related occupations— one of the most elite and influential sectors of the U.S. labor force (Hanson et al. 1999; National Center for Education Statistics 2000a, 2000b; NSB 2000; NSF 2000). This shortage of women is a reflection of the continuing bias and gender inequity in science. A focus on the science experiences of women, as an undifferentiated group, has been typical of researchers and federal data collection agencies and policymakers. It has limited the amount of information available on subgroups of women. However, researchers increasingly have come to the conclusion that not all women have the same experiences in science education and occupations and that our understanding of the unique talents, interests, and experiences of subgroups of women is heightened when we consider their experiences through a multicultural lens (Hanson and Palmer-Johnson 2000; Mau, Domnick, and Ellsworth 1995).

Preliminary research has found that young African American women are particularly interested in and engaged in science—at least during the high school years (Hanson and Palmer-Johnson 2000; Mau, Domnick, and Ellsworth 1995; National Center for Education Statistics 2000a). In spite of this interest, African American women encounter racism and sexism in the science domain and remain underrepresented in science programs and science occupations (Kenschaft 1991; Malcolm et al. 1998; NSB 2000; NSF 2000; Vining-Brown 1994).

The goal of this research is to provide information on a wide array of young African American women's science experiences at various points in the science education system. Entry into science occupations in the early adult years is also examined. Reports by the National Science Foundation (2000) and the National Science Board (2000) have gone a long way in providing increasing amounts of information on diverse populations in science education and occupations. However, the data are often provided for race groups and gender groups, but not for race and gender subgroups. Even when subgroups are mentioned, the information provided is often minimal. The focus is usually on objective measures of science experience such as scores on standardized achievement tests, course-taking (or major), and type of occupation, with few insights into young women's attitudes about and perceptions of their abilities and future experiences in science. In addition, longitudinal data are seldom included, and information tends to be presented only for certain sectors of the science pipeline (e.g., secondary school vs. post-secondary school vs. occupations).

Questions about the level of young African American women's involvement in various aspects of science (access to it, achievement in it, attitudes toward it) at multiple points in time and the extent to which these experiences differ from those of young white women are of interest here. Research questions are answered using data from the National Educational Longitudinal Study (NELS) (National Center for Education Statistics 2002). These data provide a detailed map of the science experiences of a cohort of young African American women as they move from the eighth grade (surveyed in 1988) through high school (surveyed in 1990 and 1992) to post-secondary education (surveyed in 1994) and beyond (surveyed in 2000). This paper is one of the first to look at the science experiences of women in the newest (2000) NELS panel.

Background

A Multicultural Lens

The multicultural approach to understanding gender comes out of recent work by women of color who have attempted to correct past biases in the social sciences (Collins 1990; Zinn and Dill 1996). An awareness of the white, middle-class bias of much work on gender led to a focus on the intersection of race, class, and gender (Andersen and Collins 1995; Glenn 1985; Rothenberg 1992; West and Fenstermaker 1995). This work posits that it is the combination of statuses that makes for a particular group's attitudes, position, and opportunities (Lorber 2001). Collins (1990) has been particularly effective in describing this approach. She suggests that people can be simultaneously oppressed and oppressing, privileged and penalized; no one form of oppression is primary. Rather there are layers of oppression, within individual, community, and institutional contexts. Collins also argues that each of these locations is a potential site of resistance. This approach emphasizes the unique subcultures of minority women and the role of these cultures in affecting both the structures that limit and direct women's lives and the agency that allows them to retain some control and influence within these structures (Hanson and Kraus 1998). The unique subculture of African American gender systems is important background for understanding the science experiences of young African American women.

Research has suggested that gender systems in African American subcultures might provide young women with a unique set of resources—resources that might be important for generating interest and success in science. The cultural context of the African American community is one that historically saw women working and heading families (Andersen 1997). African American women continue to have high rates of labor force participation and do not perceive work and family roles as conflicting (Collins 1987). Instead of work being in opposition to motherhood, it is seen as an important dimension of motherhood. And it is a perceived incompatibility between science careers and family pursuits that keeps many women from entering and pursuing science degrees and occupations (Matyas 1986; Ware and Lee 1988).

Historical analyses have suggested that the tradition of male dominance in white families in the United States has not been replicated in African American families. In part because of the legacy of slavery, the African American family has been typified by greater equality in family decision making and division of labor (Gutman 1976; Hill 1971; Kane 2000). As a result of these arrangements it has been suggested that gender roles are more egalitarian here than in white families (Wade 1993). Wilcox (1990) found that African American women are more dissatisfied than white women with the amount of power women have in society, and Dugger (1988) found that they were also more aware of gender discrimination. These patterns contribute to greater self-esteem, independence, and assertiveness as well as high educational and occupational expectations among young African American women (and their parents) relative to other women (Andersen 1997; Hanson and Palmer-Johnson 2000; Hill and Sprague 1999; National Center for Education Statistics 2000a). All of these characteristics have been shown to be related to success in science (Hanson 1996). In my earlier examination of the high school science achievement process (Hanson and Palmer-Johnson 2000), my colleague and I found that African American families compensate for disadvantages in some resources (e.g., socioeconomic status) by providing young women with an excess of other resources (e.g., unique gender ideologies, work expectations, and maternal expectations). And unlike white parents, they sometimes provide more of these resources to their daughters than their sons. Similarly, Higginbotham and Weber (1992) found that African American families put a greater stress on education and occupation as sources of mobility for their daughters than do white families, since they do not see marriage as a source of mobility (as white families are more likely to).

African American Women in Science

Although the last few decades have seen progress for women in science, the sciences continue to be a male domain (Hanson et al. 1999; National Center for Education Statistics 2000a, 2000b; NSB 2000; NSF 2000). The culture of science was historically, and is currently, a male culture that is often hostile to women and minorities (Harding 1986; NSF 2000; Rossiter 1982). The tendency in science research has been to talk about "women's" experiences and to focus on women or minorities without acknowledging variation within these groups.

Hence there is relatively little research on minority women in science (Bur-bridge 1991; Catsambis 1995; Clewell and Anderson 1991). Does the science system equally discourage all women from entering the sciences? There is little evidence available to answer this question. However, the literature discussed here suggests a construction of gender in the African American community which does not mirror that in the white community. The confluence of a science system that is hostile to women, minorities, and most likely minority women with an African American gender system that might provide unique resources in science is of interest here.

Some have argued that since women do less well in science than men and minority group members do less well in science than members of the majority, young African American women in science will be doubly disadvantaged (Clew-ell and Anderson 1991; Vining-Brown 1994). The argument is one of double jeopardy, assuming that the two statuses—female and African American—have an additive effect. It is true that members of non-Asian minorities and women do less well in science than do whites and men, respectively (NSF 2000), and there is some evidence that young African American women experience special barriers in science (Carwell 1977; Clewell and Anderson 1991; Hueftle, Rakow, and Welch 1983; Malcolm 1976; Vining-Brown 1994). However, there is a grow-ing (although still limited) body of research that suggests that in spite of the barriers that the science system sets up for women, minorities, and minority women, we cannot assume that members of these groups will be equally disin-terested in science. Some research shows that African American youth in general hold more positive attitudes about science than any other subgroup (Hueftle, Rakow, and Welch 1983) and that African American girls in particular are very positive about science (Creswell and Exezidis 1982), sometimes more so than their white counterparts (Mau, Domnick, and Ellsworth 1995). Interestingly, African American women are more likely than other women of color to hope to enter the field of mathematics (a male domain related to science) (Duran 1987; Kenschaft 1991).

My earlier research on African American women's science experiences during the high school years revealed that African American women do as well as (and sometimes better than) white women on a majority of science measures (Han-son and Palmer-Johnson 2000). Recent data from the National Science Founda-tion (2000) suggest that experiences in science from post-secondary school through occupations are distinctly different for women from different racial/ethnic groups. The data suggest that African American females are increasingly present in science. For example, African American women earned more than half of the bachelor's degrees in science and engineering awarded to their racial/ethnic group in 1997. And (during the same year) African Americans were the only racial group among which women earned over half of the master's degrees in science and engineering. African American women earned 46 percent of the Ph.D.s awarded to African Americans in science and engineering in 1997, while white women earned 38 percent of the degrees awarded to whites. Finally, data from the National Science Foundation (2000) suggest that African American women make up a much larger proportion of African American scientists (36%) than white women make up of white scientists (22%). It should be noted that many of these percentages were not presented in the body of the National Science Foundation report (2000) but rather were calculated from figures in appendix A. When these figures are examined with a focus on percentages of women within racial groups that are pursuing science degrees and occupations, the results often suggest a distinctly higher representation among African American women than among white women. It is important to keep in mind, however, that although the percentages are sometimes higher for African American women than for white women, the raw numbers are often very small. For example, although (within racial/ethnic groups) African American women earned a higher percentage of science and engineering Ph.D.s than did white women in 1997, the actual number of African American women and white women earning these degrees was 280 and 5,180, respectively (NSF 2000).

Although it cannot be assumed that African American women will be more discouraged in science than white women, this is not to say that these young women are able to avoid racism and sexism in their pursuit of science. Existing studies of African American women in science document considerable barriers that these women face (Kenschaft 1991; Malcolm 1976; Sammons 1990; Vining-Brown 1994). However, the potential for continued interest in and engagement with science in the white male science domain is what my research examined. In later phases of the project, the dynamics and sources of support and discouragement will be explored in greater depth.

Research Questions

A number of questions are addressed in this research. First, what are the science experiences (access, achievement, and attitudes) of young African American women in the eighth grade? A review of the reports on science education leads to the hypothesis that young African American women will start out with a high level of interest and course-taking in science. Race comparisons may show a level playing field, or maybe even that young African American women have an advantage over young white women. Young African American women will not, however, experience high levels of science achievement (as measured by standardized test scores) relative to white women.

Another question of interest is, how do the science experiences of young African American women change as they pass through high school, post-secondary school, and their early years in the labor force? How do these experiences compare to those of young white women? It is expected that as the young women progress through the white, male-dominated education system, their interest and course-taking (or likelihood of majoring or getting a degree or eventually a job) in science will lessen but, as a group, they may still be more likely than young white women to pursue science.

Data and Methods

The first goal of the study is to chronicle the science education experiences of a nationally representative group of young African American women. The National Educational Longitudinal Survey (NELS) provides an excellent data source for this activity. NELS is a nationally representative, longitudinal data set collected under the auspices of the National Center for Education Statistics (2002). The base year data for NELS were collected in 1987–88 on a nationally representative sample of 24,599 eighth graders (aged approximately thirteen) from one thousand schools. They were interviewed again in 1990 when they were in the tenth grade, in 1992 when they were high school seniors, and in 1994 when they were two years out of high school. The fourth follow-up was conducted in 2000 (when they were aged approximately twenty-five). The data include extensive information on a wide variety of science education experiences. Analyses are based on those who participated in all five survey years. Those who dropped out of high school are not included. All analyses include weights that control for sample attrition and non-response. Since NELS weights were created to project to the total U.S. high school youth population, I adjusted the weights back to sample size. The sample I used consists of 581 young African American women and 3,365 young white women.

The NELS data are a valuable tool for showing the science experiences of representative samples of U.S. youth. Their longitudinal nature makes them an excellent source of insight for transitions in the science educational pipeline and thus the development (and loss) of youth who exhibit talent in science. The data are frequently used to guide educators and policymakers. But no one has previously used this data set to focus specifically and in detail on the experiences of young African American women. With the most recent (2000) panel of information on the NELS cohort, NELS provides the richest source of information available on the science experiences of young people from age 13 (middle school) through the high school, undergraduate, and postcollege years.

Measures of Science Experiences (NELS)

NELS has excellent longitudinal data (over a twelve-year period) on the science experiences of a cohort of young people who were in eighth grade in 1988. In this research I did not include the social sciences in the definition of science. Majors and occupations involving physical science, mathematics, medical science, computer science, engineering, and technology were included. Three of the most important aspects of science experience are access to science, achievement in it, and attitudes toward it (Hanson 1996). NELS includes extensive measures of each (see table 6.1 for a full list of these measures within each survey year). For example, measures of access include enrollment in accelerated science classes (eighth grade), chemistry (tenth grade), and university/college physics classes (two years out of high school), and majoring in or earning a de-

Table 6.1. Means (and Standard Deviations) for Science Variables for
Young Women by Race: NELS

	African American Females	White Females
Science access		
1988 (8th grade)		
In advanced, enriched, or accelerated science courses (1 = yes; 0 = no)	.40 (.49) **	.23 (.42) **
1990 (10th grade)		
Coursework in chemistry (1 = one-half to two years; 0 = none)	.18 (.38)	.18 (.38)
1992 (12th grade)		
Enrolled in science classes in the last two years (1 = yes; 0 = no)	.89 (.32)	.89 (.31)
Taking a science class this term (1 = yes; 0 = no)	.43 (.49) **	.51 (.50) **
1994		
Taking/taken courses in physics last 2 years[a] (1 = yes; 0 = no)	.08 (.27) **	.12 (.32) **
Taking/taken courses in chemistry last 2 years[a] (1 = yes; 0 = no)	.29 (.45)	.29 (.46)
Science major at post-secondary institution[a] (1 = yes; 0 = no)	.25 (.43) **	.14 (.34) **
2000		
First post-secondary degree—science[b] (1 = yes; 0 = no)	.14 (.35)	.14 (.34)
Second post-secondary degree—science[b] (1 = yes; 0 = no)	.08 (.27)	.09 (.29)
Science degree expected by age 30 (1 = yes; 0 = no)	.26 (.43)**	.15 (.36)**
Science achievement		
1988 (8th grade)		
Science grades from 6th grade until now (1 = mostly A's; 0 = B's or less)	.34 (.47) **	.38 (.49) **
Science standardized score	45.12 (7.91) **	52.73 (9.11) **

(Continued on the next page)

Table 6.1. *Continued*

	African American Females	White Females
1990 (10th grade)		
Science grades		
(1 = mostly A's; 0 = B's or less)	.38 (.48) **	.43 (.50) **
Science standardized score	43.96 (7.09) **	52.06 (8.98) **
Science standardized score	43.02 (8.12) **	52.10 (8.87) **
1994		
First occupation—science		
(1 = yes; 0 = no)	.02 (.15)	.02 (.13)
Most recent occupation—science		
(1 = yes; 0 = no)	.03 (.17)	.02 (.15)
Occupation 1992—science		
(1 = yes; 0 = no)	.02 (.15)	.02 (.13)
Occupation 1993—science		
(1 = yes; 0 = no)	.02 (.16)	.02 (.13)
Occupation 1994—science		
(1 = yes; 0 = no)	.04 (.19)	.02 (.14)
2000		
Current/most recent occupation—science		
(1 = yes; 0 = no)	.22 (.42) **	.17 (.37) **
Science attitudes		
1988 (8th grade)		
Usually look forward to science class		
(1 = strongly agree or agree; 0 = disagree or strongly disagree)	.59 (.49)*	.55 (.50)*
Science will be useful in my future		
(1 = strongly agree or agree; 0 = disagree or strongly disagree)	.70 (.46) **	.65 (.48) **
1990 (10th grade)		
Often work hard in science class[c]		
(1 = almost every day; 0 = less than every day)	.64 (.48) **	.55 (.50) **

(Continued on the next page)

Table 6.1. *Continued*

	African American Females	White Females
1994		
Interested in science[c]		
(1 = important; 0 = not as important)	.42 (.50) *	.49 (.50) *
Do well in science[c]		
(1 = important; 0 = not as important)	.48 (.50)	.50 (.50)
2000		
Planned occupation by age 30—science		
(1 = yes; 0 = no)	.31 (.46) **	.24 (.43) **

* significant at .10 level
** significant at .05 level
[a] questions asked only of those in college at time of interview
[b] questions asked only of those who had attended or were attending college at time of interview
[c] questions asked only of those taking science

gree in science (including math, engineering, and technology) since leaving high school (two and eight years out of high school). Measures of achievement include (among others) grades in science and standardized science test scores during the high school years as well as science occupations in the post–high school years. Measures of attitudes include (among others) responses to questions about looking forward to science classes, feeling challenged in these classes, and being interested and doing well in science, as well as others which ask about the importance of science in the respondent's future (measured during the high school years). A question about the occupation that the respondent plans to have at age thirty (coded here as science or not science) is also included in the attitudinal questions.

Analyses

I use descriptive statistics to detail the science experiences of a cohort of young African American women (and the young white women in the comparison group) in the 1988 through 2000 NELS panels. T-tests are used to determine whether the experiences of the two groups of women are significantly different at each point in time. In order to take full advantage of the longitudinal data, odds ratios for continuing in science given earlier course-taking, achievement, and interest are calculated for the young African American women and for the young white women. Odds of continuing in an area of science given earlier success in that area (access, achievement, attitudes) as well as odds of doing well on other science indicators given success in one (e.g., odds of taking chemistry in tenth grade given positive attitudes in eighth grade) are also calculated for the two groups.[1] This analysis will provide insight about the effect of early success and interest in science (e.g., in the eighth grade) on later science experiences. Researchers are increasingly documenting the early course-taking and positive science attitudes of young African American women. The unique longitudinal data provided in NELS are important for their ability to connect these early science experiences to later ones, aiding the attempt to chronicle and explain the science experiences of these minority women from the early high school years to the post-secondary years and beyond. Later analyses will use the NELS data on families, schools, and peers as well as qualitative data being collected in a Web survey to help explain the trends found here.

Findings

Science Access. Means showing young women's access to science, by race, are reported in table 6.1. As expected, there are a number of variables on which there are no race differences and a number on which there is an African American advantage. Fewer comparisons show a white advantage than an African American advantage. Results suggest that in eighth grade, young African American women were much more likely to be in advanced, enriched, or accelerated science courses than were young white women (40% vs. 24%). In tenth grade there were no differences in the amount of coursework in chemistry. Similarly, in twelfth grade there were no race differences in the young women's enrollment in science classes over the past two years, although young white women were more likely to be taking a science class at the time of the survey. When those attending college were asked (two years out of high school) about course-taking and majors, there were no differences in chemistry course-taking.

However, young white women were more likely to have taken a physics course (12% vs. 8%). But it was young African American women who were more likely (almost twice as likely) to report a science major (at their first college or university attended, 25% vs. 14%). By the time the young women were eight years out of high school there were no race differences in their report of whether or not their first or second degree was in science. But the African American women were considerably more likely to report that they would like a degree in science by age thirty (26% vs. 15%).

Science Achievement. Means showing young women's achievement in science, by race, are also reported in table 6.1. As expected, there is considerable white advantage here. But when it is occupational achievement (not school achievement as measured by grades or standardized test scores) that is examined, this white advantage disappears. Starting in eighth grade, the young white women were more likely to get higher grades in science (in tenth grade as well) and to score higher on standardized science exams (in ninth through twelfth grades). The race difference in science exam scores was consistent across the three measurement periods. However, eight years out of high school (2000) it was young

African American women who were more likely to report that their current or most recent job was in science (22% vs. 17%). And all other job reports (measured two years out of high school) show no race difference.

Science Attitudes. Means for variables measuring science attitudes presented in table 6.1 reveal a distinctly positive attitude toward science on the part of African American women in the early years of high school. These young women, when asked in the eighth grade (1988), were more likely than young white women to look forward to science class (59% vs. 55%). In the same year, they were more likely than young white women to feel that science would be useful in their future (70% vs. 65%). Two years later (1990), in tenth grade, they were more likely than young white women to say that they often work hard in science class (64% vs. 55%). However, by the last year of high school (1992) these patterns shifted and it was young white women who claimed more interest in science (49% vs. 42%). In the same year, there were no race differences in respondent reports on how well they did in science. But the trends reverse again in the early adult years (2000), when the young women have been out of high school for eight years. Almost one-third (31%) of the young African American women reported that the occupation that they plan to have at age thirty will be in science. Less than one-quarter of young white women (24%) reported such plans.

Odds of Continuing in Science. Tables 6.2 and 6.3 show the odds for young African American and young white women of continuing in science given earlier engagement in science. Table 6.2 shows the odds of continuing within areas of science (e.g., earlier access leading to later access) and table 6.3 shows the odds of continuing within one area given engagement in another (e.g., earlier attitudes leading to later access). Findings in table 6.2 indicate that within science access, the odds of a young woman (African American or white) continuing in science courses and programs given earlier course-taking are significantly higher than for young women who did not have earlier access to science. For example, the odds that a young woman who was in eighth-grade advanced science would go on to take chemistry in tenth grade were .36 for African American women and .44 for white women. For young women who had not been in the eighth-grade advanced science class the odds of being in tenth-grade chemistry were considerably lower (.12 for African American women and .17 for white women). However, in two of the three sequences examined, the odds of continuing on were higher for African American women than for white women.

Table 6.2 also shows that in two of the three science achievement sequences, earlier achievement was significantly related to later achievement for both African American and white women. The connection between high science grades at earlier and later points was higher for young white women than for young African American women. However, the connection between having a science occupation two years out of high school and having a science occupation eight years out of high school was higher for the African American women than for the white women.

Finally, table 6.2 shows the odds of earlier positive attitudes leading to later positive attitudes toward science. Here, for the young African American women

Table 6.2. Odds of Young African American Women Having High Science Access, Achievement, and Attitudes Given Earlier Access (Odds for Young White Women Are in Parentheses): NELS

Access	Odds+	Achievement	Odds+	Attitudes	Odds+
8th Grade Advanced Science/ 10th Grade Chemistry	.36/.12*** (.44/.17)***	8th Grade Science Grades/ 10th Grade Science Grades	.64/.49* (2.23/.38)***	8th Grade: Sciences Useful/ 10th Grade: Work Hard in Science	2.66/1.07*** (1.33/1.09)***
10th Grade Chemistry/ Science Major, 2 Years post-HS	.45/.33* (.20/.15)***	10th Grade Science Grades/ Science Occupation 2 Years post-HS	.01/.03 (.02/.02)	10th Grade: Work Hard in Science/Interested in Science 2 Years post-HS	.64/.70 (1.24/.68)***
Science Major, 2 Years post-HS/ Expect Science Degree by Age 30, 8 Years post-HS	.91/.21*** (.84/.11)***	Science Occupation 2 Years post-HS/Science Occupation 8 Years post-HS	1.3/.17*** (.50/.14)***	Interested in Science 2 Years post-HS/Plan Science Occupation by Age 30	.58/.55 (.66/.26)***

*** Chi-square for association between earlier and later science experience is significant at .05 level
** Chi-square for association between earlier and later science experience is significant at .10 level (African American women only)
* Chi-square for association between earlier and later science experience is significant at .20 level (African American women only)
+ Odds of later success given earlier success/odds of later success without earlier success

Table 6.3. Odds of Young African American Women Having Success in One Area of Science Given Success in Other Areas (Odds for Young White Women Are in Parentheses): NELS

Attitudes/Access	Odds+	Access/Achievement	Odds+
8th Grade: Science Useful/Science Major 2 Years post-HS	.45/.20*** (.20/.01)***	10th Grade Chemistry/Science Occupation 8 Years post-HS	.58/.25*** (.26/.19)***
8th Grade: Look Forward to Science/Taken Physics Course 2 Years post-HS	.14/.05*** (.15/.13)	10–12th Grade Science Enrollment/Science Occupation 8 Years post-HS	.32/.13*** (.22/.12)***
10th Grade: Work Hard in Science/Taken Chemistry Course 2 Years post-HS	.50/.29*** (.44/.37)***	Taken Physics Courses 2 Years post-HS/Science Occupation 8 Years post-HS	.77/.35** (.53/.19)***
10th Grade: Work Hard in Science/Science Major 2 Years post-HS	.44/.22*** (.15/.18)	Taken Chemistry Course 2 Years post-HS/First Occupation Was in Science	.07/.00* (.03/.01)
10th Grade: Work Hard in Science/Expect Science Degree by Age 30, 8 Years post-HS	.37/.22* (.17/.19)	Taken Chemistry Course 2 Years post-HS/Science Occupation 8 Years post-HS	.60/.30** (.56/.13)***

*** Chi-Square for association between earlier and later science experience is significant at .05 level

** Chi-Square for association between earlier and later science experience is significant at .10 level (African American women only)

* Chi-Square for association between earlier and later science experience is significant at .20 level (African American women only)

+ Odds of later success given earlier success/odds of later success without earlier success (but not the white women), earlier experiences are not as strongly related to later experiences as they were in the areas of science access and achievement. Although the odds of having positive attitudes in tenth grade were higher for those who had positive attitudes in eighth grade than for those who did not, later sequences did not show this trend. These findings suggest that young African American women with positive attitudes about science early on may not continue to be as positive about science in later years. But they also suggest that an absence of positive attitudes earlier does not preclude having positive attitudes later. It makes sense that early and later engagement is less highly correlated in science attitudes than it is in access and achievement, where course-taking and knowledge tend to be cumulative along a set curriculum.

Table 6.3 shows the sequences for which the odds of having success in one area of science were significantly related to the odds of success in another area for the sample of African American women. A consideration of all possible sequences revealed significant associations in only two areas. These involved the influence of earlier positive attitudes on later access and the influence of earlier access on later achievement. In some but not all of the cases, the relation between earlier success and later success was also significant for young white women. The findings add important detail to my earlier findings that show significant science course-taking and positive science attitudes among young African American women in the early high school years. These early experiences in science are setting the stage for success in other areas of science as well. Interestingly, it is the early course-taking that seems to increase the odds of being in science occupations in later years—the area of science achievement where young African American women sometimes have an advantage over young white women.

In the research presented here, a multicultural gender framework was used to gain insight into the aspirations and characteristics encouraged among young women in the African American community. This examination suggested that simple assumptions about the mismatch between women and science are often based on the experiences of white women. In fact, in the African American community, gender is constructed in a very different way and many of the characteristics that are considered appropriate for females (e.g., high self-esteem, independence, and assertiveness, as well as high educational and occupational expectations) are not inconsistent with characteristics that contribute to success in science. It is this push into the sciences in the context of the white, male science culture, which is often hostile to those who are not white or not male, that makes young African American women's experiences an important area of study.

This examination of young African American women's science experiences from eighth grade through the early adult years (twelve years later) in three aspects of science (access, achievement, and attitudes) revealed considerable interest in and access to science among young African American women. However, it is in attitudes, even more than access, where young African American women distinguish themselves. Starting in eighth grade, and continuing into the early adult years, young African American women are often more positive about science than their white counterparts.

Findings from the NELS survey also show considerable access to science (course-taking) among young African American women. As did attitudes, this greater access continued through the post–high school years. One of the largest race differences revealed by the NELS data involved the percentage of women taking advanced, enriched, or accelerated science courses in eighth grade— 40 percent of young African American women but only 23 percent of young white women. Although the percentages were smaller in the post–high school years, young African American women still outpaced young white women on some indicators of science access. For example, the young African American women were more likely to report (two years out of high school) that they had majored in science at a post-secondary institution and (eight years out of high school) that they expected a degree in science by age thirty. Notably, a consideration of the relationship between science success in the early years and success in later years revealed that this early access to science courses and programs is critical in increasing the odds that young African American women will have science jobs in their early adult years.

It was in the area of science achievement that young African American women in the NELS sample fared the least well. When achievement in science education (as measured by grades and standardized exams) was considered, young white women in the sample always scored higher. These findings are corroborated by other research showing that young African American women score lower on standardized science exams than do young white women (Clewell and Anderson 1991; Hanson 1996). Testing biases that favor middle-class white students are most likely part of the explanation for this trend (Lomax et al. 1995). It should be noted, however, that one of the most unexpected results from the NELS survey came in achievement and it revealed an advantage for the young African American women. When they were asked as young adults (eight years out of high school) about their current or most recent job, African American women were more likely than white women to report a job in science. This finding should not come as a surprise given some of the recent data available on minority women in science occupations (NSF 2000).

Although young African American women often express more interest in science than their white counterparts, their interest nevertheless lessens over time. For example, in eighth grade, 59 percent of the young women looked forward to science class and 70 percent said science would be useful in their future. But by the time they had been out of high school for two years only 42 percent were interested in science (this is less than among white women) and 48 percent said they did well in science. And finally, in their eighth year out of high school, only 31 percent of the young African American women planned to have a job in science by the age of thirty. This "cooling out" was expected, given the fact that these young women face not just sexism, but also racism in their pursuit of science. Malcom's (1976) research on African American women scientists reveals considerable barriers, including low expectations held by teachers, especially by white teachers in integrated school systems. Other reports on African American women scientists support these claims of racism (Kenschaft 1991). Another important factor in the chilly climate that young African American women experience is that students are seldom made aware of the contributions of African Americans (much less African American women) in science (Sammons 1990; Van Sertima 1983). This invisibility, together with the small number of minority women available as teachers and mentors (Jordan 1999), creates obstacles for young African American women. Colleges that have been the most successful in encouraging minority women scientists are women's colleges and historically black colleges and universities (HBCUs), especially women's HBCUs, which have diverse faculties and give confidence to women in science (Jordan 1999).

This research revealed a large science talent base among young African American women. It constitutes one of the most careful examinations, to date, of the science experiences of these young women in both science education and occupations over time. One of the most unexpected findings revealed here was African American women's continued interest and presence in science beyond the high school years. It is important to note, however, that the number of young African American women who are interested and involved in science in the early adult years is much lower than the number who are interested and involved in the early high school years. The concern is that educational and occupational institutions are not fully developing all science talent. When gender and skin color are major factors determining who will do science, a considerable amount of science talent will be lost. The implications of this loss for scientific discovery and advancement are considerable. The implications are also great for the young people who are denied access to science, since they will not be involved in the creation of policies and technologies that will guide us through the next century.

Notes

This research was funded by grant REC-0208146 from the Division of Research, Evaluation, and Communication of the National Science Foundation. The opinions expressed do not necessarily reflect the position of the National Science Foundation. The author thanks Michelle Jiles for her analytic support.

1.   Odds of continuing within an area of science were calculated for each sequence within each area of science, e.g., from eighth to tenth grade, tenth to twelfth, etc. These odds are shown in table 6.2. In the case of the odds of continuing on in one area given earlier engagement in another area, the number of combinations was large and not all could be shown. I limited those shown in the table to the ones that were significant for the African American sample.

References

Andersen, Margaret L. 1997. *Thinking about Women: Sociological Perspectives on Sex and Gender*. Boston, Mass.: Allyn and Bacon.

Andersen, Margaret L., and Patricia Hill Collins, eds. 1995. *Race, Class, and Gender: An Anthology*. 2nd ed. Belmont, Calif.: Wadsworth.

Burbridge, Lynn B. 1991. "The Interaction of Race, Gender, and Socioeconomic Status in Education Outcomes." Working paper no. 246. Wellesley, Mass.: Center for Research on Women, Wellesley College.

Carwell, Hattie. 1977. *Blacks in Science: Astrophysicist to Zoologist*. Hicksville, N.Y.: Exposition.

Catsambis, Sophia. 1995. "Gender, Race, Ethnicity, and Science Education in the Middle Grades." *Journal of Research in Science Teaching* 32(3): 243–57.

Clewell, Beatriz Chu, and Bernice Anderson. 1991. *Women of Color in Mathematics, Science, and Engineering: A Review of the Literature*. Washington, D.C.: Center for Women Policy Studies.

Collins, Patricia Hill. 1987. "The Meaning of Motherhood in Black Culture." *Sage* 4(Fall): 3–10.

———. 1990. *Black Feminist Thought*. Boston, Mass.: Unwin Hyman.

Creswell, John L., and Roxane H. Exezidis. 1982. "Research Brief—Sex and Ethnic Differences in Mathematics Achievement of Black and Mexican-American Adolescents." *Texas Tech Journal of Education* 9(3): 219–22.

Dugger, Karen. 1988. "Social Location and Gender-Role Attitudes: A Comparison of Black and White Women." *Gender and Society* 2(4): 425–48.

Duran, Richard P. 1987. "Hispanics' Pre-college and Undergraduate Education: Implications for Science and Engineering Studies." In *Minorities: Their Underrepresentation and Career Differentials in Science and Engineering*, ed. Linda S. Dix, 73–128. Washington, D.C.: National Academy Press.

Glenn, Evelyn N. 1985. "Racial Ethnic Women's Labor: The Intersection of Race, Gender, and Class Oppression." *Review of Radical Political Economics* 17: 8–108.

Gutman, Herbert G. 1976. *The Black Family in Slavery and Freedom*. New York: Vintage.

Hanson, Sandra L. 1996. *Lost Talent: Women in the Sciences*. Philadelphia: Temple University Press.

Hanson, Sandra L., Stefan Fuchs, Silke Aisenbrey, and Nataliya Kravets. 1999. "A Comparative Study of Female Elites: Gender Attitudes among Women Scientists in Germany and the U.S." Paper presented at the annual meeting of the American Sociological Association, Chicago.

Hanson, Sandra L., and Rebecca S. Kraus. 1998. "Women, Sport, and Science: Do Female Athletes Have an Advantage?" *Sociology of Education* 71(2): 93–110.

Hanson, Sandra L., and Elizabeth Palmer-Johnson. 2000. "Expecting the Unexpected: A Comparative Study of African American Women's Experiences in Science during the High School Years." *Journal of Women and Minorities in Science and Engineering* 6(4): 265–94.

Harding, Sandra. 1986. *The Science Question in Feminism*. Ithaca, N.Y.: Cornell University Press.

Higginbotham, Elizabeth, and Lynn Weber. 1992. "Moving Up with Kin and Community: Upward Social Mobility for Black and White Women." *Gender and Society* 6(3): 416–40.

Hill, Robert B. 1971. *The Strengths of Black Families*. New York: Emerson Hall.

Hill, Shirley A., and Joey Sprague. 1999. "Parenting in Black and White Families: The Interaction of Gender and Class and Race." *Gender and Society* 13(4): 480–502.

Hueftle, Stacy J., Steven J. Rakow, and Wayne W. Welch. 1983. *Images of Science*. Minneapolis: University of Minnesota, Science Assessment and Research Project.

Jordan, Diann. 1999. "Black Women in the Agronomic Sciences: Factors Affecting Career Development." *Journal of Women and Minorities in Science and Engineering* 5(2): 113–28.

Kane, Emily W. 2000. "Racial and Ethnic Variations in Gender-Related Attitudes." *Annual Review of Sociology* 26: 419–39.

Kenschaft, Patricia C. 1991. *Winning Women into Mathematics*. Washington, D.C.: Mathematical Association of America.

Lomax, Richard G., Mary M. West, Maryellen C. Harmon, Katherine A. Viator, and George F. Madaus. 1995. "The Impact of Standardized Testing on Minority Students." *Journal of Negro Education* 64(2): 171–85.

Lorber, Judith. 2001. *Gender Inequality: Feminist Theories and Politics*. Los Angeles, Calif.: Roxbury.

Malcolm, Shirley M. 1976. *The Double Bind: The Price of Being a Minority Woman in Science*. Washington, D.C.: American Association for the Advancement of Science.

Malcolm, Shirley M., et al. 1998. "Equity and Excellence: Compatible Goals; An Assessment of Programs with Increased Access and Achievement of Females and Minorities in K–12 Mathematics Education." Report for the American Association for the Advancement of Science (AAAS-84-14).

Matyas, Marsha Lakes. 1986. "Persistence in Science-Oriented Majors: Factors Related to Attrition among Male and Female Students." Paper presented at the annual meeting of the American Educational Research Association, San Francisco.

Mau, Wei-Cheng, Margaret Domnick, and Randolph A. Ellsworth. 1995. "Characteristics of Female Students Who Aspire to Science and Engineering or Homemaking Occupations." *Career Development Quarterly* 43: 323–37.

National Center for Education Statistics. 2000a. *Entry and Persistence of Women and Minorities in College Science and Engineering Education*. By Gary Huang, Nebiyu Taddese, and Elizabeth Walter. NCES 2000-601. Washington, D.C.: U.S. Department of Education.

———. 2000b. *Trends in Educational Equity of Girls and Women*. By Yupin Bae, Susan Choy, Claire Geddes, Jennifer Sable, and Thomas Snyder. NCES 2000-030. Washington, D.C.: U.S. Department of Education.

———. 2002. *National Educational Longitudinal Study: 1988–2000*. Data files and Electronic Codebook System—Base year through fourth follow-up ECB/CD-ROM, Public Use (NCES 2002-322 CD ROM). Washington, D.C.: U.S. Department of Education, National Center for Education Statistics.

NSB (National Science Board). 2000. *Science and Engineering Indicators—2000*. Arlington, Va.: National Science Foundation.

NSF (National Science Foundation). 2000. *Women, Minorities, and Persons with Disabilities in Science and Engineering: 2000*. Arlington, Va.: National Science Foundation (NSF 00-327). http://www.nsf.gov/sbe/srs/nsf00327. Accessed February 16, 2005.

Rossiter, Margaret S. 1982. *Women Scientists in America: Struggles and Strategies to 1940*. Baltimore, Md.: Johns Hopkins University Press.

Rothenberg, Paula S. 1992. *Race, Class, and Gender in the United States.* New York: St. Martin's.

Sammons, Vivian Ovelton. 1990. *Blacks in Science and Medicine: A Biographical Directory of Black Scientists.* Washington, D.C.: Hemisphere.

Van Sertima, Ivan. 1983. *Blacks in Science: Ancient and Modern.* New Brunswick, N.J.: Transaction.

Vining-Brown, S. 1994. *Minority Women in Science and Engineering Education.* Princeton, N.J.: Educational Testing Service.

Wade, Bruce H. 1993. "The Gender Role and Contraceptive Attitudes of Young Men: Implications for Future African American Families." *Urban League Review* 16(2): 57–65.

Ware, Norma C., and Valerie E. Lee. 1988. "Sex Differences in Choice of College Majors." *American Educational Research Journal* 25: 593–614.

West, Candace, and Sarah Fenstermaker. 1995. "Doing Difference." *Gender and Society* 9(1): 8–37.

Wilcox, Clyde. 1990. "Race, Gender Role Attitudes, and Support for Feminism." *Western Political Quarterly* 43(1): 113–21.

Zinn, Maxine B., and Bonnie T. Dill. 1996. "Theorizing Difference from Multiracial Feminism." *Feminist Studies* 22(2): 321–31.

7   African Women Pursuing
Graduate Studies in the Sciences:
Racism, Gender Bias, and Third
World Marginality

Josephine Beoku-Betts

In recent years, there has been much discussion among social scientists, feminist scholars, and policymakers of the fact that, while women have made some progress in the sciences, in most societies there is still a gender gap and loss of women in science-related occupations (Etzkowitz, Kemelgor, and Uzzi 2000; Hanson 1996). In Third World societies especially, women have lower enrollments in science courses (particularly outside the biological sciences), lower achievement levels, and a negative attitude toward science disciplines, which are generally viewed as a male domain (Gender Working Group 1995; Harding and McGregor 1995). Studies of women's labor market participation in scientific occupations also suggest that while women are being hired, they are absent from specifically high-level talent pools (Hanson, Schaub, and Baker 1996; Rayman and Brett 1993). For example, women constitute less than one-third of the faculty in most universities and are significantly less likely than men to hold top-level administrative or professorial positions in the academic hierarchy (Currie, Thiele, and Harris 2002, 35).

In identifying factors that influence these trends in women's science education and careers, scholars have focused mainly on the gender-differentiated experiences of White, middle-class, North American and European scientists (Harding 1991, 1998). The main argument has been that men tend to have better access than women to resources and that this fosters the maintenance of gender-based systems of inequality in the sciences (Hanson 1996). Discrimination on the basis of race, ethnicity, and class are also additional problems women of color from Europe, North America, and Third World societies face in gaining entry to, and promotion in, scientific fields. However, very little of the comparative research has addressed the needs and experiences of women of color in these diverse contexts (Beoku-Betts 2000; Harding 1991; Jordan 1999; Mukhopadhyay 1994; Sands 1993; Subrahmanyan 1998; Sur 2001).

Harding (1991) views the lack of representation and neglect of the experiences of Black women in discourses on gender and science as a consequence of racist practices that have silenced people of African descent in the practice of science. This has made it difficult to examine ways in which people in diverse social contexts have interacted and responded to Western scientific and technological traditions and also to determine what people of European descent can learn about themselves from the experiences of others. One of the consequences is that only partial and distorted accounts can be produced of American and European experiences as well as the experiences of historically dominated and marginalized societies in the Third World. Regarding this issue, Collins also points out that "Gender analyses that rely too heavily upon the experiences of middle-class White women in the West inadvertently replicate structures of racial privilege while claiming to dismantle gender privilege" (1999, 271).

This chapter examines the experiences of African women as graduate students in scientific disciplines. These women studied in universities in North America and Europe between the 1960s and 1990s and are currently in academic positions in their respective countries. I focus on how issues of race and gender bias and marginality as Third World subjects affected their experiences as graduate students in scientific disciplines. I also examine the extent to which the women were aware of how particular educational contexts such as mentoring, peer group interaction, and the conflicting demands of marriage and graduate school affected their lives and shaped their ways of coping. I argue that the obstacles Third World women of color (specifically from African societies) face in succeeding in scientific disciplines are due to multiple factors that go beyond gender bias within the patriarchal structure of science education.

African women face struggles for legitimacy in the scientific community as a result of the particular material and political conditions of their societies. As Subrahmanyan states, "these women are marginal Third World scientists" (1998, 41). For example, the increasingly poor quality of science training and shortage of good faculty in many African universities is one significant factor affecting opportunities for more women and men to succeed in science in these societies. In many African countries, problems of economic recession and structural adjustment since the 1980s have led to an overall decline in educational budgets (Beoku-Betts 1998). On many university campuses, classes are overcrowded and students have to stand outside lecture halls or report early to their classrooms in order to secure a seat (Assie Lumumba 1995, 1). Studies also suggest that the academic qualifications of faculty at most universities have declined to a level where many of the faculty do not hold Ph.D.s (Enos 1995; Sawyerr 2002). These problems are not only typical of African societies but are also shared with many other countries in the Third World (Altbach 1987; Subrahmanyan 1998).

Racial bias is another significant factor affecting the representation of African women and their struggle for legitimacy as Third World women of color in the scientific community. Studies have shown that many international students consider racial discrimination in host countries to be an important factor affecting their experiences, allowing for variations by country of origin (Kenway and Bullen 2003). Rosser (1999) also illustrates this point when she describes the chilly environments women from Third World countries encounter in pursuing graduate work in science, engineering, and mathematics in the United States. She argues that North American scientists and engineers tend not to accept these women as qualified scientists, questioning their ability to undertake graduate work or professional scientific employment.

Although these conditions are constantly undergoing negotiation and change in local and global contexts, including the scientific community, their analysis provides a more complex understanding of the varying factors that shape the academic goals and outcomes of women who pursue graduate training in scientific disciplines. In the case of African women, the experiences of race and gender bias are consistent with feminist discourses regarding the recruitment and retention of women in science. However, their location as Third World women of color from marginalized societies in the global economic system broadens our understanding of the pervasive ways in which Western hegemony has continued to affect cultural relations throughout the global system (Kenway and Bullen 2003).

According to Collins (1999), issues of gender, race, ethnicity, and class, along with other identities, have compound consequences for women's experiences in a variety of areas. Women's experiences as graduate students in scientific disciplines must therefore be examined in ways that reflect these various conditions of their lives, the power relations which structure them in various locations, and the strategies of negotiation and resistance they employ to address their concerns.

Study and Methods

The findings I report in this paper are selected from a larger study I conducted on the perspectives and experiences of African and Caribbean women in academic and administrative careers. I draw on a small sample of fifteen doctoral-level scientists in research and academic institutions in the English-speaking sub-Saharan region of Africa. All had earned their undergraduate degrees in scientific disciplines in their home countries and their graduate degrees in other, mostly Western, countries, and had taught or done research in their own countries.

I found study participants by participating in a conference on women in science, as well as through key individuals. As a result of these contacts, I was able to gain access to women scientists through the use of the snowball technique and by earning the trust and confidence of those who were willing to be interviewed. As an African researcher, I was also able to gain the women's trust and participation by explaining my interest in bringing the voices of African women to the discussion of women in scientific disciplines and careers.

I conducted semistructured interviews to give the participants the opportunity to voice their opinions and relate their experiences on their own terms. The interview schedule allowed ample opportunity for the study participants to elaborate or to introduce issues they considered relevant. Each interview was completed in one and a half to two hours, although in some cases subsequent interviews were necessary. Interviews were conducted face-to-face or by telephone and were tape-recorded. After transcribing the interviews, I searched for general themes and sorted them by relevance and order of importance. My purpose was to provide a descriptive map to show how African women perceived particular factors affecting their ability to succeed in graduate training in the sciences while studying abroad. The main themes identified in discussions about these experiences were mentoring, peer group relations, and coping with graduate work and marriage and family life.

Characteristics of the Sample

Study participants ranged in age from thirty-seven to sixty-seven. The fifteen women came from various countries in sub-Saharan Africa: seven from Ghana, three from Nigeria, three from Sierra Leone, and one each from Cameroon and Zimbabwe. They included three professors, two associate professors, four senior lecturers/research fellows, four assistant professors/research fellows, one postdoctoral fellow, and one advanced-level Ph.D. student. The latter two were completing their studies at prominent universities in the United States. The fields of study were represented by two physicists, one chemist, one mathematician, one plant pathologist, one plant physiologist, one horticultural scientist, one pharmacologist, three biochemists, two nutritionists, one infectious diseases specialist, and one zoologist. All of the study participants studied overseas for the Ph.D., seven in the United States or Canada and eight in Europe. Among those who studied in Europe, six were trained in the United Kingdom and two were trained under "sandwich" programs arranged between their home universities and universities in Norway and Germany.

Among the fifteen women, four had at least one parent who had never been to school and seven had at least one parent who had attended only primary school. In most cases, it was the mothers who either had not been to school or had only a primary school education. Five women had at least one parent who had finished secondary school, usually the father. Of parents with university degrees, five were fathers; only one woman had parents who both held such degrees. Eight of the women were married, four were divorced or separated, and three were single, but hoped to get married and have families while pursuing a career. Nine women had two or more children, two had one child, one was pregnant, and two had no children. Finally, the majority of study participants (twelve women) were also responsible for raising the children of less privileged members of their extended family in their homes. All names used in this paper are pseudonyms.

Findings

"I lost my identity. I became the Black girl": Experiences of Racial Bias

One of the consequences of the failure of African governments to provide adequate funding for their higher education institutions is that after undergraduate education, many science graduates who have the opportunity or the means pursue further studies in Europe or North America. Among those who pursue graduate studies abroad, the experience of racism is a very distinctive factor affecting Third World women of color. In some ways, their experience of racial bias is similar to that experienced by other people of color in these societies (Jordan 1999; Kenway and Bullen 2003; Sands 1993). For example, the majority of the women in the study commented on the significance of racial bias during their graduate training. Specifically, they discussed how prejudice and discrimination against them as Black women had affected their experiences in graduate school. Among the issues they mentioned were White professors' doubts about their ability to do the work, feelings of exclusion, lack of support, assumptions that they were unable to speak English, and negative perceptions of African societies. The following statements are representative of those made by study participants:

> My graduate experience was very difficult. Being in a White institution, it was like, what is this girl doing here? I lost my identity. I became the Black girl. A lot of them assumed I couldn't understand English or [that I was dumb]. After the first semester, their attitudes changed. (Dr. Kona Bouya, Assistant Professor of Chemistry)

> In the department, there was a bit of hostility from White students, but I am a mature person and my Christianity enables me to deal with it. It wasn't a woman thing, as there were White women students, but it was more racial. (Dr. Bola Kelfa, Senior Lecturer, Plant Physiology)

> In graduate school it was a Black and White thing more than gender. X State University was the best land-grant university for the state; over twenty-four thousand students and just three hundred Black students. I always wore my [traditional African clothes]. The head of the department asked me to do a number of remedial courses in biochemistry, even though he was aware of my background. I got all A's and he was rather embarrassed. It was a "can she really do it" attitude. In these White schools, you were expected to prove yourself. When you do, they become supportive. A lot of it was more racial than anything else. I am a confident person. Your thinking that I don't measure up does not mean that I don't measure up. If I were a Black American, I would have succumbed, if all throughout they have been putting you down, but I was sure of myself. (Dr. Effie Dogu, Professor, Biochemistry)

These comments reflect the complex ways in which the women experienced racial bias as graduate students. While none of them claimed to experience overt racism, they were aware that being asked to take remedial classes and comments on their "accents" and language skills were consistent with racial stereotyping of Black people. This is clearly reflected in the comments of Dr. Dogu regarding the psychological effects of racism on African Americans and how her self-affirmation and self-validation as an African (overtly displayed through her dress) would not allow her to succumb to that.

In addition to how they felt they were perceived as Blacks, study participants were also sensitive to the nuances of being both Black and of African identity and the compounding effects on the perceptions and racist behavior of members of the host culture.

> I did my Ph.D. in Canada. That was a different story. It was an extremely cold environment. The people were as cold as the weather. For me, I was usually the only Black person in the class. That was a problem in itself. Most of the time they looked at you when you walked into class as if you were making a mistake coming in. Most times you go into a class and you are the only one. In almost all the courses I took, I was the only Black female graduate student. The perception was always bad. There was this perception that a Black woman in any aspect of science just cannot make it. It wasn't true for all the professors, but it was true for 90 percent of the professors I worked with. . . . In the sciences I think we have two problems as African students, irrespective of whether you are male or female. Because of where we have come from, they often think it is impossible for you to do good lab research. If you manage to cross the course-taking hurdle, you have that to deal with. . . . You have to prove yourself. You have no choice. When you prove yourself, you are taken more seriously. (Dr. Abiana Zimba, Postdoctoral Fellow in Pharmacy)

While racial discrimination was experienced mainly in the academic setting, a few women felt more socially isolated outside of the university setting than within it. Whether because of race or language or perceptions of their country of origin, it was therefore difficult for many to adapt and to make friends. One woman said that this probably made it difficult for graduate students like herself to succeed, and she knew some women who had returned home because they could not cope. Many, however, did cope, either by drawing on their own self-confidence or spiritual beliefs or by relying on the comradeship of other international students.

"These things are subtle": Experiences of Intersection of Race and Gender

While most of the women considered racial bias a critical factor affecting how they experienced their exclusion or felt differentiated from other graduate students, several were also aware of how the interconnections between their racial identity as Black and their gender identity as women positioned them as "outsiders." Several women were cognizant of having to grapple with the intersecting effects of the two as they tried to articulate how they experienced particular situations. For some, it was not always clear whether a particular situation was due to a race or a gender effect, or to both. Consequently, in order to make sense of what they were encountering, some found ways of determining the causes of these effects by asking whether White women had the same experience. An assistant professor and a Ph.D. candidate explained this difficulty in this way:

> It wasn't immediately obvious. These things are subtle. And I was a capable student, so, part of the flack, you know, was covered by that. But you still feel that if you had been some other person, things would have gone better in certain issues, you know. So, it's subtle, you can't pinpoint it. . . . You know, you don't get praised as much and you know you are doing a good job, just as well as the other person, other people. And you openly see them getting praised for doing equal amounts of work, or even less than that, you know. It probably was racial. Okay, sometimes it was gender, sometimes it was race, and sometimes you can't separate the two—until you say, White women are suffering, then you know it's a gender thing. (Dr. Fatma Sago, Assistant Professor of Plant Pathology)

> I believe my contribution to the field is not tied to my being a woman. It has to do with my being an African woman. In the academic setting where I am, there are adequate opportunities for women to be involved in research. For me what is important are the contributions I would make as a Black person. It's a different ballgame altogether, whether I am able to contribute in the capacity of a Black woman, not a woman. It takes time to establish yourself. It comes with commitment and hard work. (Ade Bodu, Ph.D. candidate in Infectious Diseases)

Dr. Sago and Ade Bodu were aware, if sometimes implicitly, of how societal and institutional processes can function to negate and exclude Black women from science and the production of scientific knowledge. This is consistent with Barr and Birke's study (1998) of how nonacademic women perceive science and the scientific community. In their study, they show that institutionalized racism and sexism have a negative impact on the lives of Black women, who are made to feel excluded and to feel that they do not have the capacity to do science. Both narratives also indicate some element of "the burden of representation" (Brah 1996) that these women carry as they struggle to affirm and validate their ability as scholars. With this burden of representation comes a need to identify with something larger than the self and to dispel misrepresentations of their capabilities as Black women. This misrepresentation of Black women and non-Western women of color as intellectually backward in comparison to their White counterparts is well documented in the works of postcolonial feminist and Black feminist scholars such as Avtar Brah (1996), Patricia Hill Collins (1999), Trinh Minh-Ha (1988), and Chandra Mohanty (1988). The comments of study participants about their awareness and perceptions of these gendered racial misrepresentations are also consistent with Wendy Luttrell's (1993) findings on the intersections of race and gender.

"I was always viewed as a Third World person": Third World Marginality

While race and gender bias and the confluence of the two can be interpreted as experiences shared by different categories of women graduate students, the context of Third World marginality is a distinctive factor differentiating the experiences of African women graduate students in science. For example, Kenway and Bullen (2003) draw on Hage's (1998) identification of the category "Third World–looking people" (TWLP) to differentiate between non-Western people of color and people of non-English-speaking backgrounds (NESB). They explain that "when white people who embrace the white nation fantasy look at a migrant, what they differentiate between are not those who are NESB and those who are not, or those who are European and those who are not, but those who are Third World looking and those who are not" (12). The differentiating factor in this case is the race-specific nature of looking non-White and the legacy of colonization and postcolonial relations of domination that validate the position of those categorized as White.

The dominant model of education and research training in most African countries has remained European or American. This is because major issues and paradigms are conceived and developed in these societies and only marginal issues and problems are addressed in Third World contexts, such as African societies (Goonatilake 1993; Third World Network 1993). International students in the educational institutions of host countries are therefore likely to encounter assumptions about what the future roles of people who are "Third World–looking" are likely to be, in the global community of science and elsewhere. Several of the study participants, at all levels of seniority, encountered these assumptions at some point in their studies. However, those who graduated after the 1980s were more aware of the negative stereotyping and low expectations that were embedded in the comments made to them. These are typical reports:

> I remember when the grades came out. One professor came up to me and said, "Oh, for a person who comes from the Third World, you've done pretty good." So it was like no matter what, I was always viewed as a Third World person, you know. (Dr. Fatma Sago, Assistant Professor of Plant Pathology)

> Also [there were] people's perceptions. If they met you with your African clothes, they asked you, "what are you studying?" And if you say science, they'll say, "oh, home science?" I was doing biochemistry and my room-

mate, who was Nigerian, was doing chemical engineering. It's like "these girls, they can't be serious." Nobody would believe people like us could be in those fields. (Dr. Effie Dogu, Professor of Biochemistry)

Even my landlord asked what I was doing. When I said biochemistry, he said, "wow, biochemistry, I thought you were here for home economics." Even at my church, they felt that as an African woman my place would be in home economics. (Dr. Ama Conteh, Associate Professor of Bio-chemistry)

It wasn't my first time being in England, but as a student it was the first time. I was surprised at their attitude towards foreigners, especially Blacks. I was surprised that a lecturer who taught and worked with me in the lab would ignore you in the streets. But I knew what I was there for and I focused on that. Also, my husband was in Canada and so I had to work hard to finish and go. I had also been encouraged to work hard right from secondary school, so it wasn't difficult for me. Because of this, I was able to finish in the three years I was given an award for. You had to work hard to make sure they accepted you. My supervisor made it clear that he preferred to work with the English students. His attitude was that the African brain was not as sharp as the European brain, so we had to work hard to prove that we could make it. (Dr. Sade Chad, Senior Lecturer, Horticultural Science)

Some scholars may rightly consider the grouping of these African study participants who represent diverse nation-states in a large continent to be essentialist. They might argue that more focus should be placed on the complexities of their national histories and trajectories, so as not to collude in the establishment and perpetuation of power differentials (Kenway and Bullen 2003). However, the experience of colonization under the British and the postcolonial legacies of economic and political marginality that African countries continue to experience in the global economy provide some space for an interrogation of their shared marginality. This location of marginality in the global economic system, and particularly the scientific community, permits us to use that shared membership to explore some of the fundamentally different understandings Third World women of color bring to the analysis of graduate education in the sciences in Western societies. The narratives of the women in the study make clear that while abroad, they saw themselves as African, as opposed to, for example, Nigerian or Sierra Leonean. This may be a consequence of their continent's historical and geographical location in the global community and their perceptions that people in their host country saw Africa as a "country" rather than a continent. Such scholars as Pratt (1999) and Bhabha (1994, quoted in Kenway and Bullen 2003, 10–11) argue that a strategy of negotiation and survival can evolve in certain contexts whereby these international study participants may develop hybrid identities. Claiming such a hybrid identity would thereby enable them to adapt to the dominant culture by representing themselves in ways that engage with the colonizers on their own terms.

The women were prepared to struggle with these multiple layers of negative perceptions in their host countries before they ever arrived in them. They learned from their national histories, from the ongoing trajectories of their societies in the global economy, and from their daily interactions and struggles as women in patriarchal societies, including those that are African. Their ability to survive these conditions is clearly articulated in their determination to work hard and keep focused on their ultimate goal, and in their confidence in their ability to accomplish these goals. Ade Bodu captures this determination very clearly:

> In a nutshell, I don't think I'm a quitter. I have learned a lot of that from my mother. She has a lot of drive and tenacity that I have inherited. At first, I would spend hours in tears wondering what I was doing here. I just had to try not to give up and to find other ways of dealing with my problems. If you show people you are worried, they'll move away from you. I had to move on because I could not set myself up for failure of any sort. For every five incidents that were negative, there was always one that was positive. I also realized that I was not alone. Other people were going through it and so I wanted to survive. I gravitated towards people who were helpful and who understood the constraints I was going through. These were all sorts of people and they encouraged me to just keep pushing forward. (Ade Bodu, Ph.D. candidate in Infectious Diseases)

I now focus on ways in which study participants experienced, articulated, and negotiated patterns of exclusion on the basis of their race and gender and as Third World women of color situated in their African identity. I examine their experiences and perspectives in selected educational contexts, including mentoring, peer group interaction, and juggling graduate studies and marriage.

"They are okay once they realize you are as intelligent as their other students": Mentoring

All study participants were asked, "Who were your mentors and role models in graduate school and how did they help you?" Most of the responses were consistent with the findings on the influences of mentoring in Europe and North America (Chandler 1996). Eight of the fifteen women said they had been mentored, although the mentors ranged from major advisers to supportive faculty in other departments. The women who said that they had not been mentored had relied more on self-motivation or support from other graduate students. Most of the older and more professionally senior study participants had been dependent on mentors during their graduate studies abroad, particularly those who studied in the United Kingdom. This may be due to the structure of the English educational system, which expects a graduate student to be largely self-directed. For example,

> My supervisor was helpful. Most of the time he was away consulting with the FAO [Food and Agriculture Organization]. When he came back he would set a timetable for me and I always tried to meet that deadline. In England, you had to do your own manual work without much help. I had to do everything without much help. If you are a lazy person, you can't do it. Most of the people I worked with were [other] international students. My supervisor encouraged us to write for publications and to present papers. (Dr. Sade Chad, Senior Lecturer, Horticultural Science)

The women also discussed other types of mentoring they had received. These included opportunities to attend conferences and advice on publishing and on how to conduct research, as well as general encouragement and motivation to persevere, as the following narratives show.

> I owe him [my adviser] a lot. Then at X State University, Blacks were very few and he was the only Black American on the staff. We talked every day and planned my work and I had to report every day about my results. My supervisor wouldn't allow you to go your own way. If you were on the wrong track he would correct you, so as not to waste time. He gave me support to enable me to present at four international conferences. He would always read my first draft of publications. I really owe gratitude to him and even now we still communicate. (Dr. Ama Conteh, Associate Professor of Biochemistry)

> Mentor, you are definitely not talking about the supervising? I [did have] another mentor. Actually, they were a couple. They were good. They were faculty there, but I didn't work with them. He was a really good sounding board and he had great ideas and he understood the situation in [my country]. So, I got a lot of advice in terms of directions. Like the timing of flora and other local plants. He would reinforce how important and really, you know, point out where science was going and in what direction. So, I liked talking to him, all those great ideas. And [his wife, also a faculty member,] was very supportive too and a really very talented woman. So, they were very influential and supportive. (Dr. Fatma Sago, Assistant Professor of Plant Pathology)

> My own supervisor for my Ph.D. was a wonderful lady, with a lot of experience working in Ghana and Nigeria. She had insight into the African way of life and working with Africans. One thing that influenced the way we related was I realized she was a lady who did not suffer fools gladly. She would give support if it was needed. If you stayed away and came to see her with something worth showing, she was very supportive. This has in-

fluenced me a lot. I need to know that my work is well done. (Dr. Yinka Yasin, Senior Research Fellow, Nutritional Science)

These narratives indicate the importance of mentoring and the variety of ways in which it fosters academic achievement in the sciences, regardless of where a person comes from. African women clearly benefited from this mentoring even when they felt that they were being stereotyped or marginalized, distanced from positive mentoring relationships their advisers developed with other students. From these accounts, it seemed that those advisers who had prior international experience were more helpful in supporting and mentoring their African students. Those women who lacked strong or willing mentors in their departments relied on self-motivation or found alternative ways of coping with the isolation, such as working with senior graduate students or finding faculty support in other departments.

"They just didn't think I could make it": Peer Group Interaction

Peer groups have also been found to influence the academic achievement of students in science, especially at the undergraduate level. However, women in male-dominated science courses are more likely to experience an unwelcoming environment (Seymour and Hewitt 1994), particularly if they are from Third World societies (Rosser 1999). Such experiences, though annoying or constraining, have not necessarily been found to cause women to leave the sciences (Astin and Sax 1996). In my study of the experiences of African women as graduate students in the sciences, I found that the majority did not have positive peer group experiences or collegial support from other graduate students. Nonetheless, no matter how isolating this lack proved for those who experienced it, it did not cause them to leave the sciences but rather strengthened their determination to accomplish their goal and to resist the negative racial stereotyping of their African identity. Some aspects of peer group interaction are expressed in these narratives:

At times we had to go to the main university for seminars. Most of them [colleagues] had cars, but sometimes they would be reluctant to give me a ride and prefer that I take a bus. Sometimes I would be using the lab and they would want me to leave what I was doing for them to use. They would say negative things about Africa and I would want to correct them and they wouldn't be happy about it. For example, when we had civil wars in Africa they would want to attribute it to the fact that we were uncivilized and uneducated. But they too had their own social problems. (Dr. Bola Kelfa, Senior Lecturer, Plant Physiology)

Americans tended to see you and not think you can do anything. Even fellow graduate students would make comments that implied you were no good and [not] capable of being there. They just didn't think I could make it, and I even qualified as a Ph.D. candidate before some of them. (Dr. Balu Dabo, Associate Professor of Biochemistry)

In England, there were six of us working in the lab. Again, there was one Nigerian in this lab. The color brought us together and he was very helpful. Initially, there were [some] White boys in the lab who assumed that we didn't know much, but again with time, they did realize that we did know a lot and their attitude changed. But they were very conservative and wouldn't even say good morning to anybody. I didn't know if that was their system. Social life was almost zero. It was sort of the Nigerian and myself. I think there was another Nigerian on another floor. So, it was sort of all the Blacks together and all the Whites together. (Dr. Kona Bouya, Assistant Professor of Chemistry)

With the Ph.D. and when I was a postdoc the kind of respect that you expect as a postdoc sometimes didn't come. The recognition even by, say, the graduate students in the lab that you are a postdoc and the respect that it afforded other postdocs, for instance, wasn't immediately there. And I was, I am, an independent person in the lab. (Dr. Fatma Sago, Assistant Professor of Plant Pathology)

These statements about experiencing a chilly laboratory or classroom climate are typical of most of the interviews undertaken in this study. While the complaints were more likely to come from women who studied during the 1980s and later, a few senior women shared similar sentiments about their graduate school experience. Implicit in many of these encounters between these women and their counterparts was the undertone of Western hegemonic systems of domination and subordination that continue to pervade relations between people of European descent and Third World people of color. The women were consciously aware of these power differentials and how they function to exclude people like them from the community of science. They were also aware that these situations are not necessarily based on gender, as they mention their male African counterparts in their commentaries. Their descriptions of these encounters and of how they responded to the negative representations of their African identity suggest that these women were also voicing resistance.

"Where is your husband?": Juggling Marriage and Graduate School

Another aspect of graduate training that holds similar but distinctive constraints for African women as graduate students in science is the need to manage the demands of marriage or family and graduate school. This is one experience shared worldwide by most married women who choose to pursue a scientific or professional career path. The distinct factor for African women, however, is that the majority of them are married with families, unlike their European or North American counterparts at the same stage in their careers (see, e.g., Hanson 1996, 78). In most African societies, men are privileged and do not expect to contribute to domestic labor and child-care. While this might be less of a problem for these women in their own countries, given the extended family support system and common practice of child fostering, it has adverse implications for women who study in overseas institutions. Reluctance on the part of male spouses to share the domestic workload means that women are more likely to take longer to complete their studies or may even have to drop out of their programs before completion, when it is time to return home. When asked how they coped with the demands of marriage and graduate studies in a foreign country, several of the study participants commented on emotional costs, loss of opportunities, the burden of domestic responsibilities, neglect of spouse and children, and shortchanging of their own leisure and study time.

> Soon after my first degree I got married. And so, there were recommendations for awards. But you take what you get in our part of the country. So, I was recommended for a Fulbright award to come to the U.S. So, we went to the interview. The chairman of the interview in the usual probing that goes on at interviews asked me, "Where is your husband?" And I said, "He's studying in the U.K." And he said, "Well, look, young lady, you are just going to be compounding your problems by taking a scholarship to the U.S. when your husband is in the U.K. If you see me later, there are some short courses in the U.K. which you could go on." So, although I was really first on the interview panel, and this he told me several years later, although I was first, he didn't give me the Fulbright award. He gave it to somebody else, a male colleague who then came to the U.S. for his Ph.D. (Dr. Rugi Turay, Associate Professor of Physics)

> I could not perform to my maximum, because I was a mother and wife. My husband was supportive, but once he enrolled in school, things changed a bit. At the beginning of one academic year, I drew up a timetable to share household chores. I asked him what he would be able to do. He flared up and was very angry. He felt a woman should not dictate to her husband. He decided on his own volition to do the laundry, etc. I had to study hard, because with all the negative perceptions about me, I wanted to prove them wrong. I finished my Ph.D. in four and a half years. (Dr. Balu Dabo, Associate Professor of Biochemistry)

> I got married as a student in Canada. The only reason why I was able to go through is because my husband wasn't with me. After that, when I got back, I thought I was dealing with someone who had an ego problem. He felt that because I had a Ph.D. that he needed to do one too. As an African woman, to be in the sciences, for you to be able to be successful in your career, you should have a husband who is understanding. (Dr. Abiana Zimba, Postdoctoral Fellow, Pharmacology)

These narratives indicate that being an African woman and pursuing graduate studies at an institution outside of one's own country holds particular challenges, especially for women who are married or have children. The demands they face, such as being bypassed for certain scholarships because they would be separated from their spouses, are problems not often encountered by their male counterparts. The women in this study were very conscious of the constraints and barriers they had to face as graduate students juggling marriage and family life. At some level, this strain almost seems like a necessary sacrifice they had to make to hold up their part of the marriage and parental contract, which in most African cultures is still strongly endorsed. This is clearly demonstrated by the fact that only four of the fifteen study participants were divorced or separated and that all three single women were planning to get married and to have children while pursuing a scientific career. As some feminist scholars have pointed out (Harding 1991; Rosser 1990), the difficulties of managing marriage, family, and a graduate education in science are likely to discourage more women from entering and remaining in science, if more gender-sensitive policy initiatives are not globally fostered and instituted.

Although feminist scholars increasingly note the need to develop greater awareness of diversity issues in the analysis of women's lives, this dialogue has not received much attention in discussions of women in science (Barr and Birke 1998; Birke and Whitworth 1998; Harding 1998; Rosser 1999). This study of the educational experiences of African women scientists as graduate students in universities in Europe and North America shows that we cannot assume that the road to success in graduate training is similar for all women. It reveals that the educational process for women in science is complex and requires a more nuanced understanding of how women are differently positioned according to the varied conditions of their lives, the power relations which structure their various locations, and the strategies they employ to address these concerns. The impact of racial and gender bias and negative perceptions of African societies, as a result of the marginality of these societies in the global system and international scientific community, has created very distinctive educational experiences for African women who undertake graduate studies abroad.

Racial and gender bias and the ways in which they intersect are experiences African women share with their counterparts of color in Western societies. For example, almost all of the study participants discussed racism's effect on their experiences in graduate school. In some cases it was the lack of good mentoring, in others it was a belief that they were unable to do the work, and in still others it was the social isolation they experienced in their interaction or lack thereof with White counterparts in their host institutions. Similar experiences are reported in studies addressing the experiences of African American women in science (Jordan 1999; Sands 1993).

An experience shared by the African study participants and their Western female counterparts in science was the challenge of juggling the demands of graduate studies and family. As women, they were expected to be the primary caregivers in the home, regardless of the fact that they were also students. As married women with families, many felt guilty for not being able to meet the demands of their spouses and children. While emotional costs such as divorce were not as prevalent among the African women as among their Western counterparts, four of the women in the study were divorced or separated from their spouses. These experiences are similar to those of women in science in North American and European societies. For example, a study of Finnish women in science found that one-quarter of the scientists had conflicting feelings or a bad conscience because they felt they neglected their children and families, regardless of how many hours they devoted to them (Luukkonen-Gronow and Stolte-Heiskanen 1983).

Both race and gender bias situate women of color, irrespective of who they are, in less powerful positions relative to their male or White counterparts. This affects how they acquire scientific knowledge and their lived experiences. Their structural position and the exclusionary practices they encounter in the scientific community are shaped by a racialized and gendered social order, and supported by institutional systems and processes. What distinguishes the situation of African women, as international graduate students in science, from that of their Western counterparts is the historical legacy of subordination arising from the colonial experience and the marginal position of their societies in the global economic system, including the international scientific community. This "peripheral" status in international science, as described by Altbach (1987), has particular implications for the representation of African women as graduate students in the scientific community. African women in this study were very conscious of the negative stereotyping they received within the university communities by virtue of their dark complexions. The negative assumptions about and misrepresentations of Africa and its peoples were evidenced in the low expectations that were embedded in comments made to them. The women were also cognizant of the lack of respect they received from many of their peers, as well as their isolation from key social interactions.

As well as the many challenges and the "burden of representation" borne by the women in this study, it is important to take note of their ability to resist and survive their hostile environments. Many of the women clearly articulated their determination to work hard, to keep focused on their objectives, to be self-confident, and to draw on spiritual beliefs or support from other international students as strategies for survival. The legacies of their national histories, and struggles to overcome patriarchal systems of domination in their own societies provided them with the framework to validate their own abilities to succeed in these educational environments.

In conclusion, this study indicates that an understanding of the multiple locations of struggle within intersecting relations of racial and gender bias and location as Third World women of color is fundamental to any analysis of the factors that influence the educational success of women in science. Furthermore, a more complex understanding of the varied educational contexts and hierarchical systems of power relations within which women are trained to become scientists is relevant to the development of educational initiatives to increase the numbers and retention of women in scientific careers.

Note

This research was funded partly by a grant from the Division of Research at Florida Atlantic University. I want to thank Kathleen Slobin, Jill Bystydzienski, and an anonymous reviewer for their helpful comments on the chapter. My thanks also to the African women scientists who participated in the study.

References

Altbach, Phillip J. 1987. *Higher Education in the Third World.* New York: Advent.

Assie Lumumba, N'Dri T. 1995. "Demand, Access, and Equity Issues in African Higher Education: Policies, Current Practices, and Readiness for the 21st Century." Background paper for the joint colloquium "The University in Africa in the 1990s and Beyond," Lesotho, January 16–20.

Astin, Helen S., and Linda J. Sax. 1996. "Developing Scientific Talent in Undergraduate Women." In *The Equity Equation: Fostering the Advancement of Women in the Sciences, Mathematics, and Engineering,* ed. Cinda-Sue Davis et al., 13–25. San Francisco: Jossey-Bass.

Barr, Jean, and Lynda Birke. 1998. *Common Science? Women, Science, and Knowledge.* Bloomington: Indiana University Press.

Beoku-Betts, Josephine A. 1998. "Gender and Formal Education in Africa: An Exploration of the Opportunity Structure at the Secondary and Tertiary Levels." In *Women and Education in Sub-Saharan Africa: Power, Opportunities, and Constraints,* ed. Marianne Bloch, Josephine A. Beoku-Betts, and B. Robert Tabachnick, 157–84. Boulder, Colo.: Lynne Rienner.

———. 2000. "Living in a Large Family Does Something for You: Influence of Family on the Achievement of African and Caribbean Women in Science." *Journal of Women and Minorities in Science and Engineering* 6(3): 191–206.

Bhabha, Homi K. 1994. *The Location of Culture.* London: Routledge. Quoted in Kenway and Bullen 2003.

Birke, Lynda, and Rhonda Whitworth. 1998. "Seeking Knowledge: Women, Science, and Islam." *Women's Studies International Forum* 21(1): 147–59.

Brah, Avtar. 1996. *Cartographies of Diaspora.* London: Routledge.

Chandler, Christy. 1996. "Mentoring and Women in Academia: Reevaluating the Traditional Model." *NWSA Journal* 8(1): 79–100.

Collins, Patricia Hill. 1999. "Moving beyond Gender: Intersectionality and Scientific Knowledge." In *Revisioning Gender,* ed. Myra Marx Ferree, Judith Lorber, and Beth B. Hess, 261–84. Thousand Oaks, Calif.: Sage.

Currie, Jan, Bev Thiele, and Patricia Harris. 2002. *Gendered Universities in Globalized Economies: Power, Careers, and Sacrifices.* Lanham, Md.: Lexington.

Enos, John L. 1995. *In Pursuit of Science and Technology in Sub-Saharan Africa: The Impact of Structural Adjustment Programs.* New York: Routledge.

Etzkowitz, Henry, Carol Kemelgor, and Brian Uzzi. 2000. *Athena Unbound: The Advancement of Women in Science and Technology.* New York: Cambridge University Press.

Gender Working Group, United Nations Commission on Science and Technology for Development (UNCSTD). 1995. *Missing Links: Gender Equity in Science and Technology for Development.* New York: International Development Research Center and UNIFEM.

Goonatilake, Susantha. 1993. "Modern Science and the Periphery: The Characteristics of Dependent Knowledge." In *The "Racial" Economy of Science: Toward a Democratic Future,* ed. Sandra Harding, 259-67. Bloomington: Indiana University Press.

Hage, Ghassan. 1998. *White Nation: Fantasies of White Supremacy in a Multicultural Society.* Annandale, N.S.W., Australia: Pluto.

Hanson, Sandra L. 1996. *Lost Talent: Women in the Sciences.* Philadelphia: Temple University Press.

Hanson, Sandra L., Mary Ellen Schaub, and David P. Baker. 1996. "Gender Stratification in the Science Pipeline: A Comparative Analysis of Seven Countries." *Gender and Society* 10(3): 271-90.

Harding, Sandra. 1991. *Whose Science? Whose Knowledge? Thinking from Women's Lives.* Ithaca, N.Y.: Cornell University Press.

———. 1998. *Is Science Multicultural? Postcolonialisms, Feminisms, and Epistemologies.* Bloomington: Indiana University Press.

Harding, Sandra, and Elizabeth McGregor. 1995. *The Gender Dimension of Science and Technology.* New York: UNESCO.

Jordan, Diann. 1999. "Black Women in the Agronomic Sciences: Factors Influencing Career Development." *Journal of Women and Minorities in Science and Engineering* 5(2): 113-28.

Kenway, Jane, and Elizabeth Bullen. 2003. "Self-Representations of International Women Postgraduate Students in the Global University 'Contact Zone.'" *Gender and Education* 15(1): 5-20.

Luttrell, Wendy. 1993. "Women's Ways of Knowing: Effects of Gender, Race, and Class." *Sociology of Education* 2: 33-46.

Luukkonen-Gronow, Terttu, and Veronica Stolte-Heiskanen. 1983. "Myths and Realities of Role Incompatibility of Women Scientists." *Acta Sociologica* 26: 267-80.

Minh-Ha, Trinh. 1988. "Not You/Like You: Post-colonial Women and the Interlocking Questions of Identity and Difference." *Inscription* 3(4): 71-77.

Mohanty, Chandra Talpade. 1988. "Under Western Eyes: Feminist Scholarship and Colonial Discourses." *Feminist Review* 30: 61-88.

Mukhopadhyay, Carol C. 1994. "Family Structure and Indian Women's Participation in Science." In *Women, Education, and Family Structure in India,* ed. Carol C. Mukhopadhyay and Elaine Seymour, 103-32. Boulder, Colo.: Westview.

Pratt, M. L. 1992. *Imperial Eyes: Travel Writing and Transculturation.* London: Routledge.

———. 1999. "Apocalypse in the Andes." *Americas* 51: 38-47.

Rayman, Paula, and Belle Brett. 1993. *Pathways for Women in the Sciences.* Wellesley, Mass.: Wellesley College Center for Research on Women.

Rosser, Sue V. 1990. *Female-Friendly Science: Applying Women's Studies Methods and Theories to Attract Students.* New York: Pergamon.

———. 1999. "International Experiences Lead to Using Postcolonial Feminism to Transform Life Sciences Curriculum." *Women's Studies International Forum* 22(1): 3-15.

Sands, Aimee. 1993. "Never Meant to Survive: A Black Woman's Journey—An Interview with Evelynn Hammonds." In *The "Racial" Economy of Science: Toward a Democratic Future,* ed. Sandra Harding, 239–48. Bloomington: Indiana University Press.

Sawyerr, Akilagba. 2002. "Challenges Facing African Universities: Selected Issues." Paper presented at the annual meeting of the African Studies Association, Washington, D.C., September.

Seymour, Elaine, and Nancy M. Hewitt. 1994. *Talking about Leaving: Factors Contributing to High Attrition Rates among Science, Mathematics, and Engineering Undergraduates.* Final Report to the Alfred P. Sloane Foundation on an Ethnographic Inquiry at Seven Institutions. University of Colorado, Boulder.

Subrahmanyan, Lalita. 1998. *Women Scientists in the Third World: The Indian Experience.* Thousand Oaks, Calif.: Sage.

Sur, Abhar. 2001. "Dispersed Radiance: Women Scientists in C. V. Rahman's Laboratory." *Meridians* 1(1): 95–127.

Third World Network. 1993. "Modern Science in Crisis: A Third World Response." In *The "Racial" Economy of Science: Toward a Democratic Future,* ed. Sandra Harding, 484–518. Bloomington: Indiana University Press.

8 Gendered Experiences in the Science Classroom

Molly J. Dingel

For the past twenty years considerable effort has been made to get more women to pursue scientific careers. Numerous university science departments have set up mentoring programs (Clark et al. 2000; Etzkowitz, Kemelgor, and Uzzi 2000) or weekly peer-led workshops (Ligata and Adamczeski 2000), have received grants to educate science and math faculty about research in teaching science, and have implemented feminist frameworks in scientific fields (Rosser 1990, 1997; Wenneras and Wold 2001; Jacobs and Becker 1997). The K–12 environment has sought to change curriculum and pedagogy (Flores 1997; Jones and Scantlebury 2001, 138). Despite these efforts, women are still severely underrepresented in many science disciplines.

In this chapter, I investigate this discrepancy by examining the possibility that gendered interactional practices in science classrooms play an important role in students' sense of self as potential scientists when choosing a major in college. I wish to answer the question of whether we are "doing gender" (West and Zimmerman 1987) in science classes and, if so, how it interferes with women's likelihood of success in science. A study of women's, and men's, experiences in an introductory-level science course suggests three possible manifestations of gendered classrooms. First, the "normal" scientist is perceived to be male; women are defined as an exception. Second, women are less likely than men to be recognized as authorities in the science classroom. Finally, women's resulting insecurities drive them to achieve unnecessary perfection in their work. These gendered expectations and behaviors serve as a barrier to women's full participation in science by increasing their chances of dropping out of science fields.

In the Classroom

Research suggests that gendered interactions start early and may undermine girls' education (Eccles 1984; Renzetti and Curran 1995; AAUW 1999; Seymour and Hewitt 1997; Orenstein 1994; Mickelson and Smith 1995; Martin 1998). For example, boys are more likely than girls to receive precise teacher comments about scholarship and conduct—comments that are essential to improve student learning and self-esteem (AAUW 1999, 62–63; see also Martin 1998).

Women's and men's different educational experiences may be even more pronounced in the science classroom. The sociology of gender suggests a classroom dynamic that may discourage women interested in science: doing science is doing gender. Social constructionist theorists argue that gender is the product of social interactions that rationalize a gendered division of society (West and Fenstermaker 1995, 21). From this perspective, gender is not the expression of individual tastes and dispositions so much as it is the result of our being aware of and holding one another accountable for gender-appropriate performances (West and Zimmerman 1987, 135–36). For example, in most situations we expect men to be dominant and women to be passive (Ridgeway 1993). People can choose not to conform to gendered expectations, but those who violate them generally experience negative consequences (West and Zimmerman 1987, 137; Carli 1990); for instance, passive men may have their masculinity called into question and assertive women may face social disapproval (West and Zimmerman 1987, 146). Further, because these actions and interactions seem like natural and normal aspects of social life, they legitimate power relationships and make invisible the mechanisms though which power is maintained (West and Fenstermaker 1995, 22). Given that this gender accountability is pervasive in our lives, we may then question whether "doing science" might also be a way of "doing gender." In other words, perhaps the appropriate way to act and react in science—to "do science"—is also the appropriate way to "do masculinity" but not the appropriate way to "do femininity" (West and Zimmerman 1987). Thus, in the science classroom, women could be held accountable for not acting as women when doing proper science, and for not acting as scientists when doing their gender. This possibility is reinforced by the work of those who find such gendered expectations existing in other professional fields, for example Pierce (1995), who illustrates this process in law schools and law firms.

Being a Minority

There is evidence that women students see themselves as "outsiders" in science. Women often find they cannot take for granted that they belong in their field or that they will be equitably rewarded (Kubanek and Waller 1996, 109; Seymour and Hewitt 1997). Interviews with minority group members and token women indicate that they display "less assertiveness in asking for what they needed; less inner-strength to cope with set-backs; and more dependence on others for reassurance," and face estrangement, loneliness, and a higher rate of attrition (Seymour and Hewitt 1997, 242; see also Kanter 1977, 207).

Men's numerical predominance and perceived greater authority give them a "status shield" in science (see Hochschild 1983). Men in science can use their status shield to deflect certain pressures in science classes. For instance, women in the sciences tend to spend much more time on homework than do men (Hanson 1996, 83) and seem more dependent on grades for affirmation of their ability (Seymour and Hewitt 1997, 108; Kubanek and Waller 1996, 38–39). This need for affirmation may stem from and, in turn, add to a type of "performance pressure" (Kanter 1977, 212–21) or "stereotype threat" (Spencer, Steele, and Quinn 1999, 6; Aronson et al. 1999), in which there is additional pressure for women to perform well that makes them more vulnerable to negative feedback and grades. For example, Rosabeth Moss Kanter argues, "the token does not have to work hard to have her presence noticed, but she does have to work hard to have her achievements noticed" (1977, 216). More recently, Virginia Valian has built upon this framework by showing that a woman's education and human capital seem to "buy her less than a man's buys him" (1998, 255), and that lack of recognition has negative effects on women's earnings, rates of promotion, and status. In other words, perhaps women spend more time on homework in order to get their achievements recognized. Those who receive B's may be less likely to be noticed or pushed by teachers and parents to pursue science than those who receive A's.

The pursuit of science as a career entails many stages and there are some crucial points at which women drop out of science or struggle to be recognized as equals with men. These include the sophomore year in high school (Hanson 1996), the choice to pursue an undergraduate and an advanced degree (Shauman and Xie 1999), the processes of getting a postdoctoral fellowship, getting a tenure-track job, and obtaining research funding, promotion, and tenure (Sonnert and Holton 1995), and being chosen to manage large, prestigious projects (Schiebinger 1999).

Study and Methods

I chose to investigate one crucial step in the trajectory of a science career: the beginning of the college science experience. I used a multimethod strategy, employing a survey and in-depth interviews. I administered a brief survey to all students enrolled in a calculus course at a major research university in the Midwest. This course satisfies a requirement for majors in biology, chemistry, engineering, physics, and mathematics, and is the point at which most students begin their coursework for a science degree.

Quantitative data were used to describe the sample and to identify interviewees who were representative, who "looked" like the average male and female science student and who seemed to have the ability to succeed in science. To determine typicality, the survey asked students about the math and science classes that they had taken in high school and college, their sex, year in school, major, age, race, and parents' educational backgrounds and occupations. In order to get a sense of what drives students to pursue science, I also asked what attracted the students to science; how much they were encouraged to pursue a science major by teachers, parents, and peers; and how they felt about their science classes (see Fowler 1993, 2).[1] At the end of the questionnaire, students were asked to signal their willingness to be interviewed by giving contact information.

Though the survey data can give a sense of the whole sample, they cannot provide a nuanced vision of the interactional dynamics the students experienced in the science classroom (Strauss and Corbin 1990, 19). I thus conducted in-depth qualitative interviews with twenty-one of the eighty students who had indicated a willingness to be interviewed.[2] The survey and interviews were ideal methods for this study because they allowed both a broad understanding of who initially desired to pursue a science degree and a richer understanding of the challenges that these students faced.

To select interviewees with a good chance of surviving the academic challenges of college science classes, I chose eleven women and ten men science majors who had taken at least three science classes in high school and had received A's or B's in those classes. The interviews were loosely structured to cover topics like sources of influence and support, future plans, and comparison of science with nonscience classes. I asked students how they thought their gender and, if they were a member of a racial or ethnic minority group, their race/ethnicity affected their path through or experiences with science. I transcribed each interview and then grouped together quotes from different interviews to identify common themes or patterns, a process called "open coding" in grounded theory (Strauss 1987, 28). I then searched the interviews for quotes that seemed to counter the themes I had created. In this chapter, I include samples of both supportive and seemingly nonsupportive quotes.

Survey Data

The survey provides a snapshot of the average first-year science student. Of approximately three hundred students enrolled in the calculus course, 267 (about 90 percent) completed a questionnaire. As table 8.1 shows, the vast majority of these students were in their first year (86 percent), were male (71.5 percent) and white (83.5 percent), and intended to be science majors (83 percent). Few of the survey items show major differences between men and women. Nor do the survey data show if or where doing gender conflicts with doing science.

Table 8.2 shows that women and men answered above the midpoint of the scale on three of the five questions about reasons for liking science. Both men and women were less interested in the service potential of careers in science—helping others and teaching. They tended to like their science teachers, their science peers, and the pace of their science classes. Students in this sample also received quite a bit of encouragement from teachers and parents, which is not surprising since science is a prestigious and profitable pursuit. Table 8.3 shows that students also were generally satisfied with their experiences in science.

Despite these similarities, table 8.3 indicates that women were more pleased than men with both the study groups in which they participated in high school and the way those study groups were utilized in their classes. Table 8.2 shows that women received more encouragement from teachers than did men. Since most of these students were in their first year, it may be assumed that the bulk of teacher encouragement was experienced during high school. Both men and women reported being significantly less pleased with their college teachers'

Table 8.1. Composition of Calculus Class in Fall 1999 at a Large Midwestern University

Variable	n	Percentage	Variable	n	Percentage
Sex			Major		
Female	75	28.2	Engineering	128	47.9
Male	191	71.5	Computer science	28	10.5
			Physics	10	3.7
			Biology	8	3.0
			Science major[1]	222	83
Year in School			Ethnicity[2]		
First year	232	86	White	223	83.5
Sophomore	21		Black/African American	5	1.9
Junior	4		Asian	20	7.7
Senior	7		Latino/Hispanic	3	1.1
			Other	8	3.1
College Science Experience (out of 5 classes)[3]			High School Science Experience (out of 6 classes)[4]		
No classes	47	26	Four or fewer classes	30	13.9
One class	63	35	Five classes	57	26.4
Two classes	52	29	Six classes	129	59.7
Three or more classes	18	10			

1 Includes engineering, physics, math, biology, chemistry, computer science, biochemistry, premedicine, and other science majors or double majors.

2 According to university data for 1997, the student body is 51 percent female, 3 percent Black, 1 percent Native American, 2.9 percent Asian, 2.3 percent Hispanic, and 79.9 percent White non-Hispanic.

3 The five classes asked about were precalculus, engineering, computer science, physics, and chemistry.

4 The six classes asked about were geometry, algebra, precalculus, calculus, chemistry, and physics.

Table 8.2. Reasons Students Like Science and Sources of Encouragement
(0 = not at all important, 5 = very important)
(0 = no encouragement, 5 = very encouraging)

	Female Mean (SD)	Male Mean (SD)	Difference Female-Male
Reasons Students Like Science			
Solving problems is challenging	3.27 (1.23)	3.35 (1.21)	−.08
Satisfaction of solving problems	3.93 (1.02)	3.87 (1.14)	.06
Figuring out how things work	3.87** (1.20)	4.16** (0.91)	−.29**
Prepares me for a career	4.08 (1.22)	4.00 (1.20)	.08
Prepares me to teach others	1.71 (1.38)	1.91 (1.42)	−.20
Prepares me to help others	2.82 (1.38)	2.53 (1.37)	.29
Encouragement from peers	2.33 (1.64)	2.23 (1.66)	.1
Sources of encouragement			
Teachers	4.03* (1.01)	3.79* (0.98)	.24*
Parents	3.48* (0.96)	3.70* (0.86)	−.22*

*p<.1   **p<.05   ***p<.001 (two-tailed t-tests)

teaching abilities and personalities, and with the relationships they had with those teachers, than they had been with their high school teachers (see also Seymour and Hewitt 1997). In addition, though both men and women were less happy with the pace of their college classes than with the pace of their high school classes, women were significantly less pleased than were men (see table 8.3).

The combination of the increased pace of classes and decreased support from teachers may create an atmosphere that many students, especially women, find uninviting and may therefore be a site where doing gender conflicts with doing science. Though this finding is significant, it deserves more in-depth analysis. In addition, the large number of gender similarities suggests that quantitative methods are inadequate for providing a complete understanding of why women are more likely than men to drop out of a science major.

Table 8.3. Things Students Like/Dislike in Science Classes
(0 = do not like at all, 5 = like a lot.)

	Female Mean (SD)	Male Mean (SD)	Difference Female-Male
Relationship with high school teachers	3.92 (1.06)	3.77 (1.15)	.15
Relationship with college teachers	2.49 (1.18)	2.75 (1.17)	−.26
Personality of high school teachers	3.86 (0.93)	3.86 (1.12)	0.00
Personality of college teachers	3.45 (1.12)	3.50 (1.15)	−.05
Other students in high school science classes	3.61 (1.15)	3.47 (1.22)	.14
Other students in college science classes	3.31 (1.08)	3.07 (1.21)	.24
Teaching ability of high school teachers	3.56 (1.05)	3.65 (1.11)	−.09
Teaching ability of college teachers	3.53 (1.03)	3.41 (1.27)	.12
Pace of high school classes	3.63 (1.25)	3.43 (1.16)	.2
Pace of college classes	2.67* (1.27)	3.00* (1.11)	−.33*
High school study groups participated in	2.81** (1.51)	2.26** (1.59)	.55**
College study groups participated in	2.58 (1.48)	2.43 (1.45)	.15
Use of study groups in high school	2.64** (1.59)	2.20** (1.52)	.44**
Use of study groups in college	2.27 (1.56)	2.02 (1.54)	.25

*p<.1   **p<.05   ***p<.001 (two-tailed t-tests)

Interview Data

In interviews with twenty-one students, eleven women and ten men, I was able to gain a fuller understanding of the experiences of first-year science students. These interviews illuminate how students fit into science as a gendered structure or organization—there is a cultural expectation of what the ideal student is, and that student is not a woman (Acker 1990). Acker argues that though organizations are presented as gender-neutral, assumptions abound that women do not have the technical skills, the leadership ability, or the stamina to handle the demands of traditionally masculine jobs (Acker 1990, 146–47, 152; see also Valian 1998). Three themes that emerged from the interviews support Acker's arguments: 1) women are constructed as an exception by their teachers and peers; 2) women are less likely to be seen as an authority in the science classroom; and 3) women's insecurities drive them to achieve perfection in their work.

Women as the Exception: "I can't believe there are this many women!"

About half of the students I talked with, both women and men, noted that it was generally considered strange for women to be in science. When the women noted this attitude, they always indicated that teachers, peers, and parents held this view, but they, themselves, rejected it. Men, however, did not dismiss this attitude so quickly. Brett, a first-year mechanical engineering major, stated, "there's a couple girls who work on [a team to build a working racecar] and it's just kind of, like, I think, 'What do they know about cars?' And I'll even do that, I'm guilty of it." Chris, another first-year engineering major, also noted the lack of women in the sciences: "If I was a girl, I probably wouldn't have gone towards engineering as much. . . . it just seems to be a male-dominated major, and I just don't see myself doing that if I were a female." These students recognized that women are exceptions to the rule, that the "normal" science student is male.

Women talked about the experience of being seen as the deviant "other" in science education. Amanda, a sophomore architectural and electrical engineering double major, commented that "a lot of times you'll get into a classroom, in my major, and my teachers are like, 'Wow, I can't believe there are this many women.'" She pointed out that the class of seventy students included only ten women, which shows how a small proportion of women can be perceived as a crowd.

Amanda also talked about peer discouragement, saying that "I know a lot of guys are like, 'Oh, you're a double major. You're crazy. You're not going to make it through all of those courses. And the science and physics, there's no way. You're going to drop, you're going to drop.' I find it more from guys than women." In this case, men seem skeptical that a woman can make it through the rigors of these classes. In many cases, women tried to justify these feelings of being "other." Jenny, a first-year physics and astronomy double major, stated, "a lot of people say women aren't supposed to do good in math or physics, but I've always been the top student at my high school. . . . there's always the macho guys in high school who think they're better than everyone, but they're still way back in the remedial math class." She dismissed the guilty parties as people who do not know what they are talking about, lacking adequate experience. Since many of her peers in her calculus class also thought women did not belong in science, Jenny may soon have had a harder time dismissing this attitude.

April, a first-year student who switched out of math, also recognized herself as an oddity in mathematics. Her high school science teacher offered this advice: "'Well, you're good at [math], but I don't know if you'd really like it.' Which is kind of what everybody said." In this case, the "advice" was not that women weren't good enough, but that they did not enjoy mathematics. A woman who enjoyed mathematics was certainly an exception.

Being defined as "other" is damaging to these women because it evokes a stereotype threat: they begin to see themselves as "other," outsiders in their own field. Implicit in the statement "I can't believe there are this many women" is the message that it is unusual for women to be in science. In other words, the "normal" science student is male. Feeling like an outsider may cause these women to question whether they really want to major in science, or it may make it easier for them to give up on their desire to be in science at all.

Authority: "Everyone comes to me"

Though the students I interviewed acknowledged that there were intelligent females in the sciences, about 25 percent of both men and women noted that males were more likely to be seen as a resource or authority in the classroom. Two of the men I talked with voluntarily identified themselves as "helpers" in the class. Brett noted that in high school chemistry "I was always helping out. I'd get my lab done and be able to help, just because I enjoy doing that." Andy, a first-year biochemistry major, noted that in his college chemistry lab, "when my TA is busy or I'm the closer one, everyone starts coming to me and asking me about what's going on because I'm always the first one done." These men not only are students but also take on a paraprofessional role, acting as authorities in the classroom for other students. This greater authority reinforces at least two messages: that the smartest students in science are men, and that these men have value in the classroom that women lack.

Brett, who works on a Formula One racing team, also noted his authority at another time. He told of his expertise in cars and noted that he was one of the few first-year students the upperclass students actually paid attention to or respected. It is clear he believed that his authority came in part from his "natural" authority as a man. He said, "I have a feeling that people would have looked down upon a girl playing with cars like they do. I think they get some type of criticism to them. There's just definitely a stereotype to it. And even I have that stereotype."

Women science students are aware that men in their midst have authority. Emily, a first-year computer science major, said, "A lot of the guys in the [college] programming class seem to be, like, computer professionals. They're just doing this [class] as a formality, they already know this stuff." These perceptions play into women's lack of authority in the classroom, which can cause men and women to develop different expectations for success in different subjects and cause women to feel unwelcome in science classrooms (see also Eccles 1994).

Though it seems that men are more likely to hold authority than women, there are some variations in this theme. Two of the women I interviewed thought people looked up to them or came to them for advice, but they spoke about this differently than did the men. Abbie noted that "people have always looked up to me by my school work or just anything. I've always just worked myself to my utmost ability." However, when questioned further she backed away from this claim. She said that people looked up to her in high school, but not so much in college; furthermore, "a lot of people I went to high school with are pregnant and married and not going to college or anything. So it's not like there was a high class that looked up to me or anything." When I asked her for more specifics about helping people, she switched from talking about math and science to giving more personal advice and admitted that in study groups, "usually I'm the one that needs help."

Bronwyn, a first-year biochemistry major, also often helped out other people. Bronwyn was in the honors program, and clearly a very intelligent woman. She noted that people often asked her for help; they even sat around her in calculus class. However, she was mystified by these actions. "Sometimes it's frustrating, because I'm like, 'Why are you asking me? I'm in the same class you are. I don't know any more than you do.' But, it's OK. It's kind of a mutual thing, though, because a lot of times when one person doesn't know it, a lot of people don't know it." Though students clearly looked up to her, she did not interpret this in the same ways as the men who were helpers. Bronwyn saw helping others as "a mutual thing." In another instance, she noted that when the teacher failed to clearly explain a concept she knew from high school, her fellow students sought her out:

> People were coming and asking questions, and they were like, either the teacher missed something or made it really ambiguous, so it just sort of relies on a bunch of us sitting there and being like, "OK, this is what . . . ," kind of figuring it out together. A lot of teamwork goes into the sciences.

Though she seemed to be the only one who knew how to do these problems, she still spoke of figuring out the problems together. In this way, she denied having more authority or expertise than anyone else in the room.

Fellow students recognized all of these students—Brett, Andy, Abbie, and Bronwyn—as having expertise; nevertheless, these students responded in gender-specific ways to recognition by their peers. The men interpreted helping hierarchically: as a sign of their authority. Brett, for example, spoke of himself as a sort of surrogate teaching assistant. The women, however, did not translate their peer respect into an individual sense of authority. If commitment to a field of study is based on a sense that one has strengths in that area, then this gendered attribution pattern should prompt rational women to opt out of science.

Insecurities: "I feel bad just reading the book"
vs. "That class is a waste of my time"

While it was clear that both the men and women interviewed worked hard at their science careers, three-fourths of the interviewees made statements that supported the notion that men allow themselves a greater "margin of error" than do women. In other words, men were more likely to shrug off the importance of attending class and learning everything meticulously, while women felt they had to do everything perfectly. April drew my attention to this point when she said,

> If [guys] know how to do [a problem], they'll do it. But if they don't know how they won't spend a lot of time trying to figure out how. All the other girls that I, I mean, they'll try and get it. Even if it doesn't come really easily, [guys] just assume it will someday and leave it alone.

In another case, Emily noted her distress when, in her high school chemistry class, "it seemed like everybody was catching on more than I was. Which wasn't, well, obviously not the best feeling." Feeling as though she was falling behind her classmates was stressful: she believed she was not as good as many of the other students in the class. No male students described a parallel experience.

Women also spent a lot of time talking about how they would speak to teachers during their office hours and use other resources in trying to figure out problems, even when they didn't like the teachers. However, men were less likely to report doing these things. Charles, a first-year mechanical engineering major, noted that "my strategy in high school was to do the least amount of work and get the best grades." Tim, a first-year chemistry major, stated that "in chemistry you can really slack off." Chris stated that his engineering class was

> so stupid, it really doesn't matter. . . . I just think that class is a waste of my time. The stuff that we do in there . . . so I don't really go to that. And I don't go to chemistry, just because it's an eight-hundred-person class. I went to my TA once before a test, actually, but that's about it.

Steve, a first-year computer engineering major, noted that even though his chemistry teacher tried to crack jokes, "I just wanted to leave because I'd be tired or something. . . . I just wanted to take my notes and leave." And even though Kevin admitted that he wanted "to do everything perfectly," he was very pragmatic about the time he was willing to spend on any one problem: "[I'll] go through my book, and if that doesn't help I'll just wait until the next class." These statements may be efforts to appear confident and may say very little about the hard work that these students do; even so, they show that males do not believe they must do everything perfectly in order to make it in science. Men students allow themselves a margin of error.

No woman expressed such relaxed confidence. Instead, the women talked about the classes and study sessions that they attended regardless of the actual help they provided. Amanda said, "I went to the study sessions . . . but a lot of times it wasn't what [the teacher] was doing. It wasn't specific examples. I know it's not all going to be given to you, but it helped just a little, but it didn't really help that much." Jenny emphasized her commitment to perfect attendance in her science classes when she stated, "In science classes you get behind one day, you can't do anything. You gotta be nondistracted. I mean, I was, in high school I was freaked out if I missed a single day of school. 'I'm behind in physics, I'm not going to know anything!'"

The notion of a status shield, discussed above, and the impact of not having one may shed some light on the survey finding that women disliked the pace of their classes more than did men. Since women lack a status shield, they feel they need to understand every detail and complete every problem. This may explain why women were more dissatisfied with the pace of their college classes, and suggests that greater teacher and parental encouragement would be necessary to help women overcome their lack of a status shield. We thus see why women would report that they were encouraged more than men; lacking encouragement, women simply would not choose a science major. Lack of encouragement in college could be extremely detrimental to women's resolve to pursue a science career, and it could be a reason women are more likely than men to opt out of a science career.

A few women talked in ways that made it appear they did not need to do everything perfectly. For example, Erin, a first-year double major in civil and environmental engineering, stated that if she cannot understand chemistry she just says, "'forget it, I'm never going to need chem again.' I know I can't do any better than a 'B' in this class." Nevertheless, later she stated, "You don't need a [chemistry] lecture, you can just read the book. But I go to lecture anyway just because I feel bad just reading the book; I might as well go." The telling fact here is not what she said, but what she did. Despite the fact that she disliked chemistry and did not plan to use it in the future, she still felt she had to attend class regularly.

Emily said she did not take honors classes in high school: "It was a little bit of laziness. . . . I like higher grades and I work really hard to get them, and I wasn't sure if I could get an A in honors courses, so I didn't challenge myself as much." This statement seems on par with some of the male students' comments that they wanted to do the least amount of work for high grades. However, there is also a difference. Like other women I talked with, Emily seemed wrapped up in grades as a measure of worth. In this case, fear of failure kept Emily out of honors classes, while fear of work was Charles's motivation.

Jonathan, a sophomore astrophysics major, even noted (in his own way) women's fear of failing. He said he thought teachers in high school were nicer to girls than to boys and, when questioned further, thought this differential treatment stemmed from the fact that women "bitched" more. He thought that men "just kind of deal with [problems] on their own, they don't bitch about it. Girls, on the average, not all of them, a lot of them don't, but will just complain until the teacher does something about it." His comments show, first, that he perceives women as more concerned about their problems with the material or with their science classes and more willing to voice these concerns than are males, and second, that this behavior is less acceptable. It is also interesting to note that he used a gendered insult to describe these women's actions.

Students do gender in the science classroom. By expecting women and men to conform to traditionally masculine or feminine behaviors, such as recognizing men but not women as leaders, these students reinforce the notion that science is a masculine pursuit. Further, these gendered expectations constitute a barrier for women. Those women who choose to be scientists learn important lessons about who "belongs" in science when they are sanctioned by their fellow students for expressing concerns, are discriminated against, or are not recognized by themselves or others as an authority.

Students in a science track in college have already successfully passed through many hoops in science education, and it is therefore not surprising that men and women science students are quite similar in background, preparation, and ability. The women in my study had more positive experiences with teachers and study groups in high school and seemed to be more dissatisfied than men with the college weed-out process. These two differences could be indications that there are some gender differences in patterns of social support in high school and could be warning signs that the local culture of science education makes a difference in fostering the continuance of women students. The differences between high school and college teaching suggest that while the transition to college may be difficult for all students, the challenges facing women may be particularly difficult. Despite significant findings in the survey data, the numerous similarities between men and women indicate a need to use qualitative measures to better understand women's reasons for leaving science majors.

All three of the gendered dynamics that emerged in the interviews—being defined as "other," failing to recognize one's competence, and working harder and more inefficiently because of grade anxiety—are logically interrelated. Women's feeling that they need to be perfect has both positive and negative results. First, it pushes women to work hard, to understand the material completely, and to achieve good grades. However, this drive also causes women to expend unnecessary time engaging in activities, like some study sessions, that are not necessarily helpful. In addition, feeling as though they need to be perfect may increase their expectations of themselves to unreasonable levels. Even minor failures may seem huge to these women and they may interpret grades that are by most standards good (such as a B) as not good enough and therefore a sign that they should not be in science (see Orenstein 1994). Meanwhile, in the college science environment men are not only legitimate, but also more likely to be seen as an authority in science, which further reinforces the notion that men are somehow "naturally" better at science. Women's emphasis on cooperative and collaborative approaches to learning, while good in some respects, undercuts their claims to the kind of authority awarded to men.

Being culturally defined as belonging in science courses, and seeing their male peers (even if not themselves) recognized as authorities, gives men a greater "margin of error" in courses. They are not as vulnerable to being set back by a few disappointing performances. On the other hand, without these cultural supports for their legitimacy and authority, women feel pressed to do everything perfectly. Much more is at stake in every challenge for women, and since they are likely to hold themselves to an impossible standard, it seems likely that they will "learn" that they do not belong in science and drop out, blaming only themselves. These findings give clear support to the proposition that doing science is a way of doing gender.

Notes

1. These data were analyzed using cross-tabulations and t-tests in the SPSS statistical program. Cross-tabulations were used to help gain a sense of similarity and difference between men's and women's answers by showing frequencies of answers when broken down by sex category. T-tests were used to test the statistical significance of the differences between the average scores of males and females (Vogt 1999).
2. I used a grounded theory approach, in which theory is developed in an "intimate relationship" with data to analyze interviews (Strauss 1987, 6). Though theory is an important component of research, in a grounded theory approach it is essential that the data, rather than deductive reasoning, shape the processes and products of research throughout the research process (Charmaz 1983, 110).

References

AAUW (American Association of University Women). 1999. *Gender Gaps: Where Schools Still Fail Our Children.* New York: Marlowe.

Acker, Joan. 1990. "Hierarchies, Jobs, Bodies: A Theory of Gendered Organizations." *Gender and Society* 4(1): 139–58.

Aronson, Joshua, Michael J. Lusina, Catherine Good, Kelli Keough, Claude M. Steel, and Joseph Brown. 1999. "When White Men Can't Do Math: Necessary and Sufficient Factors in Stereotype Threat." *Journal of Experimental Social Psychology* 35: 29–46.

Carli, Linda L. 1990. "Gender, Language, and Influence." *Journal of Personality and Social Psychology* 59: 941–51.

Charmaz, Kathy. 1983. "The Grounded Theory Method: An Explication and Interpretation." In *Contemporary Field Research: A Collection of Readings,* ed. Robert Emerson, 109–25. Boston: Little, Brown.

Clark, C., I. Howard, S. E. Lazare, and D. A. Weinberger. 2000. "A Peer Mentoring Program for Underrepresented Students in the Sciences." In *Women Succeeding in* the *Sciences: Theories and Practices across Disciplines*, ed. Jody Bart, 149–67. West Lafayette, Ind.: Purdue University Press.

Eccles, Jacquelynne S. 1984. "Sex Differences in Achievement Patterns." In *Nebraska Symposium on Motivation: Psychology and Gender*, ed. Theo B. Soneregger, 97–132. Lincoln: University of Nebraska Press.

———. 1994. "Understanding Women's Educational and Occupational Choices." *Psychology of Women Quarterly* 18: 585–609.

Etzkowitz, Henry, Carol Kemelgor, and Brian Uzzi. 2000. *Athena Unbound: The Advancement of Women in Science and Technology*. Cambridge, Mass.: Cambridge University Press.

Flores, Alfinio. 1997. "*Si Se Puede*, 'It Can Be Done': Quality Mathematics in More than One Language." In *Multicultural and Gender Equity in the Mathematics Classroom: The Gift of Diversity*, ed. Janet Trentacosta and Margaret J. Kenney, 81–91. Reston, Va.: National Council of Teachers of Mathematics.

Fowler, Floyd J., Jr. 1993. *Survey Research Methods*. 2nd ed. London: Sage.

Hanson, Sandra L. 1996. *Lost Talent: Women in the Sciences*. Philadelphia: Temple University Press.

Hochschild, Arlie Russell. 1983. *The Managed Heart*. Los Angeles: University of California Press.

Jacobs, Judith E., and Joanne Rossi Becker. 1997. "Creating a Gender-Equitable Multicultural Classroom Using a Feminist Pedagogy." In *Multicultural and Gender Equity in the Mathematics Classroom: The Gift of Diversity*, ed. Janet Trentacosta and Margaret J. Kenney, 107–14. Reston, Va.: National Council of Teachers of Mathematics.

Jones, Leslie S., and Kathryn Scantlebury. 2001. "Feminist Leadership in the Academy: Innovations in Science Education." In *Feminist Science Studies: A New Generation*, ed. Maralee Mayberry, Banu Subramaniam, and Lisa H. Weasel, 138–44. New York: Routledge.

Kanter, Rosabeth Moss. 1977. *Men and Women of the Corporation*. New York: Basic Books.

Kubanek, Anne-Marie Weidler, and Margaret Waller. 1996. *Confidence in Science: Interpersonal and Institutional Influences*. Ste.-Anne-de-Bellevue, Quebec: John Abbott College.

Ligata, Nermana, and Madeleine Adamczeski. 2000. "Perspectives from a Female Undergraduate Student on Successfully Integrating Learning and Researching Science with Leading Organic Chemistry Workshops." In *Women Succeeding in the Sciences: Theories and Practices across Disciplines*, ed. Jody Bart, 139–48. West Lafayette, Ind.: Purdue University Press.

Martin, Karin A. 1998. "Becoming a Gendered Body: Practices of Preschools." *American Sociological Review* 63: 494–511.

Mickelson, Roslyn Arlin, and Stephen Samuel Smith. 1995. "Education and the Struggle against Race, Class, and Gender Inequality." In *Race, Class, and Gender: An Anthology*, ed. Margaret L. Andersen and Patricia Hill Collins, 2nd ed., 289–304. Boston, Mass.: Wadsworth.

Orenstein, Peggy. 1994. *School Girls*. New York: Anchor.

Pierce, Jennifer L. 1995. *Gender Trials: Emotional Lives in Contemporary Law Firms*. Berkeley: University of California Press.

Renzetti, Claire M., and Daniel J. Curran. 1995. *Women, Men, and Society*. 3rd ed. Boston: Allyn and Bacon.

Ridgeway, Cecilia. 1993. "Gender, Status, and the Social Psychology of Expectations." In *Theory on Gender/Feminism on Theory,* ed. Paula England, 175–97. New York: A. de Gruyter.

Rosser, Sue V. 1990. *Female-Friendly Science: Applying Women's Studies Methods and Theories to Attract Students.* New York: Pergamon.

———. 1997. *Re-engineering Female Friendly Science.* New York: Teachers College Press.

Schiebinger, Londa. 1999. *Has Feminism Changed Science?* Cambridge, Mass.: Harvard University Press.

Seymour, Elaine, and Nancy M. Hewitt. 1997. *Talking about Leaving: Why Undergraduates Leave the Sciences.* Boulder, Colo.: Westview.

Shauman, Kimberlee A., and Yu Xie. 1999. "Gender Differences in the Attainment of a Science/Engineering Bachelor's Degree." Paper presented at the annual conference of the American Sociological Association, August 6–10, Chicago.

Sonnert, Gerhard, and Gerald Holton. 1995. *Who Succeeds in Science? The Gender Dimension.* New Brunswick, N.J.: Rutgers University Press.

Spencer, Steven J., Claude M. Steele, and Diane M. Quinn. 1999. "Stereotype Threat and Women's Math Performance." *Journal of Experimental Social Psychology* 35: 4–28.

Strauss, Anselm L. 1987. *Qualitative Analysis for Social Scientists.* New York: Cambridge University Press.

Strauss, Anselm L., and Juliet Corbin. 1990. *Basics of Qualitative Research: Grounded Theory Procedures and Techniques.* London: Sage.

Valian, Virginia. 1998. *Why So Slow? The Advancement of Women.* Cambridge, Mass.: MIT Press.

Vogt, W. Paul. 1999. *Dictionary of Statistics and Methodology: A Nontechnical Guide for the Social Sciences.* 2nd ed. Thousand Oaks, Calif.: Sage.

Wenneras, Christine, and Agnes Wold. 2001. "Nepotism and Sexism in Peer-Review." In *Women, Science, and Technology: A Reader in Feminist Science Studies,* ed. Mary Wyer, Mary Barbercheck, Donna Geisman, Hatice Orun Ozturk, and Marta Wayne, 46–52. New York: Routledge.

West, Candace, and Sarah Fenstermaker. 1995. "Doing Difference." *Gender and Society* 9(1): 8–37.

West, Candace, and Don H. Zimmerman. 1987. "Doing Gender." *Gender and Society* 1(2): 125–51.

Part Three.

Feminist Study of Scientific Practice

9 The Construction of Sexual Bimorphism and Heterosexuality in the Animal Kingdom

Kirsten Smilla Ebeling

> Like transgendered humans, vaginaless female hyenas may seem strange. But life shows far more variety in the sexual realm, both within and among species, than our straight-and-narrow view of normalcy might suggest. (Margulis and Sagan 1997, 199)

To most people in Western industrial cultures, the difference between male and female humans, animals, and plants is self-evident. In the same binary fashion, biologists categorize different forms of reproduction as sexual and asexual. In biology classes students may learn about hermaphrodites, like the earthworm, which have male and female reproductive organs. But despite the acknowledgment of animals that are both female and male, the sexes are considered dichotomous and heterosexual modes of reproduction are regarded as predominant.

In this chapter, I describe animal species and forms of reproduction that do not fit into the bipolar classification system of sex and reproduction. Additionally, I question whether the primacy of sexual dimorphism and gonochorism[1] can be maintained. I discuss this question in the context of interactions between human sex/gender relations and the production of biological knowledge. I point out that discursive processes involving negotiations about gender relations play an important role in biological theories about sex and reproduction.[2]

I use only examples from the animal kingdom, although an analysis of the binary system of sex and reproduction in the plant kingdom would be promising as well. For instance, as discussed in chapter 11 of this volume, the majority of plant species are not dioecious[3] but hermaphroditic. Though I have encountered no comprehensive quantitative surveys of all plant species, Sabine Riewenherm's (1996) overview of more than 1400 European flowering plants found that 92 percent were hermaphrodites and only 3 percent were dioecious.[4] Others have examined biologists' anthropomorphic heteronormative perspectives on plants (see chapter 11 of this volume; Schiebinger 1993). These are important avenues of research. Here, I limit my focus to the animal kingdom.

I selected my examples mostly from popular biological texts that were published in German or in English (and sometimes in both languages) from the 1950s to the 1990s. While I have included texts from each decade of this period, my selection does not constitute a representative sample.

Animal Species That Do Not Fit into the Dichotomous Sex Structure

The study of hermaphroditism can be traced to the ancient world. Biologists describe simultaneous hermaphrodites, which are beings that have both female and male reproductive organs, and consecutive hermaphrodites, which have female and male reproductive organs serially. One consecutive hermaphrodite is the protogynous fish species Bluehead wrasses (*Thalassoma bifasciatum*), whose habitat is the Atlantic Ocean. Among this species all individuals are born female, but some change their sex as soon as their bodies reach a certain size. This sex change—as one biology textbook explains—is "socially controlled" (Krebs and Davies 1984). If there are only a few large males and many females, some of the females turn into males.

Other hermaphrodites undergo this process the other way around. For instance, all individuals of the protandric species Anemonefish (*Amphiprion akallopisos*) are first male and may turn into females after reaching a certain body size. Again, this sex change is socially controlled: the *Amphiprion akallopisos* live as monogamous couples, and if the female of a couple disappears, the remaining male turns into a female and is joined by a new small male (Krebs and Davies 1984).

Compared to these binary classifications of simultaneous and consecutive protandric and protogynous hermaphrodites, the Thaliacea are more complex. These sea dwellers are small pelagic tunicates that have a stable but flexible body covering called a tunic.[5] Biologists describe a complicated alternation of generations among these animals: first, an egg develops into an individual that multiplies asexually by budding into four individuals; these, in turn, produce a colony through budding (Remane, Storch, and Welsch 1980). The hermaphrodites of such a colony are categorized into three types according to age. The oldest animals of a colony first produce sperm and later one egg. The middle-aged individuals produce female and male gametes simultaneously, whereas the youngest animals of a colony first produce an egg and afterward sperm. Hence, this animal species includes four different types of simultaneous and sequential hermaphrodites, as well as protandric and protogynous hermaphrodites.

In the class of the Turbellaria even more types of hermaphrodites are identified. For example, biologists describe species in which the males and the females change their sex simultaneously. Another so-called extreme form of hermaphroditism is described in lung snails (*Pulmonata*): they have an androgynous reproductive organ that produces eggs and sperm at the same time (Siewing 1985). Other animal species reveal sexually intermediate stages, as the butterfly *Morpho didius,* for instance, one side of which is completely female and the other completely male (Insect Company 2002). This type of hermaphroditism is found among birds, rays, butterflies, and bees (Remane, Storch, and Welsch 1989). Some animal species, such as some species of roundworms (*Nematoda*), are hermaphroditic in only certain generations, switching between gonochoristic (sexually differentiated) and hermaphroditic generations. They serve to illustrate the diversity of animals that do not fit into the conventional sex structure. I use these examples to point out that the sex of animals is not a fixed feature, but rather one that may vary. In the biological textbooks I examined, only eight (33 percent) of the twenty-four animal phyla were gonochoristic (sex changes were possible). A large portion of the other sixteen were hermaphroditic. Because the number of species in animal phyla varies, these calculations are only approximate. In order to obtain a more accurate count, animals need to be scrutinized on the species level.

Thus far, I have pointed out that biologists describe a diversity of animal hermaphrodites besides gonochoristic animals, and that sex is a feature that is not always fixed. All types of hermaphroditism mentioned here build directly on a binary structure. But not all animal species are easily organized into the binary sex structure, since some combine only single features of both sexes. For example, the female spotted hyena (*Crocuta crocuta*) has not a vagina but a penis that is used both for intercourse and to deliver offspring. Such a female is described as "masculinized" and her penis is referred to as a "female penis" (Margulis and Sagan 1997). Similarly, among seahorses the females deposit their eggs into the breeding pouches of the males, where the eggs are fertilized and where gestation takes place, and the males give birth (Siewing 1985). Biologists describe, in addition, all-female species in which males have never been sighted and are therefore assumed not to exist.[6]

How can we classify these animal species within the dichotomous sex model? Are they exceptions or deviations from the norm of gonochorism and heterosexual reproduction, or do they challenge the conventional concept of sex? They indicate, it seems, that the bipolar structure of sex and reproduction is not as predominant in the animal kingdom as is usually assumed.

Forms of Reproduction That Do Not Fit the Binary Concept of Sex

There are many examples of species in the animal kingdom that do not fit the conventional idea of sexual reproduction. My examples of hermaphrodites have already shown that some animals change their sex in the sequence of generations. In the group called Cnidaria, sexually and asexually reproducing organisms alternate. Here, the mobile jellyfish are heterosexual; their fertilized eggs develop into asexual polyps, which are sessile. The polyps produce jellyfish asexually. Taking into account the existence of all-female species and starting from the conventional concept of heterosexuality, one is bound to ask how these females procreate without males or sperm. The biological term for this type of reproduction is "parthenogenesis." Accounts of "pure female reproduction" are found in early Latin, Greek, Egyptian, and Christian sources, and biologists have studied parthenogenesis since the middle of the seventeenth century. The term "parthenogenesis" was coined by Richard Owen in 1849. There are many different forms of parthenogenesis, their common feature being that an egg develops without fertilization.

One type of parthenogenesis is gynogenesis. Here, females copulate with males from a different but closely related species. The sperm of these males activates the development of the eggs without fertilizing them, and only the female's genome is passed on to its offspring. According to biological descriptions, all-female species are not the only ones to reproduce parthenogenetically; heterosexual animal species do as well. For example, aphids and daphnia—depending on their environmental conditions—reproduce either parthenogenetically or heterosexually. Parthenogenetic reproduction produces only females, while sexual reproduction produces both males and females. Another type of parthenogenesis is geographical parthenogenesis. Here, one species, such as the woodlouse (*Trichoniskus provisorius*), reproduces in one geographical region parthenogenetically, and in another heterosexually.

These different forms of parthenogenesis show that biologists describe forms of reproduction that do not easily fit into the dichotomous paradigm of sexual and asexual reproduction. Accordingly, biologists debate whether the different forms of parthenogenesis are sexual, unisexual, or asexual types of reproduction. But since only the female genome is passed on to the next generation, they are often regarded as female reproduction.

In the following section, I describe in detail how all-female species and parthenogenesis are depicted in biology textbooks. I provide three examples of the impact of the idea of the "preponderance of sexual reproduction" (Gerritsen 1980, 718) and of human sex/gender power relations on the production of biological knowledge.

Depictions of All-Female Species and Parthenogenesis in Biology Textbooks

My first example is the group called the Bdelloidea. The species of the Bdelloidea are microscopic small rotifers, which live in lakes, rivers, and water ponds. They reproduce by obligatory parthenogenesis. They are estimated to be thirty to forty million years old and are the oldest known parthenogens (Voigt 1957).

In the biological descriptions of the Bdelloidea two metaphorical expressions are particularly salient: in popular science it is said that the "maleless" life of the Bdelloidea represents an "answer to the Women's Liberation Movement" (Margulis and Sagan 1997, 142), while some scientists regard it as an "evolutionary scandal" (Smith 1986). Thus, the Bdelloidea females are linked to the women's liberation movement and are perceived either negatively or positively, depending on the authors' (and readers') opinions of women's liberation. Margulis and Sagan's comment evokes stereotypes about feminists, namely that they seek a life without men. On the other hand, Smith sees the Bdelloidea as an "evolutionary scandal."[7] This metaphor bears the fundamental assumption that the different forms of parthenogenesis are evolutionarily disadvantageous in the long term compared to heterosexual reproduction. For instance, parthenogenetic species are said to persist only for a short period of time. However, the advanced age of the Bdelloidea species seems to undermine this assumption. Nevertheless, rather than treating the Bdelloidea as a reason to question long-standing assumptions, their existence is considered "scandalous." Seen as a role model for the women's liberation movement, the evolutionary success and persistence of the Bdelloidea females causes uneasiness and anger. Clearly, all-female species would not be depicted as "scandalous" were it not for the supposed predominance of sexual dimorphism.

Another example of an all-female species is *Poecilia formosa*. These are small Mexican fish that resemble the guppy, a common pet fish. The females reproduce by gynogenesis. In biological descriptions of these females some metaphors are striking. The English common name of *Poecilia formosa* is "Amazon Molly." The Amazon metaphor links the fish females with the Amazons of Greek mythology, who were depicted as an all-female population, a bellicose, "man-hating army" who violated men and threatened the Greeks (Sir Galahad 1987; Cancik and Schneider 1996).

One interpretation of the "Amazons" metaphor concerns its structural effect on scientific reasoning. Ingo Schlupp, in the beginning of his dissertation on the behavior and ecology of *Poecilia formosa* (1995), uses the Amazons metaphor. His description of *Poecilia formosa* not only includes the species' colloquial name, but also its reference to the Greek myth: "The English and German colloquial names of *P. formosa*, 'Amazon Molly' . . . and 'Amazonenkärpfling,' refer to the tribe of exclusively female warriors of the Greek myths" (Schlupp 1995, 10).[8] Schlupp describes here and elsewhere the similarities between the mating behavior of the female fish and the myth of the Amazons, and in doing so intermixes sociocultural conditions and the biological descriptions of the fish.

Like the mythical Amazons, the *Poecilia formosa* females are described as aggressive and dangerous to the males of closely related species: the females interrupt the copulation of the males with their own females, chase away the females, and "seduce" or "take advantage of copulation with" the males. Schlupp's ethological reasoning suggests that these "wrong copulations" have "catastrophic" outcomes for the males, whereas the *Poecilia formosa* females do not experience any "costs." Therefore, he regards the relations between *Poecilia formosa* and the males with which they mate as parasitism. However, the assumption that males gain no advantage from their copulation with *Poecilia formosa* does not mesh easily with the ethological presumption that every behavior results in a benefit, and leads Schlupp to ask, "Why do males of the heterosexual species copulate with *P. formosa* females?" (1995, 113). Schlupp goes on to ask whether the males are not able to distinguish between females of their own species and others, but his subsequent experiments show that the males are able to discriminate very well between the females. Schlupp therefore assumes that males will obtain an advantage from the supposedly "wrong copulations" with *Poecilia formosa*. He concludes that these fish practice "mate copying," in which females prefer to mate with males that have already succeeded in mating. Since the females of the heterosexual species observe the males' copulations with the *Poecilia formosa* females, the males gain a higher copulation rate with the females of their own species, which in turn leads to a higher reproduction rate. Therefore, one implication of Schlupp's experiments is that there is no parasitism, but mutualism.

To sum up, at the beginning of Schlupp's study the metaphor of Amazons refers to a situation in which "maleless" females harm other species' males with their aggressive mating behavior. And though Schlupp explains the reproductive system of gynogenesis (i.e., the necessity for copulations with males from different species), and invokes fundamental ethological presumptions that predict that the males' behavior will be in their own interest, he frames the females as "sexual parasites." On the landscape of ethological theories, the hypotheses of mate copying and mutualism appear as surprising news. I argue that Schlupp's findings could only have been unanticipated given the logic of the Amazon metaphor. This metaphor combines images of all-female human and animal groups that, for reproductive purposes, need "foreign" males. The active mating behavior of *Poecilia formosa* females is regarded as aggressive and harmful. This corresponds to human sex/gender relations in the sense that women's active sexuality is often considered aggressive and seen negatively. In addition, the characterization of the relationship as parasitism constructs the males as victims; the benefits males may draw from the interaction, as well as the costs the *Poecilia formosa* females may incur, are omitted from the picture.

Other studies by Schlupp and his colleagues further highlight the contradiction in framing Amazon mollies as "parasites" even though male sailfin mollies (*P. latipinna*) apparently derive some benefit from their association with Amazon mollies (Schlupp, Marier, and Ryan 1994; Schlupp, McKnab, and Ryan 2001). They conclude in one study, for example, that Amazon mollies' association with male sailfin mollies is deleterious because the smaller male sailfin mollies "sexually harass" Amazon mollies (and female sailfin mollies) by attempting forced copulation, thereby reducing the Amazon mollies' feeding time. They nonetheless continue to speak of Amazon mollies as "parasites" when describing their relationship to "their hosts in nature" (Schlupp, McKnab, and Ryan 2001, 279). And they add, further, that because reduced feeding time may also decrease fecundity, Amazon mollies try to position themselves close to larger male sailfin mollies that "court" rather than "harass" females, giving Amazon mollies (and female sailfin mollies) more time for feeding.

Thus, likening gynogenetic females to Amazons is a strategy riddled with contradictions. In modern science, a myth is regarded as superstition and fiction, as opposed to scientific rationality. And the Greek myth of the Amazons, if taken to represent any sort of reality, describes a defeated people whose ultimate destiny is death. This implies that the "Amazon Molly"—following the thread of the metaphor—should become extinct. Hence, the *Poecilia formosa* females are likened to an unverifiable superstition and it is implied that they are doomed to extinction. Schlupp's framing of *Poecilia formosa* as parasites that harm the male population of their "hosts in nature" (even though the males that mate with *Poecilia formosa* derive some benefit and *Poecilia formosa* themselves pay a cost in decreased feeding time and fecundity) suggests that while *Poecilia formosa* appear to have eluded destiny, they have not escaped "natural" relations of power, according to which males are essential and superior to females. *Poecilia formosa* not only need a "host" that is male, they seek out the larger males for protection.

Interestingly, the *Poecilia formosa* females, like the Bdelloidea, have also been said "to have brought the battle of the sexes to an end" (Berill 1954, 43) and to now live in a "feminist paradise" (Glaubrecht 1998, 32). Thus this all-female species is situated just as the other is. The reproduction of the females is seen in the context of women's liberation and feminism, and the elimination of men is said to be the goal of the women's liberation movement.

Casting *Poecilia formosa* as mythical and unreal, parasitic and destined for extinction, and as harmful to but simultaneously dependent on males, scientists force the species into a dualistic, asymmetrical model of sex and sexual reproduction that frames females negatively as weak but also harmful, and males as strong but also victimized by parasitic females. This example reveals a connection between the depiction of parthenogenetic all-female species and the gender relations of humans, and demonstrates the impact of the "primacy of sexual reproduction." This framing of *Poecilia formosa,* in turn, reinforces common assumptions in the human social world that any form of reproduction and any "female" that does not fit into the "normal" model of sex and sexual reproduction is "unnatural" and a "threat."

Even the word "parthenogenesis" itself bears the idea that parthenogens do not actually exist. It is derived from a Greek term whose literal translation is "virgin birth." It thus evokes images of the Virgin Mary and chastity. Corresponding to these images, I found the following metaphors in evolutionary biology texts: "celibacy" (Wuethrich 1998, 1981), "chastity" (Butlin, Schön, and Griffiths 1998, 314), "virgin birth" (Catton and Gray 1985, 106), "virgin females" (Rasch, Prehn, and Rasch 1970, 31), "virgins" (Margulis and Sagan 1997, 85), and "immaculate conception" (Wendt 1970, 190), just to mention a few. In a popular science magazine, one author writes about a butterfly, "The virgin's name is Solenobia. . . . As an imago [adult butterfly] it is wingless and stands pure and demure in its shell as if in a convent. There it lays eggs and dies as a virgin, without ever leaving it and without copulation" (Forsyth 1987, 213; my translation). In Schlupp's dissertation, the *Poecilia formosa* females, colloquially called Amazon mollies, are often referred to as "virgin females" (Schlupp 1995).

This metaphor associates animal reproduction with unmarried women and the Virgin Mary. Female insects, rotifers, and fish whose eggs develop without fertilization are described as "untouched," "pure," and "innocent."

Although the phrase "immaculate conception of the Virgin Mary" actually refers to the Catholic doctrine that Mary herself was conceived without original sin, the only human since Adam and Eve to be "immaculate" (unstained) in that way, it is often misunderstood to refer to Mary's conceiving Jesus by the Holy Spirit, without having sexual intercourse. Since, according to this belief, the Christian God passed his spirit through Mary, who served merely as the vessel for God's son, the metaphor of immaculate conception, misunderstood in this way, leads to the conclusion that parthenogenesis, rather than being a purely female type of reproduction, is actually an entirely male reproductive process. Parthenogenesis in the sense of a pure female reproduction, therefore, does not actually exist. Moreover, parthenogens are explicitly linked to Christian doctrine. As a consequence, this form of reproduction is moved outside the scientific realm entirely.[9]

The biological descriptions of the two all-female species and of the forms of parthenogenesis imply many different meanings, including those that represent or evoke women's liberation and feminism, feminists who seek life without men, and religious and other myths that have no scientific basis. And finally, as demonstrated in the following quotation, parthenogens are represented as nonexistent in yet another way—they disappear: "Asexual creatures save for the wearying courtship, the copulation, and the often arduous search for partners . . . [t]hey have no need for long and colorful feathers, which hinder them during flight; for antlers, which get them caught in the underbrush; or for a Porsche, which they may overturn on the freeway. After all, there is no other sex they have to impress" (Miersch 1996.) Here Michael Miersch describes "asexual creatures" in stereotypically "masculine" terms, even though he is talking about what most scientists refer to as females that reproduce without fertilization. The relations are twisted. If there were a missing sex in parthenogenetic groups, it would be not the female but the male. In Miersch's text, however, parthenogenetic females are basically "unarticulated." This example, like the previous ones, represents a perspective that sees heterosexual reproduction as normative.

My third example is the all-female species *Cnemidophorus uniparens*, a species of lizard that lives in the U.S. Southwest. Its members reproduce by parthenogenesis and mate with each other, presumably in order to support the maturation of their eggs. According to biologists, the mounting female shows "male sexual behavior." The biologist David Crews and his team studied how this behavior is triggered. Their research suggests that the female hormone progesterone, among others, causes the females to mount. Thus, a "female sex-hormone" causes a "male sexual behavior" (Crews, Grassman, and Lindzey 1986; Crews 1993, 1994b; Moore et al. 1985; Crews and Fitzgerald 1980). Since these results do not fit into the concept of binary sex, and since biologists have to work with the predetermined structure of sexual dimorphism, they encountered language troubles. They tried to get around the problem by constructing terms such as "male-like behavior," "female-typical sexual behavior," and "pseudosexual." They also put the words "female" and "male" in quotation marks. The authors applied concepts that are coherent only for heterosexual reproduction, and, according to their own statements, not for *Cnemidophorus uniparens* females. Their language refers to the heterosexual structure although the unisex animals and their unisexual type of reproduction do not fit into that structure. This example demonstrates that the conventional paradigm of binary sex is not abandoned, even in the descriptions of animal species that do not fit it—it is perpetuated instead.

Furthermore, the example demonstrates classical attribution of male and female features: mounting female lizards are said to perform an active, male-like behavior, whereas a mounted female is said to perform a passive, female-like behavior. Here, two aspects of sexual reproduction—the mounting behavior and the maturation of eggs—are described as simultaneously active and passive: a female may be regarded as active when it comes to reproduction (i.e., laying eggs), at the same time as she may be regarded as passive (i.e., being mounted). Because the authors most often speak of "active" and "passive" in connection with mounting, this behavior seems to be more important than the actual reproduction: laying mature eggs. In other words, even in evolutionary theories, where reproduction is the "essence of life," the echo of human sex/gender relations seems to be more important than the act of reproduction. Yet in both cases the attributions are questionable, since from an ethological perspective, being mounted can be regarded as an active behavior. Similarly, the mounting female, which stimulates the deposition of eggs by the other, will be mounted in a future interaction and thereby will lay eggs itself. Thus, in both of these cases, the attribution of activity and passivity appears to serve no purpose other than to render "active" equal to "male" and "passive" equal to "female," thereby symbolically expressing the gender order of Western industrial cultures.

Interestingly, David Crews and his colleagues conclude that there are no bipolar sex differences in *Cnemidophorus uniparens*, that every individual is bisexual. They suggest a continuum as an alternative model and plead for a focus on the similarities of the sexes (Crews 1994a). Hence, they question the concept of sex as bipolar. However, in their description of the *Cnemidophorus uniparens* females, they continually draw on the paradigm of sexual dimorphism and heterosexual reproduction. For example, they refer to this group of unisexual parthenogenetic females not in terms of one sex (i.e., the female) but in terms of heterosexuality, thereby failing to abandon the concepts of sexual dimorphism and heterosexuality. Note that it clearly does not make sense to talk of a "male sexual behavior" that is performed by a female and caused by a female sex hormone. In addition, the differentiation of male and female hormones becomes questionable when a "female sex hormone" triggers a "male sexual behavior." Finally, does it still makes sense even to speak of females if the second part of the binary classification—the male—is missing?

Andro- and Gynocentric Perspectives on Unisexual Reproduction

Biological descriptions of the different forms of parthenogenesis stress the idea that males do not contribute to these forms of reproduction. In the following quotation, this fact of a "maleless or pure female reproduction" is mentioned six times: "And again all offspring are female. Males are lacking. They are superfluous—the eggs develop without them; parthenogenesis is the rule. How it happens that the eggs of Daphnia are able to develop by themselves, without help from spermatozoa, is—as far as I can see—beyond our imagination today" (Berill 1954, 41). This passage can be understood as an expression of either admiration or fear. Since many biological descriptions of parthenogenesis express not only the elimination of males but also of men in general as a negative thing, I argue that findings about superfluous males cause considerable fear. One scientific paper, for example, described the absence of males as "dramatic" (Smith 1992, 662), while in a popular science text parthenogenetic locusts were called "man-haters" (Buddenbrock 1953, 53).

This fear is comprehensible only in a predominantly male and heterosexual context, characteristic of an anthropocentric and androcentric perspective. Such unreflective, male-centered views can be found in various biological texts. The title of a chapter of a popular science book on the issue of parthenogenetic invertebrates serves as an illustrative example: "The abolition of men (parthenogenesis)" (Buddenbrock 1953, 50). The authors of another popular text write about the Bdelloidea, "Might human males, like rotifer males, become evolutionarily redundant? Certainly, the cloning of women's eggs could, in principle, circumvent our two-parent sexual cycle. But it is doubtful that meiosis and fertilization in such women can be entirely abandoned. Rather, these hardy mothers will be self-fertilizing. Their haploid eggs will probably require a fertile boost of self-fusion—the egg nuclei fertilized by an egg equivalent such as another haploid nucleus from the female's own body. Indeed, this is what occurs today in all-female rotifers" (Margulis and Sagan 1997, 142). In these texts, animal parthenogenesis evokes the idea of superfluous men and all-women societies. Similarly, parthenogenetic vertebrates are said to "shake the male self-confidence" and to represent the "final fate of men" (Berill 1954, 35–36).[10]

In evolutionary theory, reproduction is a major function of life. Therefore, the contribution of the sexes to reproduction has crucial meaning. From an evolutionary viewpoint, an organism that does not reproduce goes against nature. In cases of obligatory parthenogenesis, and also in all-female species, only females pass their genes on to their offspring. Reproduction without males can be seen as an exclusion of males. But the assessment of the inutility of males as "ominous" (Buddenbrock 1953, 50) or even as "dramatic" (Smith 1992, 662) is comprehensible only in the context of a dominant notion of heterosexuality (or compulsory heterosexuality). These descriptions of parthenogenesis show that a "maleless" reproduction mode evokes the idea of the abolition of men and in this context also triggers fear. In light of this androcentric fear, the Bdelloidea might be called "scandalous" not only because they contradict evolutionary theory, but because they make males (and perhaps men) superfluous and devoid of evolutionary purpose.

Tekla Reimers, a woman biologist, connects the different forms of parthenogenesis to women and judges them positively. Reimers writes, "In her utopian novel *Herland* Charlotte Perkins Gilman describes the social structure and history of a parthenogenetic all-women society. Quite correct in the details of biological possibilities, she composes a fascinating social network on the basis of the general potency of autonomous mothers. . . . A second chance to recreate one's life—who wouldn't like that? The possibility might indeed arise for women to start their life through a 'twin daughter' once again" (1994, 217). Reimers identifies human parthenogenesis as "a form of individual transcendence of death, an incalculable number of new chances for the single woman and for female socialization" (1994, 218). Moreover, she regards parthenogenesis among humans as "a further cultural development of the natural evolution [through which] human culture realizes a potential that is naturally given in the biology of mammals" (Reimers 1994, 218). Thus, parthenogenesis can be used not only in an androcentric interpretation of animal reproduction, but also in an explicitly gynocentric perspective. Both the andro- and the gynocentric perspectives support my argument that sex/gender power relations have a strong impact on biological theories about different modes of reproduction. Altogether, the forms of parthenogenesis are judged very differently: male biologists describe them as threatening whereas women biologists tend to view them positively, as valuable possibilities for women.[11]

What I find important here are the negotiations about a unisexual type of reproduction, which are expressed in different ways. First, biologists discuss and dispute the existence of parthenogenesis differently. For example, it is controversial whether the different forms of parthenogenesis are to be understood as sexual reproduction. Certain biologists consider them to be forms of asexual reproduction, because they do not involve the combination of two parental genomes. Another group of biologists regard them as sexual reproduction, since gametes (i.e., eggs) develop into an organism. A third group of biologists define the meiosis during egg maturation as the main feature differentiating sexual and asexual reproduction. They consider only the meiotic parthenogenetic forms of reproduction as sexual ones. Second, many evolutionary biologists depict the different forms of parthenogenesis as methods of reproduction that will necessarily result in the extinction of parthenogens, although individual parthenogens are supposed to be far more reproductively successful than individual sexually reproducing organisms. At the same time, evolutionary theory cannot explain the existence of parthenogenesis.

Such biological reasoning can be seen as another version of the notion that parthenogens do not actually exist. Accordingly, numerous stories and metaphors in biological texts imply this nonexistence. To support this interpretation, I introduce yet another metaphor, the lottery, a metaphor used frequently in scientific texts. Assuming that parthenogens will become extinct relatively quickly, the field of visual images offered by the word "lottery" is vast. It conveys that parthenogens buy tickets with identical numbers, whereas animals reproducing heterosexually buy tickets with different numbers. Consequently, the parthenogens have an extremely small chance of winning. They cannot compete effectively with sexual species and therefore have little chance of survival (see, e.g., Smith 1976, 1986; Bell 1982). Other metaphors, such as "weeds" (Maslin 1971, 377), "monsters" (Maslin 1971, 366), and "disease" (Klärner 1997, 3), describe parthenogens as harmful. They are to be eliminated or shunned. In the case of the all-female species *Poecilia formosa* and *Cnemidophorus uniparens*, biologists try to produce males artificially, in order to reestablish the heterosexual mode of reproduction.

My review of biological theories and descriptions of sex and reproduction suggests overall that the existence of parthenogenesis and all-female species is negotiated both on the level of biological reasoning and in the metaphors of biological texts. The evolutionary discourse on animal parthenogenesis is related to negotiations about the possibility of human unisexual reproduction. These negotiations about unisexual reproduction, in turn, reveal that reproduction and power are not separate entities, but entirely interwoven. For example, the linkage of reproduction and power becomes apparent in texts from different scientific disciplines, such as folklore, anthropology, psychology, history, and modern reproductive technologies, all of which engage with the issue of human parthenogenesis and make statements about Western culture. These texts reveal, in their descriptions of both female "reproductive capacity" and male "potency," the interrelations between the creation of life and power. Some selected quotations demonstrate the connection between reproduction and power:[12]

> Of the two mysterious basic forms . . . the female one is older, more powerful, more ethnic, since even in the animal kingdom parthenogenetic creation prevails. . . . Moreover: the original female in the animal female not only propagated itself, but it created the male by itself; the male never created anything without the crone. (Sir Galahad 1987, 11)

> [M]en imagined reproduction of their kind as a type of parthenogenesis and thereby they acknowledged the tremendous power of their female partners to create life. (Badinter 1988, 46)

> There is doubt that during the earliest epochs of human history the magical power of the woman was regarded as the same type of wonder as the universe. Therefore, the woman obtained a wondrous power and it was a main objective for the male portion of the population to break, control, and use it for their own purposes. (Joseph Campbell, quoted in Corea 1988, 274)

> The father was acknowledged as procreator. He introduced the idea of fatherly descent. . . . The omnipotence of the father replaced the omnipotence of the mother. The power of procreation is now his alone (Badinter 1988, 91)

I do not intend to discuss or acknowledge these statements' claim of truth; rather I want to point out a connection they make between power and the creation of life, which is depicted in the descriptions of both the female power of reproduction and the male power of procreation. In these statements, the question of the exclusive ability of one sex to reproduce structures the design of sex/gender roles. The sex which is able to reproduce by itself is said to have a more powerful position than the other. Hence, the dominance of one sex constitutes itself by continuous discursive processes of negotiation over the capacity for unisexual reproduction. In regard to animal parthenogenesis, since reproduction is connected to power, and the ability of one sex to reproduce by itself serves as a decisive element in the constitution of power relations between the sexes, the forms of parthenogenesis described in biology present a very suitable discursive field of negotiations of power relations. In this cultural-historical context, given interactions between scientific theories and societal circumstances, the biological descriptions of parthenogenesis in the animal kingdom appear to be part of the negotiations about a pure female type of reproduction. Yet animal parthenogenesis is connected not only to a position of power, but also to the existence of one sex, namely the male sex.

Meanwhile, feminist scholars of science studies point to naturalizations of sex/gender relations in many biological theories. Bonnie Spanier, for example, indicates that the conventional bipolar sex concept appears to be a cultural pattern. She describes in detail attributions of sex/gender to bacteria in biological texts, even though biologically this group of organisms cannot be differentiated by sex (Spanier 1995).[13] An example of sex/gender constructions in vertebrates is found in the ethological studies of mountain sheep by Wolfgang Wickler (1969) and Valerius Geist (1971). Ruth Hubbard, in the 1980s, questioned the attributions of sex/gender stereotypes to mountain sheep (Hubbard 1990; Ebeling 2002). Yet these accounts of sex/gender constructions in the animal kingdom are individual cases found by chance (Hubbard, personal correspondence). More systematic analyses of the biological concepts of sex and reproduction are still needed. Such analyses are important because biological knowledge provides a blueprint for interpretations of human sex/gender relations. In the view of both naturalization and the linkage of gender-specific reproduction and power, animal parthenogenesis represents an exemplary field in which to study negotiations of power relations.

Biology describes many animal species and types of reproduction which do not fit into the binary classification of sex and reproduction. I have pointed out that animal species can alter their type of reproduction and that individuals can change their sex. Moreover, the assumption of a predominant sexual dimorphism has a powerful role in the production of biological knowledge. As I have demonstrated in many examples, heterosexuality and sexual dimorphism are perpetuated even in animal species and types of reproduction that do not fit the binary concept of sex. These examples indicate that sexual dimorphism in the animal kingdom is not as predominant as is commonly believed and that it needs to be scrutinized more systematically.[14]

Metaphors have different effects and functions in the evolutionary discourse about reproduction. They include interactions between human sex/gender relations and biological theories about animals. Metaphors of a "feminist paradise" and "women's liberation" clearly reveal the mutual influence of societal and biological theories and display circular reasoning: once the asymmetrical sex/gender power relations are integrated into biological knowledge, and once they are strengthened by the biological context, they reinforce societal values and serve to maintain existing sex/gender power relations. In other words, in biological texts metaphors tell stories about animals and about humans. Hence, metaphors further the ways that biological theories about parthenogens serve as models of human reproductive capabilities. For example, all-women societies can be interpreted from an andro- or a gynocentric perspective. The exclusive ability of one sex to reproduce is also negotiated via metaphors in the evolutionary discourse of parthenogenesis. Since these negotiations about unisexual reproduction are connected to power, sex/gender power relations are negotiated simultaneously. In the same way, evolutionary discourse simultaneously expresses and strengthens the culture's dominant pattern of interpretation in terms of norms and deviation from them. Heterosexual reproduction and sexual dimorphism are arranged as the norm, in opposition to deviant parthenogenesis and all-female species. Thus, even in unisexual, parthenogenetic animal groups heterosexual reproduction serves as a fundamental structure.

The metaphors examined here serve as blueprints for multiple interactions of societal and biological discourses that concern sex/gender power relations which assume a central role in the production of biological knowledge. They structure the pattern of the evolutionary discourse of reproduction in multiple ways. My analysis of biological theories about sex and reproduction provides only one example of the interconnection between biology and society. To understand and to advance the production of biological knowledge, biologists should include the social context of biology in their scientific work and regard biology not as separate from but as part of society. Biological and societal/cultural understandings of sex/gender and reproduction are interconnected—the latter is inscribed in the former—and the biological theories about sex/gender and reproduction serve as one important source of societal understanding of sex and gender.

Notes

1. Gonochorism means that individuals are either female or male and not both.
2. See Ebeling 2002, in which I systematically analyzed different types of biological texts about sexuality and reproduction from the period 1950–2000: biology textbooks, biological journals, and articles in popular science publications. Biologists scrutinize parthenogenesis in various subdisciplines. For example, some work on artificial parthenogenesis in order to analyze the functions of sperm. I chose evolutionary theory, because evolutionary biologists compare parthenogenesis to sexual reproduction involving two sexes.

3. In dioecious species an individual plant carries either female or male flow-ers; in monoecious species an individual plant carries both female and male flowers.
4. The distinction between monoecious and dioecious plants makes the situa-tion even more complex in the plant kingdom; since monoecious and dioe-cious plants combine in various ways with hermaphroditic flowers, both categories are subdivided even further. In addition, these combinations may vary within a population.
5. Tunicin is a cellulose-like material.
6. The acknowledgment of these all-female animal groups as species is still controversial in biology. The biological concept of species, which is mostly defined by Ernst Mayr (1967), refers to reproductively isolated groups in which the individuals mate. In this view, it is not the similarity of the indi-viduals that is crucial, but the reproductive isolation of the group. Therefore this concept bears a heteronormative perspective and cannot encompass uni-sexual species.
7. For examples of similar metaphors see Hurst, Hamilton, and Ladle 1992, 145; Cockburn 1995, 80.
8. All translations from German sources are my own.
9. While I provide only one interpretation of the metaphor of immaculate con-ception here, there are many more. For example, one could focus on the belief that "immaculate conception" may mean sinless heterosexual intercourse.
10. The meaning and effects of the metaphors used in a decade such as the 1950s differ from those of other decades. However, some ideas are persistent. Berill's association of parthenogenesis and men's uselessness can, for example, still be found today.
11. I did not systematically compare the use of metaphors and language by men and women biologists. However, most biology texts I analyzed were written by men.
12. The texts from which I quote date mainly from the 1970s, 1980s, and 1990s. They provide a review of epochs of the past in the sense that they describe old myths (such as matriarchal deities) and the beliefs of ages past (such as the Middle Ages).
13. For additional examples, see Anne Fausto-Sterling (1985, 2000), Donna Haraway (1991), Ruth Hubbard (1990), Evelyn Fox Keller (1992), and Londa Schiebinger (1999), just to mention a few.
14. In my current project on the construction of sexual dimorphism in zoology and botany, I examine the development of basic biological concepts concern-ing sex and reproduction, such as sexuality, copulation, and procreation, using a cultural/historical perspective.

References

Badinter, Elisabeth. 1988. *Ich bin Du.* (I am you.) München: Piper.
Bell, Graham. 1982. *The Masterpiece of Nature: The Evolution and Genetics of Sexuality.* Berkeley: University of California Press.

Berill, Norman J. 1954. *Natur und Geschlecht.* (Sex and the nature of things.) München: Biederstein Verlag.

Buddenbrock, Wolfgang von. 1953. *Das Liebesleben der Tiere.* (The love life of animals.) Bonn: Athenäum-Verlag.

Butlin, R. K., I. Schön, and H. I. Griffiths. 1998. "Introduction to Reproductive Models." In *Sex and Parthenogenesis: Evolutionary Ecology of Reproductive Modes in Non-marine Ostracods,* ed. Koen Martens, 1–24. Leiden: Backhuys.

Cancik, Hubert, and Helmuth Schneider, eds. 1996. *Der neue Pauly. Enzyklopädie der Antike.* (The new Pauly: Encyclopedia of the ancient world.) Vol. 1. Stuttgart: Metzler.

Catton, Chris, and James Gray. 1985. *Sex in Nature.* London: Croom Helm.

Cockburn, Andrew. 1995. *Evolutionsökologie.* Stuttgart: Gustav Fischer. (Orig. pub. 1991 as *An Introduction to Evolutionary Ecology.*)

Corea, Gena. 1988. *The Mother Machine: Reproductive Technologies from Artificial Insemination to Artificial Wombs.* London: Women's Press.

Crews, David. 1993. "The Organizational Concept and Vertebrates without Sex Chromosomes." *Brain, Behavior, and Evolution* 42(4–5): 202–14.

——. 1994a. "Constraints to Parthenogenesis." In *The Differences between the Sexes,* ed. R. V. Short and E. Balaban, 23–49. Cambridge: Cambridge University Press.

——. 1994b. "Geschlechtsausprägung bei Wirbeltieren." (The sex characteristics of vertebrates.) *Spektrum der Wissenschaft,* March: 54–61.

Crews, David, and Kevin T. Fitzgerald. 1980. "'Sexual' Behavior in Parthenogenetic Lizards (*Cnemidophorus*)." *Proceedings of the National Academy of Sciences* 77(1): 499–502.

Crews, David, Mark Grassman, and Jonathan Lindzey. 1986. "Behavioral Facilitations of Reproduction in Sexual and Unisexual Whiptail Lizards." *Proceedings of the National Academy of Sciences* 83(24): 9547–50.

Ebeling, Kirsten Smilla. 2002. *Die Fortpflanzung der Geschlechterverhältnisse. Das metaphorische Feld der Parthenogenese im evolutionsbiologischen Fortpflanzungsdiskurs.* (The reproduction of sex and gender relations: The metaphorical field of parthenogenesis in evolutionary theory.) Mössingen-Talheim: Talheimer-Verlag.

Fausto-Sterling, Anne. 1985. *Myths of Gender: Biological Theories about Women and Men.* New York: Basic Books.

——. 2000. *Sexing the Body: Gender Politics and the Construction of Sexuality.* New York: Basic Books.

Forsyth, Adrian. 1987. *Die Sexualität in der Natur. Vom Egoismus der Gene und ihren unfeinen Strategen.* Munich: Deutscher Taschenbuch Verlag.

Geist, Valerius. 1971. *Mountain Sheep.* Chicago: University of Chicago Press.

Gerritsen, Jeroen. 1980. "Sex and Parthenogenesis in Sparse Populations." *American Naturalist* 115(5): 718–43.

Glaubrecht, Matthias. 1998. "Auch Fisch-Amazonen gehen mitunter fremd. Aber der Seitensprung dient nur der Jungfernzeugung." (Even fish-Amazons cheat on their partners: But the infidelity serves for parthenogenesis only.) *Der Tagesspiegel Berlin,* January 29: 32.

Haraway, Donna. 1991. *Simians, Cyborgs, and Women: The Reinvention of Nature.* New York: Routledge.

Hubbard, Ruth. 1990. *The Politics of Women's Biology.* New Brunswick, N.J.: Rutgers University Press.

Hurst, Laurence D., William D. Hamilton, and J. R. Ladle. 1992. "Covert Sex." *Tree* 7(5): 144–45.

Insect Company. 2002. "Oddities and Rarities: A Collection of Gynandromorphs and Other Strange Deformities." http://www.insectcompany.com/main/ oddities.shtml. Accessed February 16, 2005.

Keller, Evelyn Fox. 1992. *Secrets of Life, Secrets of Death: Essays on Language, Gender, and Science.* New York: Routledge.

Klärner, Diemut. 1997. "Wenn Jungfernzeugung ansteckend wird." (When partheno-genesis becomes contagious.) *Frankfurter Allgemeine Zeitung,* July 30: 3.

Krebs, John R., and N. B. Davies. 1984. *Einführung in die Verhaltensökologie.* (An intro-duction to behavioral ecology.) Stuttgart: Georg Thieme.

Margulis, Lynn, and Dorion Sagan. 1997. *What Is Sex?* New York: Peter N. Nevraumont.

Maslin, T. Paul. 1971. "Parthenogenesis in Reptiles." *American Zoologist* 11: 361–80.

Mayr, Ernst. 1967. *Artbegriff und Evolution.* (Animal species and evolution.) Ham-burg: Parey.

Miersch, Michael. 1996. "Wozu taugt der Mann?" (What good are men?) *Die Zeit,* October 4, p. 35.

Moore, Michael M., Joan M. Whittier, Allen J. Billy, and David Crews. 1985. "Male-Like Behavior in an All-Female Lizard: Relationship to Ovarian Cycle." *Ani-mal Behavior* 33: 284–89.

Rasch, Ellen M., Lillian M. Prehn, and Robert W. Rasch. 1970. "Cytogenetic Studies of *Poecilia* (Pisces). II. Triploidy and DNA Levels in Naturally Occurring Popula-tions Associated with the Gynogenetic Teleost, *Poecilia formosa* (Grigard)." *Chromosoma* 31: 18–40.

Reimers, Tekla. 1994. *Die Natur des Geschlechterverhältnisses. Biologische Grundlagen und soziale Folgen sexueller Unterschiede.* (The nature of gender relations: Basic principles of biology and social consequences of sex differences.) Frank-furt: Campus Verlag.

Remane, Adolf, Volker Storch, and Ulrich Welsch. 1980. *Systematische Zoologie.* (Sys-tematic zoology.) Stuttgart: Gustav Fischer Verlag.

———. 1989. *Kurzes Lehrbuch der Zoologie.* (Short textbook of zoology.) Stuttgart: Gustav Fischer Verlag.

Riewenherm, Sabine. 1996. "Wenn Pflanzen zu sehr lieben." (When plants love too much.) *22. Kongreß von Frauen in Naturwissenschaft und Technik. Dokumenta-tion,* 278–94. Darmstadt: Frauen in der Technik—FiT—Verlag.

Schiebinger, Londa. 1993. *Nature's Body: Gender in the Making of Modern Science.* Boston: Beacon.

———. 1999. *Has Feminism Changed Science?* Cambridge, Mass.: Harvard University Press.

Schlupp, Ingo. 1995. "Biologie und Ökologie der Weibchenart *Poecilia formosa* Grigard 1859 (Poeciliidae, Teleostei)." (Biology and ecology of the all-female species *Poecilia formosa* Grigard 1859 [Poeciliidae, Teleostei].) Ph.D. diss., Hamburg University.

Schlupp, Ingo, Cathy Marier, and Michael J. Ryan. 1994. "Benefit to Male Sailfin Mol-lies of Mating with Heterospecific Females." *Science* 263: 373–75.

Schlupp, Ingo, Ryan McKnab, and Michael J. Ryan. 2001. "Sexual Harassment as a Cost for Molly Females: Bigger Males Cost Less." *Behaviour* 138: 277–86.

Siewing, Rolf, ed. 1985. *Lehrbuch der Zoologie.* (Textbook of zoology.) Vol. 2. Stuttgart: Gustav Fischer.

Sir Galahad [Bertha Eckstein-Diener]. 1987. *Mütter und Amazonen. Liebe und Macht im Frauenreich.* (Mothers and Amazons: Love and power in women's world.) Frankfurt am Main: Ullstein Sachbuch.

Smith, John Maynard. 1976. *The Theory of Evolution.* Cambridge: Cambridge University Press.

——. 1986. "Contemplating Life without Sex." *Nature* 324: 300–301.

——. 1992. "Age and the Unisexual Lineage." *Nature* 356: 661–62.

Spanier, Bonnie. 1995. *Im/Partial Science: Gender Ideology in Molecular Biology.* Bloomington: Indiana University Press.

Voigt, Max. 1957. *Rotatoria. Die Rädertiere Mitteleuropas. Ein Bestimmungswerk.* (Rotatoria of Middle Europe: A field guide.) Berlin-Nikolassee: Gebrüder Borntraeger.

Wendt, Herbert. 1970. *Das Liebesleben in der Tierwelt.* (About animals' love lives.) Rheinbek bei Hamburg: Rowohlt Taschenbuch Verlag.

Wickler, Wolfgang. 1969. *The Sexual Code: The Social Behavior of Animals and Men.* Garden City: N.J.: Doubleday.

Wuethrich, Bernice. 1998. "Why Sex? Putting Theory to the Test." *Science* 281: 1980–82.

10 Feminism and Science: Mechanism without Reductionism

Carla Fehr

Modern science is a historical enterprise born in the sixteenth and seventeenth centuries. We all know its "fathers": Francis Bacon, René Descartes, John Locke, Galileo Galilei, Sir Isaac Newton, Robert Boyle; this intellectual honor roll goes on and on. They developed a way of producing knowledge that was astonishing in its ability to explain the natural world. Carolyn Merchant, in her book *The Death of Nature* (1980), points out that modern science developed along with, and was dependent on, a change in the metaphors used to make sense of nature. Merchant argues that modern science took Mother Nature and turned her into a machine and that this mechanistic worldview allowed for new ways to see order in the complexities of nature, fostering an attitude of control and domination toward both women and the natural world.

Historian Londa Schiebinger (1989) argues that as this new method and metaphor developed from the infant creations of the "founding fathers" into powerful social institutions, women intellectuals were systematically excluded from modern science. The nascence of modern science involved the domination of a female world and the exclusion of women. Can there be such a thing as feminist science(s)? Developing responses to this question has become a cottage industry during the last twenty years. I am adding an analysis of the relationship between mechanism and reductionism to this literature. I argue that we need to tease the concept of mechanism away from harmful and exclusionary reductionist philosophies, and embrace, rather than merely tolerate, a plurality of scientific methodologies. Dethroning reductionism and embracing a plurality of methods that engage the world at a variety of organizational levels can create space for feminist methodologies, and for a science that is more inclusive of women and those with previously marginalized perspectives. The methodological pluralism that I endorse involves a plurality of methods and approaches to science that recognize causal patterns at, and among, different levels of organization.

Reductionism, Mechanism, and Their Strengths

Reductionism is the view that things in the universe are arranged hierarchically, and that causation only occurs at the lower levels of this hierarchy.

Reductionism entails a relationship between parts and wholes such that wholes are explained in terms of their parts; hence, a reductionist scientific method focuses on describing and attempting to understand a phenomenon in terms of its parts. Reductionist views range from strong to weak depending on the level of organization to which phenomena are ultimately to be reduced. Strong reductionism assumes that there is a single, most fundamental level at which entities are real, where causation happens, on which explanations rest, and where methodologies should focus. For instance, in the biological sciences, strong reductionism holds that phenomena should be reduced to their physiochemical constituents. Weak reductionism assumes that causation, explanation, and methodology should focus on the parts of wholes but does not insist that these parts be the simplest or smallest entities possible.

Helen Longino (1990) and Robert Brandon (1996) have pointed out that there are different senses of reductionism, which are often conflated.[1] Brandon makes a distinction between ontological, explanatory, and methodological reductionism, and Longino makes a distinction between ontological (or metaphysical) and methodological reductionism. Ontologically (that is, in relation to the structure of the world), reductionism means that big things are composed of smaller things and little things are composed of tiny things and so on, and that causal relations only happen among the smallest functional parts of a whole (Longino 1990). Brandon characterizes this view of causation as "bottom up" or "inside out," meaning that "parts of a whole and their interaction cause the behavior of the whole, but never vice versa. It never happens that properties of wholes cause the behavior of their parts" (1996, 189).

Methodological reductionism, in its weak and strong sense, prescribes that we should search for the physiochemical bases of biological phenomena or that researchers should characterize and seek to understand phenomena in terms of their parts (Brandon 1996; Longino 1990). Explanatory and ontological reductionism have been tightly linked in the causal mechanical view of explanation. According to this view of explanation, we understand phenomena by showing how they are causally produced.

I work within a causal mechanical view of scientific explanation because science is in the business of finding out how things work, and this view is implicit or explicit in most scientific practice. Mechanism, as a philosophical perspective, is the general view that arose with the Newtonian concept of the universe as consisting of matter in motion. It sees nature as machine-like and explainable without referring to any supernatural entities or causes. A mechanism is simply a causal process. A mechanical explanation is a description of the causal process that produced the phenomenon in question. Mechanism is based on causation or production. If one embraces ontological reductionism, which rests on the *a priori* assumption that causation only acts at lower levels of organization than the phenomena in question, then a causal mechanical explanation must involve descriptions of processes and entities at those lower levels. Reductive mechanism is the view that real causal processes involve interactions among the smallest parts of a whole, scientific explanations are elucidations of these low-level causal processes, and scientific methodologies should search for explanations at low levels of organization. I will argue that even though reductionism and mechanism can fit together, it does not follow that they must. Brandon (1996) argues—and I concur—that a causal mechanical view of explanation can and should stand without the *a priori* assumptions about causation entailed by ontological reductionism. If we discover causal regularities at higher levels of organization, then we have examples of nonreductive causal mechanisms. Brandon gives many such examples, and I discuss the example of the role of hormones in development here.

If one is a strong reductionist, the only genuine explanations of phenomena involve descriptions of causal processes among the "least functional bits" of a phenomenon. The result is a methodological monopoly in which it is assumed that the only methods that yield genuine understanding are those that characterize phenomena in terms of those least bits. According to this view, methods aimed at higher levels of organization at best have heuristic value and at worst represent a fruitless search for nonexistent causes. If one is a weak reductionist, the result will not be a monopoly of methods, but rather a systematic privileging of methods that characterize phenomena in terms of their parts at some lower level of organization. Both types of reductionism marginalize methods that search for higher-level causes or that characterize parts in terms of wholes. This methodological monopoly, or privileging of lower-level methods, reflects a normative continuum ranging from sciences such as physics, which is stereotypically viewed as engaging the world at the lowest levels of organization, to sciences such as ecology, which is seen as engaging the world at higher levels of organization. Thus we end up with the quip that biologists suffer from "physics envy." Although women are woefully underrepresented in science generally, it is worth noting that this underrepresentation is most severe in sciences that from a reductive perspective are seen as being the "most scientific."

I do not believe that reductionist methods, which have been at the heart of modern science since its inception, are universally misguided or inappropriate and I do not want to say that all mechanistic/reductionist research need be androcentric. The use of these principles in chemistry and biology has led to major accomplishments, such as the elucidation of the structure of the atom and of the biochemical nature of DNA. These methods are powerful and are one of the reasons that science is so successful. However, I take issue with the argument that the past successes of reductive science provide a reason why nonreductionist methods should not be considered within the rubric of science. When considering past successes, one must ask, "Successful at what?" especially since the past successes of science have benefited some groups and classes more than others, and since researchers have historically paid insufficient attention to important areas of investigation, especially in biology. I want to decenter, not eliminate, reductionist approaches. Allowing for and valuing a plurality of approaches may allow a new set of successes and help us address neglected topics.

Three Feminist Concerns

Given that reductionist mechanism has been so successful, what is the justification for feminist concerns about this sort of science? There are many reasons why some feminist theorists are hostile to scientific practice and production. I want to focus on three of these worries: the biological determinism problem, the methodological/epistemological problem, and the cultural/historical problem. These problems are interrelated in complicated ways and it would be worth considering these relationships more closely in future work.

The Biological Determinism Problem

Biological determinism (in respect to gender) is the view that the properties of gender are caused by, or determined by, the properties of sex. A tight link exists between reductive mechanistic science and biological determinism. It has received extensive treatment in feminist debates concerning the relation between sex and gender (Nicholson 1994), in feminist critiques of the biology of sex differences (Birke 1986; Longino 1990; Longino and Doell 1983), and in feminist critiques of animal behavior and sociobiology (Hrdy 1986; Hubbard 1979; Kaplan and Rogers 1994). This link is due to the reductive assumption that causation acts in an upward direction from lower to higher levels of organization. This means that biological causes are considered to be the only causes, or at least the most important causes, of feminine behavior. With this sort of tight, unidirectional causal link, upper-level phenomena, such as the social characteristics connected with gender, are seen as caused by lower-level phenomena, such as genetic or hormonal factors. According to this sort of analysis, as long as the lower-level phenomena remain constant, the upper-level phenomena will be unchangeable. More particularly, a healthy woman cannot help but be nurturing, or tend to be emotional rather than rational, or have whatever gender characteristic one cares to name. This sort of reasoning is problematic not only because it implies that seeking political change in the face of biologically determined sex differences is futile, but also because this is not an accurate explanation of the mechanisms of sex and gender.

These specific concerns about reductionism and biological determinism, although focused on sex and gender as an object of scientific investigation, are also crucial to women as scientists. The results of deterministic reductive analyses of sex and gender support cultural understandings of femininity that call into question women's ability to enter and remain in the traditionally masculine world of research. Sandra Harding writes, "were these biological determinist arguments true, a 'woman scientist' *should* be a contradiction in terms" (1986, 93; italics in original). Consider the gender characteristics that are argued via reductionism to have a biological basis and hence be static, and ask which tend to be associated with women and which tend to be associated with men: mathematical ability, spatial ability, rationality, competitiveness, compassion, listening and nurturing skills, emotional awareness, and cooperation. The first four characteristics correspond not only to traditional Western understandings of masculinity but also to traditional conceptions of science as objective, impersonal, and competitive.

This conflict between stereotypical feminine and scientific virtues can be ameliorated in at least two ways. On the one hand, one can attempt to break the association between masculine values and science, showing that this association already is mythical or unnecessary. For example, Evelyn Fox Keller (1983) reports that the intimate, emotional connection between Barbara McClintock and the individual plants, cells, and chromosomes of the corn she studied allowed her to conduct her Nobel Prize–winning work on transposition. On the other hand, one may also argue against the reductionist perspective that leads to biological determinism and hence to the construction of an essentialized woman's (and man's) nature. Readdressing the masculine gendering of conceptions of science is an important task, but here I want to tackle the problem of reductionism and determinism, which leads to the calcification of cultural stereotypes of women's intellectual nature. If nonoppressive scientific investigations of sex and gender are to be conducted, scientists must address the problem of reductionism.

One tactic, which both defends science and ameliorates the effects of the stereotypical conclusions concerning the nature of women, is to argue that the science that "shows" that women, for example, lack spatial ability is simply bad science. Studies showing that women are less apt than men at tasks traditionally associated with the practice of science, such as spatial or mathematical tasks, have largely been discredited because of statistical problems and unjustified assumptions (see Longino 1990, 123–26 for an overview). It must be clear to what the "bad science response" refers. If it refers to poorly conducted statistical analyses, then presumably the research can be redone more clearly and produce better data. If, on the other hand, it refers to assumptions about the nature of causation, for example ignoring or discounting the existence of higher-level social causes, then the "bad science response" is more radical because it reveals a foundational reductive ideology. Even though there is justification for setting aside these particular results, the problem of reductive determinism still exists. As long as one holds the assumption that causation only works from the bottom up or inside out, then higher-level phenomena associated with gender will be viewed as determined by lower-level phenomena associated with sex.

The Methodological/Epistemological Problem

Reductionist mechanism leads to several methodological and epistemological feminist concerns. It puts blinders on researchers that limit their choice of problems and objects to study. Many possible sources of bias can limit the problems that researchers focus on. One such source is the assumptions of reductionistic mechanism. If one believes that scientific explanation involves revealing causal interactions that take place at low levels of organization, then one may not think to, or may consciously choose not to, investigate complex phenomena or problems that do not appear reducible to tiny enough bits. This set of assumptions leads to hostility toward holistic approaches to medical and environmental problems, and the exclusion of things recalcitrant to reductionism from the domain of scientific investigation. For example, Sue Rosser points out with regard to dysmenorrhea that "[o]nly after prostaglandins were 'discovered' was there widespread acceptance among the male medical establishment that this experience reported by women had a biological component" (2002, 231). The discovery of a possible low-level hormonal basis for dysmenorrhea accorded it the status of being biological, even though women had long been reporting bodily experiences of the phenomenon.

Reductive mechanistic science not only limits the choice of problems and the recognition of objects of investigation, it also limits methods of inquiry. Strong reductive mechanism enforces a methodological monopoly on the study of acceptable scientific problems, such that if one is investigating a phenomenon or problem that could be approached from either above or below, one must approach it from below because that is where the true causes of the phenomenon are to be found.

Finally, reductive mechanism leads researchers to decontextualize their objects of investigation. When they seek to understand a phenomenon in terms of its smallest parts, they lose sight of the relationships between the phenomenon and a greater whole. Possible higher-level causal influences that may affect the phenomenon disappear and become inaccessible to investigation, giving false support to the reductionist view that causation only happens at low levels of organization. For example, Jacquelyn Zita (1989) argues that clinical discourse creates a reductionist view of the female "medicalized body" according to which causal patterns are located within the body. With regard to research on premenstrual syndrome (PMS), she notes that researchers tend to assume that only internal variables and not contextual variables can sufficiently explain the patterns of observations related to PMS.

The Cultural/Historical Problem

The third problem concerns the historical association of reductionism and mechanism and the current effects of this association. Merchant (1980) argues that during the seventeenth century machines became models for not only the way the world was structured, but also for the best, or perhaps the only, way emerging scientists gained knowledge of that world. According to Merchant, these new metaphors make possible an ideology of control and domination of nature and of women; human beings create machines to do the tasks that they require. This machine metaphor developed into a mechanical philosophy that came to dominate European science.

Merchant describes several intertwined assumptions that linked science and machines and were constitutive of this new mechanistic worldview. These assumptions are still involved with what many consider to be "the scientific method" (Merchant 1980, 228–35). The first assumption concerns the particulate structure of nature. In the seventeenth century matter came to be seen as "corpuscular, passive and inert; change was simply the rearrangement of particles as motion was transmitted from one part to another in a causal nexus" (Merchant 1980, 102). This assumption that matter has a fundamental low-level structure and that change can be characterized at this level paves the way for reductionism. A second assumption is that the universe has a lawlike natural order. The concept of order is the heart of a mechanistic view both in the seventeenth century and today. The replacement of an organic conception of order corresponds with the replacement of Aristotelian science, which searched for the purpose of phenomena in the context of the universe as a whole, with modern science, which abandoned questions of purpose in a quest to discover how things work. The domain of modern science included those things that exhibited lawlike behavior, or, in a more general and contemporary sense, those things that exhibit causal regularities. In the modern worldview the machine was the perfect metaphor because it was an artifact. The relationships among its parts are transparent and its purpose is obvious because humans designed it to do a particular job.

A third assumption is that knowledge and information can be abstracted from the natural world. Merchant calls this the assumption of context independence and points out that it springs from the idea that "only quantities and context-independent entities can be submitted to mathematical modeling" (Merchant 1980, 230).[2] The idea behind this assumption is that the world is organized such that we can describe it in terms of discrete entities and that scientists can come to know and manipulate these entities independently of the context in which they occur. A final assumption is that "problems can be analyzed into parts that can be manipulated by mathematics (the methodological assumption)" (228).[3] The methodological assumption, as Merchant presents it, involves an ambiguity between a reductive approach to problems and a reductive approach to entities in her analysis of Descartes and Hobbes respectively.

With this list of assumptions Merchant reveals that the heart of early mechanistic philosophy in the modern period is based on the idea that the world can be understood in terms of the motions of particles engaging only in efficient, proximal causal interactions, and that this conception of nature paid little attention to higher causes. I agree with Merchant's arguments regarding the changing metaphors of nature, her general conception of early mechanistic philosophy, and her thesis that mechanical metaphors grant permission and develop a means for a science that is based on control and domination. These arguments have contemporary counterparts. For example, Emily Martin, in her 1987 book *The Woman in the Body: A Cultural Analysis of Reproduction*, reveals a strong metaphor in which a woman's body with its reproductive functions is seen as a machine and the gynecologist and obstetrician are mechanics who not only maintain the machine, but also retain tight control of the reproductive process, especially the process of birth itself. Merchant's historical arguments and Martin's contemporary analysis show the cultural and political effects of a reductive mechanistic view, and provide impetus for deconstructing this position.

Although the four assumptions that Merchant lists co-occurred in the development of a reductive mechanistic worldview, they do not imply one another. They are tied together by ontological reductionism. Once this cord is severed, context stripping and methodological reductionism can be teased away from the heart of the mechanistic view. Although the assumption of the particulate structure of nature was developed in terms of reductionism, it need not imply or prescribe ontological reductionism. Modern physicists were interested in describing observable interactions among medium-sized composite bodies (for example, billiard balls). The existence of causal interactions among particles is not an argument against the possibility of higher-level causal interactions.

Once the assumption of ontological reductionism is set aside, natural order, in the mechanistic sense, need not be confined to interactions among the least bits of phenomena. Without ontological reductionism there is no *a priori* justification for context stripping. Only if one presumes that causal interactions happen exclusively within an object, or at lower levels of organization than the object in question, can one feel secure considering an object out of its context because one presupposes that there are no contextual causal relations that may be overlooked. Similarly, unless one holds an *a priori* commitment to ontological reductionism one cannot prescribe methods that focus exclusively on illuminating the interactions among the parts of an object. One of the goals of science (many would argue the primary one) in both modern and contemporary periods is to reveal causal patterns and regularities in nature. But those regularities may well include interactions at higher levels of organization in which the context of a phenomenon plays an integral role. This search for causal order need not be limited to methods that characterize a phenomenon in terms of its parts, since order may well be found in the interactions between the phenomenon and its context. If one abandons presuppositions of ontological reductionism, it becomes clear that the assumptions Merchant identifies in modern science are historically—but not necessarily—connected.

While it is true that machines are made of parts, are based on order, and give us power over nature, the machine metaphor need not be reductionist. Brandon gives the example of a car engine:

> If we go inside the engine and look, say, at the behavior of an intake valve, reductionism fails us. The internal constitution of the valve is not totally irrelevant to its behavior. . . . But consideration of its internal parts does not begin to explain its up and down motion. To explain that we imbed it into the larger mechanical system, the engine. (1996, 195)

Merchant's important thesis about order, domination, and control is not contradicted by a nonreductionist conception of mechanism. The discovery of causal regularities allows us to intervene in nature more efficiently. Coupled with technological advances, this gives humans incredible power over nature; this power deserves immediate and continued critical attention. But focusing on nonreductive mechanism highlights the contingent relationships among the scientific assumptions that generate this power, forces us to pay attention to context, and endorses a plurality of methodologies that search for causal patterns at a variety of levels of organization. For example, research which pays attention to the ecological context of an endangered species and uses methods which engage nature at high levels of organization may help us more effectively intervene to protect that species and the environment in which it lives.

Feminist concerns about biological determinism, methodological problems (such as narrow choice of problems, privileging of methods that engage the world at low levels of organization, and context stripping), and cultural/historical problems in which the combination of reductionism and mechanism allows for a science of domination and control are in varying degrees ameliorated by removing reductive assumptions from a mechanistic view of science. Feminist critiques of reductionism need not be critiques of mechanistic science.

Philosophers on Mechanism

I previously defined a mechanism as a causal process; at this point this terse and vague definition needs to be fleshed out. I turn to two conceptions of mechanism: Wesley Salmon's, which is central to contemporary ideas of causal mechanical explanations, and that of Peter Machamer, Lindley Darden, and Carl Craver, which is recent and under development.

Philosopher of science Wesley Salmon (1984, 1989, 1994, 1998) championed a causal mechanical view of explanation in which an explanation of a fact is a description of the causal processes that produced the fact. Salmon points out two kinds of causal mechanical explanations. A *constitutive* explanation describes "the causal processes that constitute a phenomenon and exhibit the phenomenon's internal causal structure" (1984, 270). A constitutive causal explanation of a clock would involve exhibiting its gears and springs and revealing the causal processes that they undergo. Although this type of explanation is at least weakly reductive, as it is looking at wholes in terms of parts, it does not specify that those parts be least bits. This is an example of Brandon's inside-out characterization of ontological reductionism.

Salmon's second type of causal mechanical explanation is *etiological:* one explains "a given fact by showing how it came to be as a result of antecedent events, processes and conditions" (1984, 269). In this case, an explanation fits the fact-to-be-explained into an "external pattern of causal relationships" (270). This second type of explanation explains a whole not in terms of its parts, but rather in terms of its history. This sort of an explanation of a clock would reveal how the clock was made, or how winding it up starts a process that allows it to keep time. This kind of explanation may have reductive aspects if one assumes that these antecedent causes must be low-level.[4]

Salmon's view is based on a revolutionary reconception of causation in terms of causal processes and not in terms of a chain linking discrete events. There is controversy in the philosophy of science concerning what sorts of processes count as casual in this new view. Jim Woodward (1989, 2003) argues that Salmon's view applies primarily to simple cases in physics, such as the momentum and energy involved in the collision between a baseball and a window. If this is the case, causal interactions that are not based on physics will be ruled out and we are back to ontological reductionism.[5] However, Salmon (1989, 1998) employs examples of causal mechanical explanations that range from physics to psychology and anthropology. Furthermore, there is a significant body of literature in the philosophy of biology that documents higher-level causal interactions in terms of Salmon's view. For example, Brandon (1990) gives a causal mechanical account, based on Salmon's work, of explanations of evolutionary adaptation. Given the possible reductionist implications, this is an area in the philosophy of science that merits further feminist attention.

Machamer, Darden, and Craver (2000) have recently developed a more general theory of mechanism. Although Salmon's view need not be reductive, if critics are not convinced by the replies to the concerns regarding Salmon's view, Machamer et al.'s position escapes these concerns. Machamer, Darden, and Craver offer the following definition: "Mechanisms are entities and activities organized such that they are productive of regular changes from start or set-up to finish or termination conditions" (2000, 2). They still hold the general view that descriptions of mechanisms "explain how a phenomenon comes about or how some significant process works" (Machamer, Darden, and Craver 2000, 2). Both the set-up and termination conditions are idealizations that are privileged for any number of local reasons. The set-up conditions are those conditions that scientists privilege as the beginning of the mechanism. The termination conditions are the privileged end point of the mechanism and are a description of what it is that the mechanism is being used to explain. In this sort of mechanistic explanation of a clock, the entities involved are the springs and gears and the activities are what those springs and gears do. The start-up conditions include the existence of the clock and the activity of winding it up. It then regularly and expectedly carries out a series of activities that result in the clock's keeping time. Scientists in different disciplines, at different stages of an investigation, and with different goals may privilege different points as the beginning and end of the mechanism that they are attempting to describe. A mechanism may contain intermediate entities and activities that fill in the gap between the set-up and termination conditions, showing that a continuous process produces the phenomenon that scientists are trying to explain. It is a description of this productive process that makes a mechanism intelligible. Before analyzing what Machamer, Darden, and Craver mean by "intelligibility," I illustrate these two concepts of mechanism at work in an example that has received extensive treatment in feminist criticism of science.

Hormones as Causes of Behavioral Sex Differences

There has been significant feminist analysis of studies of genes and hormones as causes of the development of behavioral sex differences. Longino (1990) and Longino and Doell (1983) have analyzed the role of male bias in this field, and have used their work as a case study to show how different kinds of values influence scientific practice. Lynda Birke, in her book *Women, Feminism, and Biology* (1986), scrutinizes the problems that reductionism has raised with regard to the role of hormones in developmental biology. One example that she discusses is the study of hormonal effects on the development of sexual behavior in rats. These rodent studies are not only an interesting example of research concerning sex in the animal kingdom, but are also important because animal studies are centrally used in research on human sex differences, particularly in scientific studies of human sexual orientation. In an early model of this system, which Longino has called the "linear-hormonal model," prenatal and perinatal hormone levels are assumed to be the basis for behavioral sex differences. The assumption is that a gene on the Y chromosome triggers the development of testes, and that the hormones released by these brand new testes affect the structure of the rodent brain during a critical developmental period. The presence of testes and the hormones that they produce are assumed to cause the male brain to develop such that the male performs "stereotypical" male sexual behavior such as mounting; in the absence of testes and the hormones that they produce the female brain develops in such as way as to cause the female to perform "stereotypical" female behaviors such as lordosis (in which the female arches her back and raises her rear, presumably to solicit mounting). Longino (1987, 52) points out that in this explanatory model it is assumed that there is a "unidirectional and irreversible sequence of (biochemical) events."

Birke writes of this early model,

> Even if it is not always made explicit, the framework within which this line of reasoning has progressed is that hormones in early life are the prime determinants of adult patterns of sexual behaviour. To a large extent this assumption still holds, and the existence of sex differences in behaviour in adults is frequently attributed to perinatal hormone effects. Now it may well be true that hormones exert a large effect. The problem is not there, but with the exclusive focus on hormones from the individual's own testes or lack of them which dominated research for several decades. (1986, 96)

This is a powerful model because it fits within a reductive framework and is a case of reductive mechanism: genes cause anatomical differences, which cause hormonal differences, which cause brain differences, which cause behavioral differences. Notice that this causal structure proceeds from lower to higher levels of organization. Birke argues that the relationship between hormones and behavior is much more complicated than this tidy reductionist picture implies; an "interactionist model" is replacing the linear-hormonal model. She points out that research is beginning to embrace a much more complex picture of the causes of sexual behavior, a picture that involves interactions among the mother, the fetus, and the physiological and social environment as well as the genes, the internal anatomy, and the brain structure of the developing fetus. In this model, before birth the pups are influenced by factors such as the sex of other members of the litter, the mother's environment, and the mother's hormonal states. Birke writes, "even before birth . . . it is difficult to separate the individual pup and its hormones from a network of complex processes" (Birke 1986, 97). After birth, factors such as the pup's own hormones, maternal care (which is differential depending on the sex of the pup), the physical environment, and the other pups influence adult sexual behavior. When the pup becomes an adult, factors in its physical and social environment as well as its hormonal states and its own behavior influence its sexual behavior.

The linear-hormonal model fits both types of mechanism, and is reductionist. In terms of Salmon's view it has both constitutive and etiological components. The causal processes are inside the individual. The constitutive parts of the individual—genes, hormones, and brain structure—produce the behavior. The etiological aspect involves the causal processes that take place during the development of the individual. The causal process begins with the genes and ends with the behavior. According to Machamer, Darden, and Craver's conception of mechanism, the linear-hormonal model can be described in terms of a series of entities and activities that begins with the individual's genetic makeup and terminates with the sexual behavior in question.

The interactionist model is mechanistic, but not reductionist. In the interactionist model, there is not a simple linear causal sequence from the bottom up, from genes to hormones to behavior, all occurring within an individual. Instead, the social and physical environments also provide complicated and interacting causes. Factors at higher levels of organization, such as the ratio of the sexes among the littermates and the behavior of the mother, affect phenomena at lower levels of organization, such as the sexual behavior of the individual pup when it reaches maturity.

The interactionist model can be described as a causal mechanism in Salmon's terms. There are still constitutive causes of the behavior within the organism. But the set of etiological causes is greatly expanded to include such things as social interactions. Note that etiological explanations place phenomena in the causal nexus that makes up the world. The interactionist model provides a much more detailed and accurate description of the causal nexus in which we can locate sexual behavior. Very clear causal patterns connect maternal behavior and social environment with pup behavior, and these patterns are not limited to low-level genetic and hormonal interactions that occur during the development of the pup.

The interactionist model fits very well with Machamer, Darden, and Craver's definition of mechanism. Entities regularly and predictably go through activities, but we are not limited to looking downward along the hierarchical organization of developmental processes. The set-up conditions include not just the presence or absence of a Y chromosome or testes, but also the social environment of the pup and the mother, the behavior of the mother, and other factors. Also, this is a multilevel mechanism; the entities and activities involved are found at many levels of biological organization. The entities range from genes, to hormones, to individuals, to social and physical environments, and the activities range from biochemical to social interactions. Birke's description of the interactionist model is an example of very good, nonreductionist, and yet mechanistic science.

One may object that even though the hormonal influences on development may come from many sources—some from the fetus, some from the littermates, some from the mother, and some caused by the behavior of the mother—the action of these hormones, no matter what their source, still occurs at the micro-level of the biochemical interaction of the hormone and its receptor site. In other words, the objection is that even in this more complex view of the role of hormones in development, the action is still occurring at the microlevel and hence this view does not offer an alternative to reductionism. The problem with this objection is that it involves a misunderstanding of the object of the investigation. In this example, researchers are not trying to investigate how hormones interact with their receptors. The interaction between hormones and their receptors is a reasonable topic to investigate, and many researchers have worked on it and are continuing this work. But in the example at hand, researchers are trying to understand sexual behavior.

The biochemical and cellular interactions between hormones and their receptor sites are associated with a wide variety of both stereotypically male and female behaviors. Both males and females have "male" and "female" hormones and receptor sites. In order to associate a particular hormone/receptor mechanism with a particular behavior or small class of behaviors that tends to be linked with one sex or the other, one needs to know such things as the abundance and activity of the receptors, and the amount of hormone circulating in an individual or the individual's ability to produce a particular hormone. Details of the microlevel hormone/receptor mechanism are not sufficient to explain these varying abundances. To answer the "how much" question, one needs to refer to causal influences that are both internal and external to the individual in question.

In Machamer, Darden, and Craver's terms this objection is an example of researchers choosing to privilege different set-up and termination conditions of the mechanism in question. One who is pushing this objection is privileging the hormone receptor before it interacts with the hormone as the set-up condition, and the hormone receptor while it interacts with the hormone, or some cellular consequence of this interaction, as the termination condition. The interactionist model privileges a much larger set of set-up conditions.

The intuition underlying both concepts of mechanism is that figuring out how something comes about or how it works is crucial to—or, more strongly, is constitutive of—understanding the phenomenon in question. Machamer, Darden, and Craver argue that describing the continuous process from set-up to termination conditions renders a phenomenon intelligible.

Intelligibility

Machamer, Darden, and Craver point out that a mechanistic worldview is "a conviction about how phenomena are to be understood" (2000, 21). "Understanding" is a vague, multivocal term. They maintain that scientific understanding makes phenomena intelligible. Their use of "intelligibility" needs to be carefully read. They write that "intelligibility consists in the mechanisms being portrayed in terms of a field's bottom out entities and activities" (2000, 21). This talk of "bottoming out" may lead one to think that their conception of mechanism also relies on reductionist assumptions. In some passages the authors seem to support this sort of understanding. For example,

> 1. lower level entities, properties and activities are components in mechanisms that produce higher level phenomena. (2000, 13)

Or

> 2. mechanisms typically bottom out in lowest level mechanisms. These are components that are accepted as relatively fundamental or taken to be unproblematic for the purposes of a given scientist, research group, or field. (2000, 13)

In many cases, statement 1 will be true, but one may also read it to imply that higher-level entities do not play a role in the production of lower-level phenomena. If one accepts this implication, then explanatory reductionism follows; we understand upper-level phenomena only in terms of the activities of their lower-level parts. This is a problematic consequence, but one that doesn't necessarily follow from the core of Machamer, Darden, and Craver's conception of mechanism. Statement 1 occurs in the context of a description of the mechanisms of neurobiology and molecular biology as multilevel, meaning that entities and activities at multiple levels of organization are required to make a mechanism intelligible. They write later in the paper that "higher level entities and activities are . . . essential to the intelligibility of those at lower levels, just as much as those at lower levels are essential for understanding those at higher levels" (2000, 23). As a result, it seems that even though they state that lower-level entities and activities explain the production of higher-level phenomena, the opposite can also occur, and hence explanatory reductionism does not necessarily follow.

In statement 2 it is tempting to interpret "fundamental" to mean "lower-level." But this is an extremely truncated notion of "fundamental."[6] "Fundamental" refers to basic processes, entities, or theories. In some instances those basics may occur at lower levels of organization than the phenomenon in question and in some they may occur at levels above the phenomenon in question. It is important to note that Machamer, Darden, and Craver's work is confined to analyses of mechanism in the fields of neurobiology and molecular biology. In these fields, bottoming-out activities that occur at low levels of organization may indeed be basic or fundamental, or, in other words, low-level activities may be necessary to understanding neurobiological and molecular phenomena. But in other fields, such as developmental biology, evolutionary biology, or ecology, the activities basic to understanding of the phenomenon may exist at levels above the entity or process being investigated. In the case of the development of sexual behavior, the social and physical environments as well as genetic and hormonal states are fundamental to an understanding of the phenomenon in question.

Machamer, Darden, and Craver's conception of intelligibility is "historically constituted and disciplinarily relative." There is no logical story to be told about how these bottom-out activities, these kinds of production, come to inhabit a privileged explanatory position. "What is taken to be intelligible . . . changes over time as different fields within science bottom out their descriptions of mechanisms in different entities and activities that are taken as, or have come to be, unproblematic" (2000, 22).

They also explicitly focus on the context of the phenomenon being studied, stating that "[t]he entities and activities in the mechanism must be understood in their important, vital, or otherwise significant context" (2000, 23). This focus on historically contingent factors within scientific disciplines and on the context of the phenomenon under investigation makes their view particularly useful for feminist science studies and feminist science.

Whereas a reductive mechanistic approach to science results either strongly in a methodological monopoly, or weakly in privileging low-level methods, once mechanism and reductionism are teased apart, mechanistic science is open to a plurality of methods that can be used to investigate nature at a variety of levels of organization. The interactionist model provides a highly mechanistic explanation of the development of sexual behavior that is consistent with both Salmon's and Machamer, Darden, and Craver's views of mechanism. It reveals how the development of sexual behavior works, both in terms of the constituents of the behavior within the organism and in terms of the processes, both internal and external to the organism, by which the behavior develops. This model is pluralistic; it does not ignore or rule out the causal influence of genes and hormones, but it also does not privilege these low-level causes over higher-level causes, such as social influences. Social influences as well as genetic influences exhibit causal regularities and productive processes. Very different methodologies are required to uncover the genetic influences and social influences on behavior. The interactionist view requires methodological pluralism with respect to the elucidation of causal influence at these different levels of organization. It is an example of an approach to science that produces results that ameliorate many feminist concerns about the linear explanatory model and that, because of the explicit inclusion of higher-level causes and social factors, can be used to counteract one of the most central and important feminist criticisms of the biological sciences: biological determinism.

A view of mechanism unhindered by *a priori* reductionist assumptions allows us to address the three feminist concerns raised earlier. The problem of biological determinism is caused by the assumption that biological causes are the only or primary determinants of gendered behavior, and that these causes are low-level and unidirectional. Nonreductive, mechanistic approaches such as the interactionist model ameliorate this problem. The inclusion of environmental and social factors, and the explicit emphasis on the interactions among these factors, means that high-level phenomena are not solely determined by genetics or hormonal states. Change in those upper-level phenomena may be induced not only by genetic change, but also by environmental or social change.

The interactionist model also offers insight into how nonreductive mechanism escapes the methodological/epistemological problem. Although a reductive perspective encourages researchers to focus on phenomena that exhibit low-level causal regularities, a mechanistic approach need not do so. The interactionist model is an example of a mechanistic approach to the role of high-level causes in the explanation of complex phenomena. I have argued that nonreductive mechanism implies methodological pluralism. Studies of low-level phenomena may very well be easier to conduct, but at least part of the reason for this is that methods that engage the world in terms of its tiniest bits are central to sciences such as physics and molecular biology. Within disciplines such as these, reductive methods have been honed and taught over long periods of time. Focusing on mechanism without reductionism provides an avenue for revaluing sciences such as ecology and evolutionary biology, which have long provided mechanistic accounts of complex ecological and biological phenomena.

Mechanism without reductionism allows for a plurality of investigational styles and methodologies. Through whatever complicated interaction of causes, it may turn out to be the case that some women approach research differently from most men. One of the characteristics that may be more common in women's labs is a tendency to approach the work more contextually and holistically. This concept of mechanism takes away many claims that such work is unscientific. A plurality of methods, such as those inherent in the interactionist model, opens the door for researchers to choose an approach that best suits not only the object of study but also their political interests. The opening up of methodology that results from teasing mechanism away from reductionism may give feminist researchers the opportunity to look at gendered issues from a perspective designed to ameliorate past injustices, or to address previously overlooked areas of investigation. Finally, sciences that attract the most women tend to be seen as the "low-status" or "softer" sciences. This approach gives us an intellectual framework in which to revalue those sciences. The fact that evolution is not reductionist does not mean that it is less scientific or even less mechanistic.

The point of Carolyn Merchant's work was that a mechanistic worldview paved the way for a cultural philosophy of domination and control in which both women and the environment suffered. Separating mechanism from reductionism can go some way toward relieving her valid concerns. A reductionist stance affects the way that we see the world. If the methods that we use to learn about the world focus on breaking our objects of study into pieces, we can easily overlook the effect that changing those pieces has on the greater whole. Such methods also encourage us to overlook the context in which we do our research. This is just the myopia that accounts for much of the success of reductionist science, but that has also led feminists to worry. This stance toward the environment and toward health care in particular has led to the neglect of a more holistic, contextualized view; it has not helped us to address our present environmental crises or to develop an empathetic practice of medicine. A science that allows for higher-level and interlevel mechanisms makes it possible to conduct research that pays attention to context and complicated causal situations.

Notes

Thanks to Travis Butler, Jill Bystydzienski, Mark Ereshefsky, Jackie Foster, Margaret Holmgren, Anna Keyte, Kevin de Laplante, Jacqueline Litt, Helen Longino, Jennifer Schmahl, Tad Schmaltz, Virginia Valian, Mark Wunderlich, and an anonymous reviewer for help with various drafts of this chapter.

1. Ernest Nagel's (1961) famous work on reductionism is not within the scope of this chapter.
2. One could also focus on the role of mathematical models in this analysis.
3. Merchant (1980, 228 and 232) notes a fifth assumption that I will not discuss here.
4. There is current debate in the literature regarding this topic (see Woodward 2003).
5. There is also a serious concern regarding moves from causal explanations of events to causal explanations of regularities (Woodward 1989, 2003).
6. Thanks to Marc Ereshefsky for bringing this point to my attention.

References

Birke, Lynda. 1986. *Women, Feminism, and Biology: The Feminist Challenge*. Brighton, U.K.: Wheatsheaf.
Brandon, Robert. 1990. *Adaptation and Environment*. Cambridge, Mass.: MIT Press.
———. 1996. "Reductionism versus Holism versus Mechanism." In *Concepts and Methods in Evolutionary Biology*, ed. Robert Brandon, 179–204. Cambridge: Cambridge University Press.
Harding, Sandra. 1986. *The Science Question in Feminism*. Ithaca, N.Y.: Cornell University Press.
Hrdy, Sarah Blaffer. 1986. "Empathy, Polyandry, and the Myth of the Cot Female." In *Feminist Approaches to Science*, ed. Ruth Bleier, 119–46. New York: Teachers College Press.
Hubbard, Ruth. 1979. "Have Only Men Evolved?" In *Women Look at Biology Looking at Women: A Collection of Feminist Critiques*, ed. Ruth Hubbard, Mary Sue Henifin, and Barbara Fried, 29–41. Cambridge, Mass: Schenkman.
Kaplan, Gisela, and Lesley Rogers. 1994. "Race and Gender Fallacies: The Paucity of Biological Determinist Explanations of Difference." In *Challenging Racism*

and Sexism: Alternatives to Genetic Explanations, ed. Ethel Tobach and Betty Rosoff, 21–45. New York: Feminist Press.

Keller, Evelyn Fox. 1983. *A Feeling for the Organism: The Life and Work of Barbara McClintock.* San Francisco: W. H. Freeman.

Longino, Helen. 1987. "Can There Be a Feminist Science?" *Hypatia* 2(3): 51–64.

———. 1990. *Science as Social Knowledge: Values and Objectivity in Scientific Inquiry.* Princeton, N.J.: Princeton University Press.

Longino, Helen, and Ruth Doell. 1983. "Body, Bias, and Behavior: A Comparative Analysis of Reasoning in Two Areas of Biological Science." *Signs: Journal of Women in Culture and Society* 9(2): 206–27.

Machamer, Peter, Lindley Darden, and Carl Craver. 2000. "Thinking about Mechanism." *Philosophy of Science* 67(1): 1–25.

Martin, Emily. 2002. *The Woman in the Body: A Cultural Analysis of Reproduction.* Boston, Mass.: Beacon. (Orig. pub. 1987.)

Merchant, Carolyn. 1980. *The Death of Nature: Women, Ecology, and the Scientific Revolution.* San Francisco: Harper and Row.

Nagel, Ernest. 1961. *The Structure of Science: Problems in the Logic of Scientific Discovery.* London: Routledge and Kegan Paul.

Nicholson, Linda. 1994. "Interpreting Gender." *Signs: Journal of Women in Culture and Society* 20(1): 79–106.

Rosser, Sue. 2002. "Androcentric Bias in Clinical Research." In *The Gender of Science,* by Janet A. Kourany, 228–36. Upper Saddle River, N.J.: Prentice Hall.

Salmon, Wesley. 1984. *Scientific Explanation and the Causal Structure of the World.* Princeton, N.J.: Princeton University Press.

———. 1989. "Four Decades of Scientific Explanation." In *Scientific Explanation,* ed. Philip Kitcher and Wesley C. Salmon, 3–219. Minnesota Studies in the Philosophy of Science 13. Minneapolis: University of Minnesota Press.

———. 1994. "Causality without Counterfactuals." *Philosophy of Science* 61: 297–312.

———. 1998. *Causality and Explanation.* New York: Oxford University Press.

Schiebinger, Londa. 1989. *The Mind Has No Sex? Women in the Origins of Modern Science.* Cambridge, Mass.: Harvard University Press.

Woodward, James. 1989. "The Causal Mechanical Model of Explanation." In *Scientific Explanation,* ed. Philip Kitcher and Wesley C. Salmon, 357–83. Minneapolis: University of Minnesota Press.

———. 2003. "Scientific Explanation." *The Stanford Encyclopedia of Philosophy (Summer 2003 Edition),* ed. Edward N. Zalta. http://plato.stanford.edu/archives/sum2003/entries/scientific-explanation/. Accessed February 16, 2005.

Zita, Jacquelyn N. 1989. "The Premenstrual Syndrome: 'Dis-easing' the Female Cycle." In *Feminism and Science,* ed. Nancy Tuana, 188–210. Bloomington: Indiana University Press.

11 Across the Language Barrier: Gender in Plant Biology and Feminist Theory

Dana A. Dudle and Meryl Altman

Making difference visible does not mean making it disappear. (Barbara McClintock [Keller 1985, 163])

Feminist theory today faces no more urgent task than learning to understand and work with the powerful discourses of science and technology that shape our world, rather than simply resisting them or turning away. Crucial to that work will be reaching consensus about what some key terms—"sex," "gender," "the body," "sexuality"—can usefully mean and do, some thirty years after what is now called second-wave feminism first deployed them in the service of what was then known as women's liberation. Scientific inquiry and technological developments in the intervening decades have themselves tested the limits of that terminology, as Anne Fausto-Sterling (2000), Donna Haraway (1991), Lynda Birke (2000), and many others have shown. Yet genuine intellectual partnerships between feminist theorists and working scientists, so frequently called for, remain rare, and mutual suspicion and misunderstanding are still the norm. This chapter reports on one attempt to build such a partnership: a collaborative effort between an evolutionary biologist who studies flowering plants and a science-phobic literary critic trained in the 1980s, while we were both teaching at a small midwestern liberal arts college.

Our interchange began during a faculty meeting debate about "saving" a gingko tree in the academic quad, which was slated for demolition. Some (including Dana) found the tree beautiful and shady; others (including Meryl) found it simply smelly. Then Jim Benedix of the biology department suggested that Meryl, as director of women's studies, really ought to be supporting the gingko, since the gender of the tree was female, and the meeting broke up in general hilarity.

But Meryl and some others from women's studies remained puzzled by Jim's reference to "the gender of the tree." Surely he must have meant, not "gender," but "sex"? The sex/gender distinction has been foundational for three decades within mainstream social constructionist feminist theory, with "sex" referring to the biological givens of human existence, "gender" to the differences culture adds on or embeds, in ways that have varied from culture to culture and over the course of history. Politically, the sex/gender distinction was crucial to second-wave arguments that "biology is not destiny" and that long-standing human arrangements injurious to women could and should be questioned and changed. Despite recent "third-wave" reinterpretations and challenges, the idea that gender is culturally constructed—as expressed in Simone de Beauvoir's famous phrase "women are not born, but made" (1952, 249)—remains the central precept of feminist studies in the United States today.[1] So how could a tree have a "gender"?

As it turned out, Dana and other plant reproductive biologists have been happily pursuing research programs that use the term "gender" in a rather different way. Plant reproductive biologists employ the term "sex" to describe categories based on whether an individual produces male gametes (sperm cells), female gametes (egg cells), or both. But an individual's sex does not determine the extent to which its gametes function successfully, nor the relative number of male and female gametes an individual produces; these features constitute the plant's *gender*. A plant's gender may be influenced by its environment, including the gender (and sex) of neighboring plants. However, plant gender remains a quantifiable trait that is tightly linked to reproductive activity.

When we compared notes, we discovered that feminist theorists and biological scientists (some of whom are also feminists) speak of "gender" in ways that are very different, but not necessarily opposed, and that the ways we were misunderstanding one another were fascinating. Neither usage could be labeled "wrong"; might they be made compatible? And if so, how? We decided to pursue the question by spending a summer reading together in both fields, to see if we could begin to translate between the languages of science and feminist theory more fruitfully.

On Collaboration

The communication gap between "feminists" and "scientists," and ongoing efforts to overcome it, are widely discussed (see, e.g., Rosser 2002; Mayberry, Subramaniam, and Weasel 2001), and some ways to work in coalition (hiring more women, changing pedagogy and curricula, bringing scientific topics to women's studies core courses) are obvious. But there are real barriers to deeper intellectual collaboration which should not be glossed over. Working scientists have been understandably skeptical about "feminist critiques of science" which question the epistemological and practical underpinnings of what they do, particularly when mounted from outside science by historians and philosophers. On the other hand, feminists outside the sciences are sometimes wary of scientific claims, because scientific and pseudo-scientific explanations have been deployed to keep Woman in her place for centuries—eugenics, misapplied statistical reasoning, some forms of sociobiology are examples. Less often discussed are the deep commitments to distinct professional and intellectual *vocabularies* that can make it hard for feminist scholars and women scientists to understand one another, on individual campuses and in international debates, or (even worse) can make us believe we have understood one another when we really have not. Collaboration across fields was the ideological and practical origin of women's studies, and shared work (now called "hybridity" or "bordercrossing") is still at its heart. But in a true collaboration, each partner must be prepared not just to contribute something, but to leave something behind.

Meryl would not describe herself as engaged in the "feminist critique of science," both because that term has been a barrier and because of the ethico-intellectual principle best articulated by Bob Dylan: "don't criticize what you can't understand." Scholars like Anne Fausto-Sterling (2003) and Evelyn Fox Keller (1985) describe themselves as immigrants from one world to another, or as "intrepid travelers," but a more honest term for Meryl's position would be *tourist*: armed with a phrasebook she knows is inadequate, seeking to understand what she is seeing and hearing, alert for false cognates which sound the same but are not, prepared to endure some discomfort, but ultimately planning to go home. Embarrassing as this is, it is a reasonable figure for the ways most general education students encounter the part of the curriculum farthest from their comfort zone. So as we try to increase scientific literacy among women's studies students (and faculty), perhaps aiming to make tourists less irresponsible and more open-minded is a modest but realistic goal.

Meryl soon saw that she would have to ditch some baggage. She would have to stop saying "I don't believe in biology" and stop explaining the sex/gender system to her Introduction to Women's Studies class by saying, "sex is what the biologists study, it's natural and fixed, while gender is what we study, it's cultural and we can change it, though not easily." She would have to place provisionally on hold her absolutist investment in social construction: probably her deepest and longest-held intellectual commitment.

Meryl had other incentives to rethink social constructionism, too. As feminist theory solidified during the battles of the 1980s, identifying even a moment of "essentialism" or "biologism" in the argument of an opponent had been sufficient to discredit and dismiss her, but more recent work was taking a more nuanced view. Judith Butler (1990), for example, argues that "sex" may also be socially constructed, that in some ways gender underlies or produces sex, or even precedes sex historically (Laqueur 1990). Toril Moi points out that other forms of essentialism (for instance, religious ones) may operate just as oppressively as biological dualisms (1999), and Eve Sedgwick (1990) shows that a constructivist position may be mobilized just as easily as an essentialist position to further either a highly repressive agenda or a progressive one. Thus Meryl's self-scrutiny would parallel recent complications in feminist theory.

Meanwhile, Dana had to agree to be patient with feminist writers who use "biological" casually and who refer to one another in a highly ingrown, mysterious manner. Like the beginning science student, she had to accept the need for a difficult technical vocabulary to describe new phenomena. Perhaps hardest,

Dana had to accept that language questions were not peripheral, that the way a concept was expressed was not an afterthought, but *was* the thought. And we both had to give up debating who "owns" a particular discourse and has the "right" to say what certain words "really" mean. We had to set aside an English teacher's passion for correctness and a scientist's passion for transparency and clarity, and replace the struggle for mastery with something closer to Wittgenstein's "common sense" theory of language: "the meaning of a word is its use" (Wittgenstein 1953, 51). This is not a relativist approach: some words are more useful than others, depending on what sort of *work* one is using them to do, and on who is using them in what sort of "discourse community" or "language game." So we prepared ourselves to ask, not what words mean, but what words do.[2]

But why *plants*? Perhaps plant "gender" was simply an arbitrary signifier like "grammatical gender" in the romance languages, or a dead metaphor like the "male" and "female" electrical plugs on sale in hardware stores. What might be at stake *politically* in a discussion of how angiosperms or gingkos reproduce? Much feminist writing about science deals with topics like the human genome project, in vitro fertilization, hormone replacement therapy, and nuclear war, where political and ethical stakes and commitments are only too apparent; but it was hard to imagine large numbers of feminists getting as agitated about stamens and pistils as by, for example, Randy Thornhill and Craig Palmer's *A Natural History of Rape* (2000). We did hope that the chance to bracket angry investments and sidestep well-worn debates might help us see something about how scientific language is made and how it operates. We hoped that analogies would become more clearly visible as analogies, metaphors would stand out as metaphors, and unconscious, problematic slippages from biological explanations to biological determinism would be easier to arrest. An example from the early history of plant reproductive biology soon showed us how this sort of inquiry could make a difference.

Sex, Plants, and Metaphor: A Historical Example

Carolus Linnaeus's "Sexual System" of classification for plants, based on what he called "the marriages of the plants," was a hierarchical, arithmetical system in which genera of plants were placed into twenty-four *classes* based on the number, position, and length of stamens (male organs), and within each class the genera were assigned to *orders* based on the number of pistils (female organs). Linnaeus used strongly gendered, metaphorical language, supplemented by scientific prose, in the key to his sexual system of classification (see Stearn's introduction to Linnaeus 1957). Sometimes that metaphorical language was inaccurate in describing the species being classified. For example, the technical definition of species in Class XXIII, "Polygamia," translates as "hermaphrodite flowers, and male ones, or female ones in the same species." However, if one follows the steps in the key to classify such a species, one sees that this is a "public marriage" in which "husband and wife have separate beds," and "husbands live with wives and concubines." Aside from the inappropriate creep of human gender roles into the supposedly scientific description of plant species, and the spurious primacy placed on male reproductive organs over female reproductive organs, this metaphorical description of "polygamous" species could easily have led to inaccurate assumptions about the plants. As just one example, literal interpretation of the metaphor quoted above allows for the husbands—male flowers or male organs within hermaphroditic flowers—to engage in "extra-marital" fertilizations, but does not allow for female organs or flowers—wives or concubines—to receive pollen from multiple sources. As Linnaeus probably realized, it is common for stigmas to receive pollen from neighboring plants as well as from the anthers within the same flower. But his metaphorical schema did not permit him to describe the plants properly: not just because he used human analogies, but because he could only imagine the kind of human arrangements that were normative in his day and class. Anthropomorphic metaphor may be inevitable (see below), but in this case, we would argue, Linnaeus picked the wrong one.

If plant systematics or plant reproductive ecology had been founded on the Linnaean classification system, later scientists might have inadvertently ignored the complexities of female reproductive fitness simply because the metaphorical descriptions of the classes and orders did not allow for certain behaviors of the female floral organs. This potential danger was averted, possibly because the Linnaean system did not reflect evolutionarily meaningful relationships among plants and fell out of use as biology began to incorporate evolutionary principles, and possibly because empirical and theoretical work on female reproductive behavior in plants is usually at least a step ahead of the work on paternity in plants.[3]

We identified three problems with Linnaeus's system: he uses metaphorical language where literal description would have been clearer; he anthropomorphizes in a fairly hilarious way; and as a result, he actually gets things wrong, misses part of the story. Cultural assumptions about humans underlie supposedly objective scientific descriptions of plants. But these assumptions only become glaringly visible to us *because they are funny,* and quaint, that is, because human cultural sexual arrangements and the way we think and talk about them have changed since Linnaeus's day. What equally problematic assumptions in current discourse about plants might be invisible to us, because we are part of the culture that produced them?

The Sex-Gender Distinction in Plants and Humans

Among plant biologists today, the word "sex" describes categories defined by the morphological structures and gametes that a plant or flower produces. One way to describe sexes is by using the following categories: hermaphrodite, male, and female. In a common use of these words, a hermaphroditic plant has functional anthers (with pollen containing viable sperm) and functional pistils (with ovules containing viable egg cells), whereas male and female plants have only anthers or pistils, respectively. A crucial difference between animals (including humans) and plants regarding sex and gender is the distribution of sexes within populations. Unlike in most animal species, where a large number of individuals can be classified as male or female (based on reproductive morphology or chromosomes), more than 95 percent of flowering plants would be classified as hermaphrodites, because they produce both sperm and eggs. The label "hermaphrodite" obscures important variation among individuals in the relative number of sperm and egg cells produced, and in the sex cells used as the basis for the next generation of plants. Thus a further distinction became necessary.

David Lloyd's groundbreaking work in the mid-1970s and early 1980s identified and illuminated this tendency of the sexual labels to hide important variation among hermaphrodites, and introduced the term "gender" to plant biology as a partial solution to the problem. Contemporary plant reproductive biologists use the term "phenotypic gender" to describe the relative investment in male and female reproductive structures on a given plant at a particular time. For example, a morphological hermaphrodite whose floral display includes fifty bisexual flowers (containing both egg and sperm cells) and a hundred pistillate flowers (containing only egg cells) is considered to be "phenotypically female." A related term, "functional gender," describes the proportion of an individual's genes passed to the next generation in sperm, relative to eggs. For example, if the hermaphrodite described above fails to produce any seeds but transmits sperm (within pollen) to two hundred eggs on surrounding plants, this individual's functional gender would be 0 percent female. Conversely, if the same plant developed fifty fertilized egg cells into seeds but did not successfully fertilize any eggs with its pollen, it would be functionally 100 percent female. In hermaphrodites phenotypic gender and functional gender are both quantitative traits, with possible values anywhere between 0 and 100 percent.

Dana has been told more than once by nonscientist feminist colleagues that "gender" is not an appropriate term for the phenomena explored by biologists. But the borrowing of the word makes an important point. In this usage, both phenotypic and functional gender depend on context. A hermaphrodite's phenotypic gender depends on many environmental variables, including both abiotic factors (e.g., time in the season) and biotic factors (e.g., pollinator activity). The functional gender of a hermaphrodite also largely depends on the gender of surrounding plants.

There is an analogy to the sex/gender distinction in humans, which feminists were developing around the same time. Human gender was usually viewed as a product of environmental factors; gender was something that could change, both in individuals and populations, in a way that sex could not;[4] gender, sex, and reproduction in humans were acknowledged to be related, but not in any simple, obvious, or causally necessary way. Gayle Rubin's formulation in "The Traffic in Women" (1975, 159) is classic: "a 'sex/gender system' is the set of arrangements by which a society transforms biological sexuality into products of human activity, and in which these transformed sexual needs are satisfied." The definition of what it means to be a man, or a woman, in any given culture is then the outcome, rather than the starting point, of a set of cultural conventions and operations; and there is tremendous variation both across cultures and across individuals of the same biological sex within the same culture. In fact, one reason feminists drew the sex/gender distinction was to legitimize human variation in gender by showing that statistical norms did not match up to what was described as "normal" in the post–World War II United States—following up on the goals and the methods of both Kinsey et al. (Kinsey, Pomeroy, and Martin 1998) and de Beauvoir (1952).

Functional gender in plants is notoriously hard to measure, and can only be accurately calculated at the end of an individual's life. It is inextricably linked to sexual reproduction; an individual that does not reproduce sexually cannot be assigned a value for functional gender. In other words, sexual fitness and functional gender are almost synonymous in plants. Furthermore, sex and gender are also highly correlated: a plant that produces only pistillate flowers can never be functionally male. It is only in plants that are morphological hermaphrodites that gender can be anything but 0 or 1—but more than 95 percent of plant species fit this criterion. So, as with humans, sex and gender tend to be statistically linked, but this does not tell us very much about what actually happens in either individuals or populations. And, as with humans, the term "gender" serves to highlight variation and diversity previously masked by a simple binary opposition.

But even though plant gender is influenced by other plants, terms like "socially constructed" seem inapplicable. Perhaps the clearest sign that we are seeing an analogy, not an identity, is if the analogy works up to a certain point and then stops. Perhaps the sex/gender distinction in plants is best understood, not as an explanatory schema, but as a descriptive model, or a metaphor.

Until this point, we have been using the term "sex" to refer to a trait or quality of an individual, what feminists sometimes call "sexual difference." But of course "sex" is also used to mean "sexual activity": what people do in bed, and plants do in fields (or in labs). And here is where the analogy between the sex/gender distinction in plants and in humans really breaks down. For a plant, sexual activity and reproductive activity are indistinguishable;[5] and for people, or at least for feminists, they are not. Sharon Kinsman, for example, defines "sex" simply as "mixing DNA" (Kinsman 2001, 194) and a plant biology textbook says, "when you think of sex, think of meiosis" (Raven, Evert, and Eichhorn 1999, 169)—both being mainly concerned to distinguish the most important aspect of sexual reproduction from other sorts of reproductive activity. But those who teach introductory women's studies classes may feel equally committed to convincing students that gender is *not* reducible to baby-making.

As we explained earlier, functional gender is construed retrospectively for plants, after reproduction has occurred, so that reproductive activity (at least through sexual reproduction) and functional gender cannot really be separated. In contrast, one key reason for emphasizing the sex/gender distinction in humans was to make the political point that the physical configurations of one's body did not determine what one did, or wanted to do, with that body, and did not exercise determination over what came to be called "sexuality"—which could be further broken down into sexual activity, sexual identity, object choice, et cetera, all arguably decoupled or unlinkable from each other and from both gender difference and sexual difference. In other words, far from being a "functional" definition (as with plant gender), the creation of a new category for people was designed to avoid functionalist explanations that reduce "what it means to be a woman" to the role female bodies play in the continuance of the species. A plant's phenotypic gender is in some ways closer to human gender than functional plant gender is. But human gender can be discussed apart from human reproductive activity, whereas plant gender always refers to reproductive activity, or at least the potential for it. So to speak of male and female gender in plants is clearly metaphorical rather than strictly "literal."

Metaphor across the Disciplines

In *Blindness and Insight* (1983), literary theorist Paul De Man observes that by yoking a thing to something else which is similar but clearly nonidentical, metaphor highlights and illuminates some features of the thing previously left in shadow, and casts into darkness other features which might otherwise have been salient. For example, consider the literary tradition of describing women as delicate flowers which, if not picked, tragically wither on the vine. A pretty thought, from a certain angle; feminist poets make other choices.

In this view, the problem is not metaphor as such, but the uses to which metaphors are put. Metaphor is inevitable, since no descriptive account can be complete; human language and life would be impoverished without it; metaphor has a cognitive and not simply a decorative function. Certain accounts suggest that the activity of making metaphors, of "carrying over" words and part of words from one context to another, is basic to meaning-making activity and motivates all language change. However, the use to which metaphor is put, the specific meanings it is used to make, still can, indeed must, be subject to scrutiny, especially with "dead" or unconscious metaphors which over time have come to seem simply literal and "true."

Along these lines, we then notice that human gender is metaphorical, too, since language and cultural discourse generally reduce the myriad variability of human behavior to a choice between one of two boxes marked "M" and "F."[6] Viewing human gender as a metaphor helps us see that other metaphors, alternative interpretive frames, are possible. Eve Sedgwick elaborates this point in *Epistemology of the Closet* (1990) when she shows that male/female and homo/hetero are only two of an enormous array of binary axes around which a sex/gender system might be oriented.

But within the traditions of the sciences there is a strong suspicion of metaphor and of figurative language generally. Linguists M. A. K. Halliday and J. R. Martin trace the Enlightenment project of making a new and separate "language of science" back through Leibniz to William of Ockham but especially to Bacon, who complained about "the confusion of the literal and the figurative meanings of a word" (1993, 5). Thus technical neologisms proliferate: the risk that most readers may find the text opaque seems preferable to the danger that everyone will easily grasp a concept that is, in fact, wrong. Scientists are explicitly trained to look for language that will be both accurate and transparent, as demonstrated by Anne Sakai and Stephen Weller (1999) in their review of the terminology of plant reproductive biology. "Accurate" language would not lump things together that are not the same, or artificially create differences that do not exist in the real world, and "transparent" language would be free of misleading connotation and association. The ideal scientific prose, some have argued, should move from the writer's mind to the page to the mind of the reader (or student) as a pure signal with a minimum of "noise." We can see in a recent botany textbook that this also means with a minimum of *cultural* noise, as when the authors say certain terminology is purely conventional, like the view that a gamete is called "male" simply because it is smaller (Raven, Evert, and Eichhorn 1999, 372). When biologists take on board the idea that scientific objectivity has been compromised by the "creep" of social attitudes, incarnated in language, the first impulse is to purge. One particular category of metaphor, anthropomorphism, draws particular censure not because it is "politically incorrect," but because it is seen as unscientific.

Feminist and evolutionary biologist Marlene Zuk (2002) has recently taken on just this issue. Zuk argues against allegorizing animal behavior in *either* direction, pointing to the illogical circularity with which "[w]e both judge . . . animals by rules for human behavior and at the same time look to them as role models" (2002, 3). Zuk shows scientists applying cultural assumptions to a variety of birds, mammals, and insects, leading to misinterpretations of data, from the Victorian supposition that "mate choice" is something only male animals do (persisting well into the twentieth century in descriptions of "coy" females), to unrealistic expectations of a universal "selfless maternal instinct," to a deterministic account of a "gay gene."

Zuk's intervention is particularly interesting because she is against both sexist and feminist anthropomorphizing: her project is not simply to oppose certain stories about "nature" but to warn against storytelling itself as fanciful, to prevent the natural world from being taken as ground, source, excuse, or justification for *anything*. But it is very easy to show that a "pure" literal language is not possible: she herself uses metaphors, and her book makes science accessible by telling stories that solicit reader identification with birds and bugs. Anthropomorphic metaphors are especially common (and arguably useful) in teaching; efforts to avoid them seem doomed. Instead, we could try to use metaphor more self-consciously, reminding ourselves not to confuse the model with the world, not to mistake descriptions for causes or causes for justifications, to attend to difference as well as analogy, and so on.

In the end, we came to agree with Evelyn Fox Keller that "metaphors of gender" govern what scientists see or do not see, what questions are culturally ask-able and thus answerable (Keller 1985, 43), as in the case of Linnaeus's "mar-riages of the plants" above. But where Keller's famous article about the "master molecule" and the slime mold blamed (masculinist) metaphor for impeding ob-jectivity and setting back scientific progress, her recent view is more nuanced: some metaphors turn out to *work* better than others, and judgments about "ob-jectivity" are hard to make. In *Refiguring Life* (1995, xii–xiii) she contrasts two culturally contingent "ways of seeing" the sperm and the egg, one "evocative of the Sleeping Beauty myth" and a newer version "cast in the language of equal opportunity," concluding that "a socially effective metaphor twenty years ago has ceased to be so, in large part because of the dramatic transformation in ideologies of gender that has taken place"—and yet "scientific productivity would have to be granted to both," since each led to insightful research.

When people are exhorted to watch their language use, two different things may be meant. One is a simple surface change, for example, "stop saying 'Negro,' say 'black'" or "stop referring to 'the scientist' as 'he' or as 'the valiant war-rior battling disease'—it discourages girls from going into physics." Deborah Cameron (1995) calls this "verbal hygiene" and shows it is both inevitable and important, especially when it recognizes the claims of excluded groups.[7] But Keller (1995) is calling for a deeper examination, claiming that by changing lan-guage we can change our cognitive categories and actually see differently, see better, see more. We agree that rather than trying to wash metaphors out of discourse, scientists should seek to use them more intentionally, to be aware of both blindness and insight.

Is Plant Gender a Useful Metaphor?

What, then, is the function of the "gender" metaphor in plant biology? Is it worth keeping? One test of functional or felicitous metaphor—or indeed of any neologism—might be *does it add clarity*, does it describe something that actually exists in a way that previously could not be done? A close reading of chapter 7 of *The Different Forms of Flowers on Plants of the Same Species* (Dar-win 1877, 287–88) suggests that Charles Darwin recognized the difficulty of using labels based on morphology (sexes) to describe variable plants in natural populations. However, David Lloyd was the first to explicitly describe the con-cept of quantitative gender (particularly functional gender), without naming it, in 1972. He identifies the insufficiency of single terms for the evolutionarily im-portant variation he saw among the sexual morphology of individuals in the genus *Cotula* (Lloyd 1972, 1182):

> The examination of the functional degree of separation between the sexes in diclinous populations[8] often ends with the choice of a single word to describe the sex expression. It appears that exceptions to the neat division of the various diclinous conditions presupposed in their simple morphological definitions are usually considered to be of little significance and are dismissed as unimportant anomalies, irregularities, or "intersexes."

Here, Lloyd expresses frustration with the paucity of available terms for the variation he sees in natural populations. "Intersex," as Lloyd's predecessors had used it, was not a helpful label for an evolutionary biologist studying variation in populations because it provides no information about how plants function in populations, nor does it help predict populations' responses to natural selection. It lumps dissimilar plants that are subject to different selection pressures into a single group. Finally, this term is imprecise: the boundary between "intersexes" and other classes of plants is necessarily arbitrary in a classification system based on morphological traits that vary continuously.[9]

In 1980, Lloyd formalized his earlier ideas in an important article, "A Quantitative Method for Describing the Gender of Plants." Lloyd's method crystallized a salient argument about plant populations that biologists since Darwin (1877) had struggled to express: in plants, sexual *morphology* alone does not define the sexual *function* of an individual. Because an individual's sexual function defines in large part its evolutionary fitness, it is the most important aspect of a plant's "sexual identity" from an evolutionary standpoint. Lloyd's frustration and insight in 1972 made it obvious that new labels and metrics were necessary to describe sexual function separately from morphology. In 1980, the new lexicon he proposed enabled plant reproductive biologists to make a meaningful distinction that could not be stated previously. Notice that the theory of functional and phenotypic gender in plants arose from observation. Lloyd used data to drive his descriptions and his theory—unlike Linnaeus, who, it seems, worked the other way around.

A second test of a functional metaphor might be *is it used carefully, meaningfully, and consistently in the relevant literature?* Sakai and Weller (1999) echo David Lloyd's argument that plants need a morphology/function distinction and thus a sex/gender distinction. They indicate that the boundary of language between the two concepts should be carefully maintained in order to clarify the theoretical and experimental goals of investigators. Some scientists, including botanists and ecologists, do seem to use "gender" to mean "sex" in the scientific literature, and outside the primary literature on plant reproductive biology the slippage between the concepts of sex and gender in plants is even more common—in fact, the comment mentioned at the beginning of this chapter about the gingko tree's "gender" would more accurately have referred to the tree's "sex," since the speaker (an animal biologist) was probably referring to its morphology rather than its relative seed fitness. When the discussion moves beyond the biology department to the rest of the world, policing usage seems even harder; but the distinction remains important.

A third test of functional metaphor, following Keller (1995), might be *is it productive?* Lloyd's description of gender as a quantitative trait has allowed theory and experiments exploring plant reproductive biology to answer questions that were not even conceivable previously. By that we mean, not that the *work* was not conceivable, but that the *questions* were never asked because the available terminology in the field effectively discouraged the study of the individuals and populations whose variability defied the usual nomenclature. The key point is that the concept of functional gender allows us to predict evolutionary trajectories, and to measure evolutionary responses to environmental pressures. We now have named a quantitative trait that can respond to an array of environmental stimuli, and that reflects evolutionarily important function, so the science can focus on evolutionary processes that result in the complex patterns that we see in natural individuals and populations. For example, plant reproductive biologists have gained tremendous insight into topics such as the evolutionary stability of and transitions between plant breeding systems (McCauley et al. 2000; Dudle, Mutikainen, and Delph 2001), the natural selection of plants with different sex morphs (Charlesworth 1989; Eckhart 1992), and the different responses of plants that vary in functional gender to environmental stresses (Delph 1990; Case and Barrett 2001; Ashman 2002). This work depended on the ability of researchers to imagine and represent the potentially continuous variation of hermaphroditic plants between their male and female function. To take another measure of productivity, the *Science Citation Index* lists 285 articles published within the past ten years whose abstracts include the term "gender" in reference to plants.

"Plant gender" would seem to meet Evelyn Fox Keller's (1995) directive to seek metaphors that are productive rather than reductive, that open up new ways of seeing and thinking rather than closing them down. But we also need to remember to ask the ideological question of what is produced, what newly thrown into shadow, when we move the lamp.

Turning the Lens the Other Way; or, Can Plant Biology Change Feminism?

Anthropomorphisms of the Linnaean sort, from slime molds to baboons, make it harder to see and respect the diversity in the biological world; but the danger is more serious when the metaphors run back in the other direction: when the behavior of nonhuman organisms is used to interpret either what human animals "really" are like, or what they should be like. Unreflective use of economic metaphors of "investment," "bet-hedging," or "banking" may hide value judgments about what human (female) sexuality is really "for," or may even explicitly underwrite such value judgments to the extent that they become prescriptions, delineating some human behavior as normal, other behavior as deviant. Meanwhile authors as different as Kinsey (Kinsey, Pomeroy, and Martin 1948) and Haraway (1991) have sought to intervene in ideological debates by providing counterexamples, counter-anthropomorphisms, such as Kinsey's famous photograph of two bulls vigorously copulating with one another in a Midwestern field (Gathorne-Hardy 2000) or Haraway's comment that she "would rather develop a theory of the unconscious based on the reproductive practices of the fern rather than the nuclear family" (Goodeve and Haraway 1999, 124). Zuk's (2002) caution about the "naturalistic fallacy" seems relevant here: these examples should unsettle assumptions, not ground new norms. But when one is dealing with issues so fully laced with conscious and unconscious social assumptions for all of us, the very act of making the familiar strange—as Lynda Delph, a plant biologist, did simply by saying, "human beings are dioecious"—can be intellectually valuable and liberating. Suppose we ask (as someone in Dana's lab group did), "what is the relationship between phenotypic and functional gender in humans?" The answer must be that, as with plants, there is a statistical correlation, but not a causal imperative. Perhaps this opens up a new and productive (metaphorical) way of discussing what Adrienne Rich calls "compulsory heterosexuality" (Rich 1980).

Some "third-wave" revisions to the sex/gender divide leave us with human beings who look even more like plants. Judith Butler (1990) speaks of "doing gender" rather than "having gender"—closer to what plants do. Recent accounts that criticize the rigidity of second-wave understandings of the sex/gender binary emphasize gender as "performative" or processual, unfolding through time, and bending back recursively upon "sex," which can no longer be seen as an irreducible preexisting substrate. In Anne Fausto-Sterling's definition, "gender is a process of creating distinguishable social statuses for the assignment of rights and responsibilities. . . . as a process, gender creates the social differences that define woman and man" (2000, 250). One might recall E. P. Thompson's famous statement that "class is a relationship, and not a thing" (1968, 11) and say the same is true of gender: it is not something fixed that we *have*, or *are*, but rather something created in a context. Perhaps human gender is also usefully seen as a continuum, like functional gender in plants.

If Fausto-Sterling (2000) is right to take intersex bodies as paradigmatic, the same may be true also of "biological sex," which then becomes not just a constructed category, but a continuum as well. Fausto-Sterling documents the oppressive, unethical, even violent ways medical doctors impose value judgments on intersex or ambiguously sexed infants—intermediate, ambiguous, indeterminate, or undecidable bodies which occur more frequently than most people realize; the norm the doctors impose is that there are, or at least there ought to be, two sexes and only two, with no gray zone between them. "Functional gender" might look like an improvement on this practice.[10] At the very least, thinking about plants may remind us that the relationships between gender and sexuality, and between gender and reproductivity, remain fuzzy in many feminist accounts, even though, as Meyerowitz (2002) shows, for example, it is the (deep-rooted cultural) connections among them that locate gender with respect to social power.

Plant sex/gender metaphors might further contribute to our thinking about sex/gender relationships in humans if we consider how plant biologists actually study the interactions among the traits that plants display and the environments in which they find themselves. Phenotypic gender, functional gender, and sex (as well as water-availability status, competitive environment, etc.) are all variables with which we can identify and describe plants, but are also factors that influence each other in complicated and hard-to-predict ways, within and between individuals. Elucidating the interactions between these traits and environments is the underlying goal of plant evolutionary ecology. Since beginning work on this project, Dana has been thinking about these traits as axes of identity for plants. It is intriguing that sometimes certain "plant identities" are more salient than others, depending on what a biologist is studying—for example, if someone is interested in the effects of plant density on the sex ratio of gynodioecious populations, she is unlikely to manipulate or even measure the plants' water availability in her initial experiments. So the "water-availability" axis is not important in that study—until the investigator begins to suspect that plant density is mediated, in part, by access to water during the dry season. In other words, a good scientist is always aware that multiple "identities" of the plants might be contributing to a phenomenon, even while she is focusing on a small subset of these identities.

Similarly, some aspects of *human* identity influence and change the performance of other aspects, though unpredictably; some potential aspects of identity may go unmarked. Sedgwick (1990) brings our attention to some dimensions, or axes, of human sexuality that are not salient in contemporary American society—for example, "orgasmic/nonorgasmic, noncommercial/commercial, using bodies only/using manufactured objects, in private/in public, spontaneous/scripted"—but that may in fact be much more salient in the lives of individuals than "male/female" or "heterosexual/homosexual." Some of these dimensions might well influence or be influenced by one's gender, or sex, or genital configuration, and so on; for some, the influence might be reciprocal, while in other cases, different axes might operate independently. Feminist theory is also struggling toward nonhierarchical ways to map the mutual interactions of race and class as well as gender on human identity and experience (terms like "intersectionalities" reflect such attempts). The ideal model would also recognize a complex interaction between self-understanding and perception by others (cultural ascription vs. creative agency)—"what we make of what they make of us."

For example, Dana suggests replacing "intersectionality" with something more like the genetic concepts of pleiotropy and epistasis. "Pleiotropy" means that one gene can affect the expression of more than one trait of an individual, and "epistasis" means that a single trait can be affected by the identity and interactions of more than one gene. Provided we could bear in mind that these were in fact metaphors, not claims that race or gender identity were genetically caused, pleiotropy and epistasis could help signify the complexities of human identity described here—and remind us that geneticists do, in fact, deal in complex ways with complexity.

Similarly, physicist Neal Abraham has suggested a statistical model for beginning to understand the complexity of human identity that borrows the plant biologists' use of mathematical language and methods: he points out that, when analyzing any complex data set with many continuous or categorical variables, the relationships between the variables are rarely clear before analysis; that certain kinds of statistical analysis (for example, principal value decomposition) can illuminate which variables are redundant and which are independent; and that the relationships between variables are often surprising to the investigators.[11] And in recent work, Anne Fausto-Sterling (2003) has suggested using dynamic systems theory to map human gender, not as a permanent static quality but as a nonlinear, emergent property, involving self-organization from multiple processes both internal and external to the organism. These approaches seem very promising to us, provided researchers and writers (and those who build on and popularize their work) can avoid the slippage from description to functional explanation to deep causality to justification to prescription that has been all too common in the history of scientific inquiry into human sex and gender.

With the proliferation of these axes (or dimensions, or continua, or spectra), all interacting and affecting one another, the potential level of complexity becomes mind-boggling in short order. But this is already well known among plant biologists trying to make sense of nature: the *practice* of biologists is the opposite of reductive or one-dimensional. Zuk (2002), among others, finds it rather irritating when people fail to understand the interactional character of biological explanations. Since "[a]ll behaviors are the result of genes, developmental conditions during environmental life, and the subsequent environment where the organism finds itself . . . the absurdity of arguing over which part of a behavior, whether it is hole-drilling in woodpeckers or homosexuality in humans, is innate or cultural" ought to be apparent (Zuk 2002, 14). Ideologues looking to "biology" for simple stories about human sexuality will find instead a bewildering array of complexity and variation: as Barbara McClintock puts it, "anything you can think of, you will find" (quoted in Keller 1983, 199).

It is worth reminding ourselves that the concept of "gender" came into feminist theory (as into plant reproductive biology) to do certain specific work, which may still need doing. By showing that sex and gender could be separated, but were connected, as it were, arbitrarily—or metaphorically—Rubin (1975) and those who followed helped us see that cultural work was being done in holding them together. This made it possible to see and name a sex/gender *system,* holding male power in place through compulsory heterosexuality. Certainly we now need more nuanced and multidimensional terms of analysis, models that more clearly unhook sexual identity from sexual behavior and subjective gender role from the gender of one's object choice, as well as a less rigidly dimorphic understanding of sex. But if sex and gender are both continua or spectra—indeed, if we further break down "gender" into "gender role, gender behavior, sexuality" with "sexuality" further divided into desires, behaviors, and identities, and all these subcategories must also themselves be continua or spectra—the question becomes how to map all these different axes together so that we may continue to speak meaningfully about *power.* What *ethical and political* analysis might the next generation of feminist theorists put in place of the sex/gender system? Could feminist theory without some form of the sex/gender distinction still be recognizably feminist? How?

The dualism that currently seems most problematic to us, and most in need of elimination, is not "sex/gender" or "nature/culture," but "feminism/biology."

Locally speaking, our collaboration was successful in the sense that it generated new insights and new questions neither of us could have seen on our own; the crucial factor was that each was willing to take the other's language (both literal and metaphorical) seriously, rather than attempting to dismiss, translate, "colonize," or assimilate either field as a subset or mere object of scrutiny within the other. Our modest hope is that others will be encouraged both to undertake such cross-disciplinary collaborations and to proceed cautiously, with attention to language and with mutual respect. Future work toward a better understanding of both sex and gender, in people and other organisms, will require numerous conversations between many varieties of biologists and many sorts of feminists. We have only just begun to talk.

Notes

For insight and support we thank the DePauw University Faculty Development program; L. Delph, C. Lively, and their lab groups at Indiana University; the DePauw women's studies workshop, especially K. Ahlm, D. Geis, B. Gourley, J. Jerz, and J. Stockton; and N. Abraham, J. Benedix, D. Cameron, J. Johnson-Licon, K. Hall, K. Nightenhelser, and H. Smith-Borne.

1. The terms "second wave" and "third wave" are contentious, but seem useful here to quickly express a contrast between the position outlined by, for example, Gayle Rubin (1975) and the challenges posed by, for example, Judith Butler (1990) and Jay Prosser (1998).

2. See also another influential "commonsense" philosopher of language, J. L. Austin (1952). This approach is standard in sociolinguistics (see Cameron 1995; Cameron and Kulick 2003) and is also embraced by Evelyn Fox Keller (1995) and by such otherwise divergent theorists as Judith Butler (1990) and Toril Moi (1999). See also Altman (1990).

3. Seeds tend to stay put and are therefore easy to count, extract DNA from, or otherwise analyze. But pollen can travel kilometers before it lands on a stigma, so it is often hard to assign paternity to new plant embryos. Therefore, "female fitness" is easier to measure.

4. Here we are concerned with the basic second-wave formulation. See also Haraway (1991).

5. That is to say, in general, if plants *can* reproduce, they do. Although plants may produce more flowers than will be developed into fruit, even if all of the flowers are successfully pollinated (Sutherland and Delph 1984; Bell 1985), they do not have sex just for fun, to express intimacy, creativity, commitment. (As far as we know.)

6. Rubin (1975, 179) states, "Men and women are, of course, different. But they are not as different as night and day, earth and sky, yin and yang, life and death. In fact, from the standpoint of nature, men and women are closer to one another than either is to anything else—for instance, mountains, kanga-

roos, or coconut palms. Far from being an expression of natural differences, exclusive gender identity is the suppression of natural similarities."

7.   Such self-monitoring occurs in Raven, Evert, and Eichhorn (1999); for example, when a violet is described as making a reproductive "choice," the quotation marks express some uneasiness.

8.   According to Lloyd (1972, 1183), diclinous or "sexually dimorphic" populations "consist of two classes of plants, each occurring in considerable frequency, one of which contributes exclusively or predominantly male gametes and the other female gametes to the next generation."

9.   Lloyd's use of "intersex" is unrelated to more recent uses of that term to describe some physical configurations in humans.

10.   While space is lacking for a full discussion of feminist history, the novelty of Butler's view (the distance between two waves) is sometimes overstated. Sex as a continuum is a very old idea (see Meyerowitz 2002), and can be found in Rubin (1975), Rich (1980), and de Beauvoir (1952).

11.   Personal communication, 2003.

References

Altman, Meryl. 1990. "How Not to Do Things with Metaphors We Live By." *College English* 52(5): 495–506.

Ashman, Tia-Lynn. 2002. "The Role of Herbivores in the Evolution of Separate Sexes from Hermaphroditism." *Ecology* 83: 1175–84.

Austin, J. L. 1952. *How to Do Things with Words.* Cambridge, Mass.: Harvard University Press.

Beauvoir, Simone de. 1952. *The Second Sex.* New York: Alfred A. Knopf.

Bell, G. 1985. "On the Function of Flowers." *Proceedings of the Royal Society of London,* series B, 224: 223–65.

Birke, Lynda. 2000. *Feminism and the Biological Body.* New Brunswick, N.J.: Rutgers University Press.

Butler, Judith. 1990. *Gender Trouble.* New York: Routledge.

Cameron, Deborah. 1995. *Verbal Hygiene.* New York: Routledge.

Cameron, Deborah, and Don Kulick. 2003. *Language and Sexuality.* London: Cambridge University Press.

Case, Andrea L., and Spencer C. H. Barrett. 2001. "Ecological Differentiation of Combined and Separate Sexes of *Wurmbea dioica* (Colchicaceae) in Sympatry." *Ecology* 82: 2601–16.

Charlesworth, Deborah. 1989. "Allocation to Male and Female Function in Hermaphrodites, in Sexually Polymorphic Populations." *Journal of Theoretical Biology* 139: 327–42.

Darwin, Charles. 1877. *The Different Forms of Flowers on Plants of the Same Species.* London: Murray.

Delph, Lynda F. 1990. "The Evolution of Gender Dimorphism in New Zealand *Hebe* (Scrophulariaceae) Species." *Evolutionary Trends in Plants* 4(2): 85–97.

De Man, Paul. 1983. *Blindness and Insight: Essays in the Rhetoric of Contemporary Criticism.* 2nd ed. Minneapolis: University of Minnesota Press.

Dudle, Dana A., Pia Mutikainen, and Linda F. Delph. 2001. "Genetics of Sex Determination in the Gynodioecious Species *Lobelia siphilitica:* Evidence from Two Populations." *Heredity* 86: 265–76.

Eckhart, Vincent M. 1992. "Resource Compensation and the Evolution of Gynodioecy in *Phacelia linearis* (Hydrophyllaceae)." *Evolution* 46: 1313–28.

Fausto-Sterling, Anne. 2000. *Sexing the Body: Gender Politics and the Construction of Sexuality.* New York: Basic Books.

———. 2003. "Thinking Systematically about the Emergence of Gender." Paper presented at the Women's Sexualities Conference, Indiana University, November 13.

Gathorne-Hardy, Jonathan. 2000. *Sex the Measure of All Things: A Life of Alfred Kinsey.* Bloomington: Indiana University Press.

Goodeve, Thyrza Nichols, and Donna Haraway. 1999. *How Like a Leaf.* New York: Routledge.

Halliday, Michael Alexander Kirkwood, and James R. Martin. 1993. *Writing Science: Literacy and Discursive Power.* London: Taylor and Francis.

Haraway, Donna. 1991. *Simians, Cyborgs, and Women: The Reinvention of Nature.* New York: Routledge.

Harding, Sandra. 1986. "Gender and Science: Two Problematic Concepts." In *The Science Question in Feminism,* 30–63. Ithaca, N.Y.: Cornell University Press.

Keller, Evelyn Fox. 1983. *A Feeling for the Organism: The Life and Work of Barbara McClintock.* New York: W. H. Freeman.

———. 1985. *Reflections on Gender and Science.* New Haven, Conn.: Yale University Press.

———. 1995. *Refiguring Life: Metaphors of Twentieth-Century Biology.* New York: Columbia University Press.

Kinsey, Alfred, Wardell B. Pomeroy, and Clyde E. Martin. 1998. *Sexual Behavior in the Human Male.* Bloomington: Indiana University Press. (Orig. pub. 1948.)

Kinsman, Sharon. 2001. "Life, Sex, and Cells." In *Feminist Science Studies: A New Generation,* ed. Maralee Mayberry, Banu Subramaniam, and Lisa H. Weasel, 193–203. New York: Routledge.

Laqueur, Thomas. 1990. *Making Sex: Body and Gender from the Greeks to Freud.* Cambridge, Mass.: Harvard University Press.

Linnaeus, Carolus. 1957. *Species Plantarum.* Facsimile of the original 1753 edition, with an introduction by William T. Stearn. London: Ray Society.

Lloyd, David G. 1972. "Breeding Systems in *Cotula* L. (Compositae, Anthemideae). I. The Array of Monoclinous and Diclinous Systems." *New Phytologist* 71: 1181–94.

———. 1980. "Sexual Strategies in Plants III: A Quantitative Method for Describing the Gender of Plants." *New Zealand Journal of Botany* 18: 103–108.

Mayberry, Maralee, Banu Subramaniam, and Lisa Weasel, eds. 2001. *Feminist Science Studies: A New Generation.* New York: Routledge.

McCauley, David E., Matt S. Olson, Stacie N. Emery, and Douglas R. Taylor. 2000. "Population Structure Influences Sex Ratio Evolution in a Gynodioecious Plant." *American Naturalist* 155: 814–19.

Meyerowitz, Joanne. 2002. *How Sex Changed: A History of Transsexuality in the United States.* Cambridge, Mass.: Harvard University Press.

Moi, Toril. 1999. *What Is a Woman?* London: Oxford University Press.

Prosser, Jay. 1998. *Second Skins: The Body Narratives of Transsexuality.* New York: Columbia University Press.

Raven, Peter H., Ray F. Evert, and Susan E. Eichhorn. 1999. *Biology of Plants.* 6th ed. New York: W. H. Freeman.

Rich, Adrienne. 1980. "Compulsory Heterosexuality and Lesbian Existence." *Signs* 5(4): 631–60.

Rosser, Sue. 2002. "Twenty-five Years of NWSA: Have We Built the Two-Way Streets between Women's Studies and Women in Science and Technology?" *NWSA Journal* 14(1): 103–23.

Rubin, Gayle. 1975. "The Traffic in Women: Notes on the Political Economy of Sex." In *Toward an Anthropology of Women,* ed. Rayna R. Reiter, 157–210. New York: Monthly Review Press.

Sakai, Anne K., and Stephen G. Weller. 1999. "Gender and Sexual Dimorphism in Flowering Plants: A Review of Terminology, Biogeographic Patterns, Ecological Correlates, and Phylogenetic Approaches." In *Gender and Sexual Dimorphism in Flowering Plants,* ed. Monica A. Geber, Todd E. Dawson, and Lynda F. Delph, 1–32. Berlin: Springer-Verlag.

Sedgwick, Eve Kosofsky. 1990. *Epistemology of the Closet.* Berkeley: University of California Press.

Sutherland, Steve, and Lynda F. Delph. 1984. "On the Importance of Male Fitness in Plants: Patterns of Fruit-Set." *Ecology* 65(4): 1093–1104.

Thompson, E. P. 1968. *The Making of the English Working Class.* Harmondsworth: Penguin.

Thornhill, Randy, and Craig T. Palmer. 2000. *A Natural History of Rape: Biological Bases of Sexual Coercion.* Cambridge, Mass.: MIT Press.

Wittgenstein, Ludwig. 1953. *Philosophical Investigations.* Trans. G. E. M. Anscombe. Oxford: Blackwell.

Zuk, Marlene. 2002. *Sexual Selections: What We Can and Can't Learn about Sex from Animals.* Berkeley: University of California Press.

Part Four.   *Remedies*
          and Change

12 The Graduate Experience of Women in STEM and How It Could Be Improved

Anne J. MacLachlan

The changing position of women earning doctorates in science and engineering is described by the title of the National Research Council's 2001 report, *From Scarcity to Visibility* (Long 2001). It reflects the impact of thirty years of study, programs, and initiatives as women have grown from 8 percent of all Ph.D. recipients in science, technology, engineering, and mathematics (STEM) in 1966 to 39 percent in 2002 (NSF 2003, 38; Hoffer et al. 2003, 13). Yet women earning STEM Ph.D.s today are still largely white, as the growth in female Ph.D. attainment has not been paralleled by similar attainment among U.S. minorities, even though more earn Ph.D.s than thirty years ago. Among the growing but still low number of individuals from underrepresented groups earning STEM Ph.D.s, women remain virtually invisible. Indeed, of the 14,313 Ph.D.s awarded to United States citizens in 2002 in STEM fields, only 353 went to African American women, 103 to Chicana women, and 32 to Native American women (Hill 2003). Clearly barriers remain in STEM doctoral education for women of all ethnicities (Hollenshead et al. 1996).

This chapter examines the graduate school experience of an ethnically diverse and highly successful group of sixty-three women who earned their Ph.D.s in science and engineering fields from several University of California (UC) campuses between 1980 and 1990. They were admitted to some of the departments ranked most highly by the National Research Council (NRC), and despite the difficulties they may have had, succeeded in finishing their doctoral programs. These women of all ethnicities also overcame other obstacles in the form of sexism and racism. They all are bright and good at science, with a strong will to succeed, to obtain the skills necessary to move ahead, and to structure a professional life compatible with personal and community values. If individuals of this caliber had difficulties in getting through their graduate programs, we can be sure that others found these difficulties insurmountable and left the program. The problems the study participants had with their training should, therefore, be taken all the more seriously as they are the survivors.

There are two parts to this chapter. The first presents some of the findings of the qualitative study described below, which focused on the types of difficulty the diverse women participants experienced as they progressed through the graduate program, the role of their advisers, and the benefits, obstacles, and omissions they perceived in their training. The second part goes beyond the study to make a series of systemic recommendations about how these documented issues could be addressed to make graduate education more responsive to the needs of women and other underrepresented groups. It situates the study's findings in the wider literature on graduate education, and it uses the recommendations of the study participants as well as the many programs and practices I personally know to be successful, along with those documented in the literature, such as the programs described in *Preparing Future Faculty in the Sciences and Mathematics* (Pruitt-Logan, Gaff, and Jentoft 2002).

Description of the Study

The data analyzed here come from a four-year qualitative project entitled "A Longitudinal Study of Minority Ph.D.s from 1980 to 1990: Progress and Outcomes in Science and Engineering at the University of California during Graduate School and Professional Life." Included in the study were African Americans, Chicanos, and Native Americans, because these groups are critically underrepresented in higher education. They were augmented by Asian Americans and Hispanics with a matched white group. The match was made by selecting a white graduate student from the same lab or who had studied with the same adviser as any minority student in the study and who received her or his Ph.D. around the same time. Data were collected through telephone interviews generally lasting two hours using a standard questionnaire. Questions covered respondents' entire lives, from their family background and early schooling through graduate school and professional life. A total of 158 interviews were completed.

Of the 13,700 students who earned a STEM Ph.D. in the UC system between 1980 and 1990, 206 were members of underrepresented minorities (URM) (African American, Chicano/Chicana, Native American). The total number of women among the 206 is not known. On the Berkeley campus, the source of the majority of respondents, there were a total of 87 URM Ph.D.s in STEM, of whom twenty-three (26 percent) were women. The very small number of women Ph.D. recipients overall (even white women were only 22.6 percent of all whites) led to the inclusion of as many other women of color as could be located, so that Hispanic and Filipino women were also interviewed. Women were therefore overrepresented in the study as a whole. On the other hand, they were definitely underrepresented in their doctoral programs without respect to ethnicity, and were thus in a minority position, with all that this entails.

The Experience of Graduate School

This analysis focuses on the graduate experience of the 63 women in the study: 10 African Americans, 8 Asian Americans, 6 Chicanas, 9 Hispanics, 2 Native Americans, and 28 whites. As a group, these women did not differ significantly from male students completing STEM doctorates—entering graduate school in their early twenties, having earned a B.S. with good to excellent grades in the same field as their Ph.D. or a closely related one, and going through the program in a timely manner. It would be a grave mistake, however, to view them in only this way.

There were substantial differences among the women in family background and ethnic identity, differences which affected their experiences in graduate school. Chicanas and African American women came from relatively poor and uneducated families, while white and Asian American women came from relatively prosperous families with college degrees. Hispanic women were mostly foreign-born and came from well-educated families. The two Native American women in the study fit no pattern, as one had a father with a Ph.D. and the other had parents with only a high school education. They, along with the Chicanas and African American women, not only identified with their ethnic and cultural heritage, but were identified by it on occasion in unprofessional and unconstructive ways.

The majority of the women had been recognized as bright and capable at an early age, often by a grade school teacher. They also enjoyed very strong support systems provided by their families, including parents with virtually no education who sustained them through school. As undergraduates, the majority attended major research universities such as MIT and Cal Tech or good state comprehensives; a few attended leading liberal arts colleges. A small group attended historically Black colleges and universities (HBCUs), which made their transition to a majority institution more stressful. Their grade point averages were almost uniformly high, ranging from 3.0 to a perfect 4.0, with a heavy concentration above 3.5.

Before turning to the answers to specific questions from the study, a few general observations about graduate school are in order. How graduate school is experienced, the kind of mentoring one receives, the degree of induction into the discipline or profession, the extent to which one builds a personal network—all play a role in whether one remains in the program or leaves (Golde 1998, 55–59; Nettles 1990, 497). These factors also play a role in the choice of postdoctoral programs and subsequent career decisions. Succeeding at UC Berkeley and in other programs requires a student to develop relationships with various faculty advisers, establish personal support systems, and cultivate a general "toughness."

One question asked students what they wished they had known when they started their program. The number of different areas of vulnerability the study participants articulated is revealing.

One major theme in answers to this question was how much simply being a woman "affected absolutely everything." A woman coming from an undergraduate institution, where women were 50 percent of the students, wished she had known "how to interact with her fellow students better." She was the only female in her adviser's group, and the only female teaching assistant. As a result, she was "always on the outside in my colleagues' group." Another wished she had known "that women would interact differently from men in math. I should have formed a woman's group to study so there would be a safe environment."

Other comments in answer to this question covered almost the entire process of graduate training, from someone who wished she had known "more about what is an appropriate course of study," to another who wished she had known more about career options. Needing to be mentored and to ask for help, but not knowing how to do so, or that it was permissible to do so, was a recurring theme. Women also wished they had known about the importance of academic politics and competition as well as the need to be aggressive in asking questions and making sure their own work was recognized.

Specific lack of skills or knowledge was articulated more clearly in answer to the question "What training would you like to have received?" Thirteen women mentioned grant writing here, in addition to those who mentioned this as something they wished they had known. Nine mentioned techniques and specific training in field-related areas, while an additional seven in public health mentioned areas specific to their field. One woman mentioned the need for interdisciplinary research. The lack of guidance and the need for a mentor was mentioned by six. Career skills of various kinds were mentioned by twenty-four.

Twenty-five questions asked about advisers; twenty-one offered a choice of ranked answers, and four were open-ended. The responses to one question, "How would you rate your overall relationship with your adviser?" were used as a general measure of that relationship. This was only a rough measure, however, because the rankings of answers to more specific questions often were at odds with the general assessment. Low rankings on questions such as whether the adviser provided "teaching opportunities," "opportunities to present," "information on grant writing," "information on preparing articles for publication," "the ethics of science," or "information about lab management" were common in conjunction with a high overall ranking. Some individuals refused to answer such questions because it never occurred to them that their advisers should have taught them some of these things. Women in particular tended to blame themselves for not having received specific information and training from advisers.

Since study participants were women who got through the doctoral program, it is not surprising that on a 1 to 5 scale, with 1 the highest ranking, 72 percent of advisers received overall evaluation scores of 1 or 2. Typical comments for those giving high rankings included this from a Hispanic woman: "He was always a good role model, very supportive and gave me freedom." From a Chicana: "Gave me training that allowed me to succeed and be professional." And from a white woman: "Our relationship was phenomenal, he is very knowledgeable and willing to share his knowledge." An African American woman who ranked her adviser as a 3 still remarked, "I gained a lot of independence, self-reliance. This has been helpful in terms of initiating new activities and programs in my environment and to develop confidence in my ability." But even those who did rank advisers as a 1 or 2 made comments suggesting ambivalence, such as "He was great, but he wasn't there much, it was a problem for a couple of years . . . he didn't micro manage which I adored."

Thesis advisers were usually perceived as benign to actively supportive, their faults those of omission rather than commission. But these omissions were often very significant for the women concerned. Lack of direction, particularly in the first year, led to floundering in the program and lost time, and to initial poor choices of lab or adviser; later lack of advice led to failed experiments, unpublished results, and poor choice of employment or postdoctoral positions. Only one adviser out of this group was ranked 1 in all categories by all advisees.

Even women who rated an adviser overall as 1 almost never gave rankings of 1s in all of the more specific categories. Nearly a third also said that when they entered the program they were naive, did not know what to expect, and did not expect very much. Scores for specific faculty advisers' behaviors varied widely with no apparent pattern, but on this set of questions there were some notable low scores. These, plus the comments about what students would have liked to get from their advisers but did not, provide a clear picture of the elements left out of the students' training, especially such things as grant writing, article preparation, lab management, and the workings of the academic profession.[1]

Issues of Racism and Sexism

Problems with racism and sexism tended to originate with male student colleagues. One African American woman always felt under surveillance by fellow graduate students and stated, "Berkeley was my worst experience of racism ever," although few other women of color were so emphatic. In response to the question "Have you experienced discriminatory or racist behavior during graduate school?" 42 of the 63 women stated that they experienced some form of discrimination that was either racist, sexist, or both. Sometimes prejudice was perceived, but its form was hard to describe. While often only single events were mentioned, what was more clearly remembered was the ineffable sense of being different, being excluded in small ways, and sometimes a pervasive sense of discomfort (Ibarra 2001). There were also blatant cases of discrimination: two faculty members were known to always fail women in either the preliminary or the oral examination, and failed the two women who reported this.

Women's comments suggest that faculty were often unaware that they were treating women or persons of color differently. Yet the women reported such different treatment fairly extensively. Some women reported that an older white male faculty member likened them to a daughter (although one woman was very happy with this). Women also could sense that they were directed toward applied rather than theoretical areas, and when they pursued theory nonetheless, they encountered noncomprehension. Some sensed doubt about their overall ability. Many observed switches in behavior when they, as lone women, joined gatherings of men, whether in the lab or the hallway. No man would be congratulated on passing his qualifying exam by being kissed and having his ear bitten, as was one respondent. As discussed above, many also said that they did not receive much advice or mentoring, leading to unsuccessful experiments or poor choice of postdoctoral positions.

Women of color in the study reported faculty doubt about their abilities more often than white women, although they usually found more such doubt in their workplaces than in graduate school (see also Turner 2002). They commented on subtle changes in behavior suggesting they did not belong, that they were seen as "a" or still "the" minority, not as a student or a potential colleague. The women of color felt that they were not seen as themselves, as persons, or future scientists, but as "representatives of their race," and were scrutinized and judged on that basis. Degrees of difference were accentuated by language. If white women experienced "a thousand paper cuts" (Mason 2002), persons of color experienced "micro-aggressions" (Solorzano 1998). Is this deliberate? Even many of those affected by such behaviors often did not think so, but attributed the discrimination to "ignorance." Nonetheless, the accumulation of all of these negative experiences can form a substantial barrier to success for women of color in science and engineering graduate programs.

In summary, students expressed dissatisfaction in a large number of areas, despite the fact that all had completed their degrees. Women in graduate science programs, however, often may not know how their environment could be improved because they tend to arrive with fairly poor undergraduate experiences and low expectations. As one woman remarked, "MIT was heavily male, so it [Berkeley] wasn't hard for me." Difficulties articulated by women in the study covered a broad range of areas, but very frequently women saw themselves as the source of the problem(s) rather than extensively critiquing the system of training.

Although this is only an abbreviated presentation of the findings, it is sufficient to suggest that there are better ways of training diverse women for STEM careers. In addition to the lack of guidance, most women considered aspects of their intellectual development to be missing, along with a long list of professional development activities, from grant writing to publishing. Career advice, particularly advice on postdoctoral positions, was often lacking. In what follows, I discuss how to provide a better graduate experience for women, and all students.

Reconceiving Graduate Education

The areas of omission and hardship articulated by this diverse group of sixty-three women are similar to those examined in recent surveys of graduate students (Golde and Dore 2001; NAGPS 2000), as well as in many years of research on graduate education and in various studies by federal agencies. The particular situation of women in science has also been discussed for many years, although issues faced specifically by women of color rather less so. If the goal is to address these issues in order to make graduate education more effective, and especially to avoid the all too common negative and denigrating experiences in graduate school, then the first step is to work with faculty and staff in STEM to change the culture of departments.

Although there are STEM faculty actively engaged in making graduate education more accommodating to new kinds of students, and focused on preparing students for the complex scientific workplace, their efforts are usually not systemic. On the contrary, they often rest on the initiative of an individual scientist or department chair, and are randomly distributed throughout graduate programs in the United States. Faculty consciousness of their roles vis-à-vis graduate students and the doctoral program in their departments is all too often underdeveloped because they themselves may never have been exposed to alternative models of behavior and organization, or have given the issues much thought. This lack of consciousness can all too often be compounded by a belief that surviving a program without assistance, as the faculty commonly did (or thought they did) themselves, is actually part of the process (Katz and Hartnett 1976; MacLachlan 1996; Golde and Dore 2001). Hence faculty frequently resist changing either their own behavior or the system.

Arguments made to faculty that the low numbers of women and minority group members in so many graduate programs continue the historical legacy of exclusion and segregation largely affect only rhetoric. Faculty members unwilling to reconsider their role as graduate educators will deliver only politically correct bromides on the need for diversity. Many scientists are unconvinced that there is a national workforce crisis in STEM because the current system is not developing and utilizing the talent of U.S. women and minorities. The recent reports reiterating arguments made in the 1950s (Hollis 1950) that graduate training in science is inefficient with respect to time to degree and training for future employment appear to make little impression. Moreover, the fact that STEM graduate education has been critically if not extensively examined for the last thirty years (Katz and Hartnett 1976; Baird 1993; Bowen and Rudenstine 1992) has had little impact, as STEM faculty usually do not read this kind of literature. While the demand for better training for academic employment has generated a few *Preparing Future Faculty* programs in STEM, these are not at Research I institutions with top-ranked National Research Council (NRC) departments (Pruitt-Logan, Gaff, and Jentoft 2002).

If faculty think of graduate education as a boot camp, they will not be persuaded that this form of training wastes talent and shatters lives. All programs expect some measure of attrition, but there could be another way of doing things, even if attrition in STEM fields is substantially less than in others (Nerad and Miller 1996; Rapoport 1998). Because many do not "survive," the question also arises whether the traditional model of graduate education is an efficient use of departmental resources, since many are employed to partially educate those who leave.

Graduate education has become more than ever a complex process of socialization (Austin 2002). STEM faculty understand that they must induct graduate students into the practice of science, convey to them the values of research ethics and the nature of scientific inquiry, and teach them research methods. They understand less clearly the process of socializing students into a particular social and institutional system in which academic science is practiced. There are also cultural ramifications, since students from low educational backgrounds are often assumed to suffer from a lack of "human capital" (Nettles and Millet 1999), and are expected to learn the customs, language, and manners of the academy. This can be a painful and alienating process for the student. Such students tend to have difficulties in becoming socialized into both the graduate and the science system (Zelditich 1997). Faculty's heightened consciousness of these issues, as well as their acceptance of a broader range of learning styles (reflected in their approach to instructing students whose ethnic culture and social world is unknown to them) would make a substantial difference.

In considering the issues of diverse women in graduate school it is important to bear in mind that the pedagogical efficacy of graduate programs depends heavily on the interest and activity of faculty. Not only can the relationships faculty form with students as thesis directors or principal investigators make or break the student, general faculty engagement with graduate student training in the department can promote more effective socialization. One useful way to develop this awareness is to draw from the various activities developed for faculty training and education by those seventeen campuses which have received NSF ADVANCE Institutional Transformation Grants. The ultimate purpose of the grant program is to increase the number of diverse women on the STEM faculty by transforming departments and entire campuses. Generally the goals of programs funded by ADVANCE grants are the same as those of programs or graduate women generally; they are intended to make departments welcoming to and supportive of women, whether faculty or graduate students. However, in order to specifically address the issues articulated by women in the study reported in this chapter, interventions should focus on making faculty more effective graduate educators. The role of faculty in making graduate education successful for more students than it currently is should also be discussed, and language and behaviors that can exclude women and students of color from the mainstream of graduate education should be explained and demonstrated. Specific training in effective mentoring should be included—knowledge useful for working with undergraduates and junior faculty as well. This training would also encourage faculty to organize and participate in the various activities described below. Ideally such training would be mandatory for all faculty at all levels, whether new hires or long-tenured professors. While there are always faculty who support such work in any department, resistance can be expected from many more. For that reason such training has to be part of broader considerations about how a department wants to function, and must be based on strong leadership from the chair and the dean and supported by the top administration.

Direct external pressure on grant recipients by the large federal funding agencies, such as the National Science Foundation (NSF), the National Institutes of Health (NIH), and the Department of Energy (DOE), is highly desirable for fostering systemic change. Small steps have been taken in this direction by the structure of some grant programs, such as the NSF ADVANCE program and the Alliance for Graduate Education and the Professoriate, which provide the resources to increase the number of graduate students of color earning STEM Ph.D.s and going into the professoriate. Mandatory education components in large institutional grants are also helpful. On the whole, however, federal agencies need to be much more active (NSB 2003).

Faculty reconsideration of graduate training—however it is brought about—could lead departments to adapt the recommendations below to suit their individual disciplines and needs. These recommendations build on the graduate student study participants' comments and draw from models of such successful activities at the department level as can be identified. They are organized sequentially, following the path of the new student through a program, and are intended to provide the missing pieces which are more likely to make more students successful in their STEM graduate programs. Taken together, these programs reorganize the system of graduate education and remove obstacles which can hinder successful completion of a STEM graduate program. Since there are no adequate data on who leaves—and certainly many white men leave—such programs need to be for everyone, with special components for women and students of color. They focus on three main areas: (1) orientation to the department and its people, to the discipline, and to academic expectations; (2) mentoring to provide individual support, encouragement, knowledge, and advice, including advice on how to mentor others and build networks; and (3) socialization to the culture of the academy and the discipline, and advice on the steps beyond, which include training for life after the doctorate (professional development). In practice, these components are not so tidily separated. Moreover, the goals of these activities have been sought for at least thirty years, with only varying degrees of success. That is why the majority of faculty in any given department must be engaged for these activities to be successful.

Recommendations for a Successful Graduate Student Experience

Graduate student orientations should be distributed throughout the first and subsequent years, organized sequentially, with participation considered a part of the formal program. Incoming students in particular are often greatly overwhelmed with information as they try to learn about their new environment. Whether or not students participated in a campus recruitment visit, their initial contact with faculty and staff upon arrival should make them feel welcome. In order to be successful, students need information and guidance. They also need to be actively involved in their own success, and if they arrive without knowing what they want to get out of a doctoral program beyond just the degree, they should be encouraged throughout to think about what they want to learn, why they want to learn it, and in what kind of working environment they want to use their knowledge.

Just prior to term, the first orientation should be a two-day department re-treat to meet faculty and staff, get acquainted with peers, examine various labs and other campus facilities, and pair up with an older student for mentoring. In a well-organized presentation, faculty and some advanced students should discuss the structure of the academic program, the nature and timing of "mile-stones" such as the preliminary or qualifying exam, meeting learning expecta-tions, and what options are available to students (courses in this or other de-partments, lab rotations, available equipment, and opportunities to use off-site labs). As many of the department's faculty as possible should be present to in-troduce themselves and briefly discuss their current research projects, their current students' work, and the placement of former students. Faculty in inter-disciplinary areas who collaborate with others either on or off campus should speak to how to prepare for interdisciplinary work and describe opportuni-ties for such work in other departments and with other faculty. Interspersed throughout the day should be informal opportunities for new students to speak with faculty and the more advanced students present.

Built into the orientation should be breakout sessions so that all incoming graduate women students can meet with more advanced women and establish a formal mentoring relationship. If there is a department-sponsored women's organization, representatives should be present to describe their activities and invite all of the new women to participate. Women faculty should also be avail-able to meet the new female students and discuss resources available to women on campus, along with the resources of national organizations like the Associa-tion for Women in Science (AWIS) and any disciplinary-specific organization, such as the Women Chemists Committee of the American Chemical Society (ACS), which offers a great deal of valuable information for chemists in train-ing. At this meeting, too, it should be emphasized to the students that they need to develop a plan for their own education. Senior women should explain, from their more advanced perspective, how pitfalls can be avoided. A discussion of the real department climate would also be helpful to incoming women, along with suggestions of what to do if they encounter sexist or racist behavior. All of this can be done in a positive way so that students feel informed and pre-pared, but not frightened or discouraged.

At another time there should be a breakout session for all minority students, men and women, to discuss the same themes as in the women's and general ori-entations. Since it is all too likely there will be few or no minority faculty mem-bers in the department, this session should be coordinated with other STEM departments so minority students have the opportunity to meet one another and the faculty, minority or not, who are interested in their welfare. Since most leading graduate school faculty are white (Nelson 2004), most minority stu-dents in STEM will have to find a white adviser. If there are minority faculty in STEM on campus, they should attend; some might even be invited from neigh-boring institutions. More advanced minority students from several departments should also be invited and encouraged to become official student mentors for the new students. If there are minority student organizations on campus, their coordinators should be introduced and describe the goals and activities of the organizations. Advanced students or faculty could also discuss minority organizations such as the Society for Advancement of Chicanos and Native Americans in Science (SACNAS), the National Society of Black Physicists, the National Association of Mathematicians, and Mathematicians of the African Diaspora, and minority sections of larger organizations, such as the Women Chemists Committee of the American Chemical Society and the Minority Scientists Network of the American Association for the Advancement of Science (AAAS). The minority and the women's groups should continue to meet regularly both socially and to discuss graduate issues as they arise.

All of these initial programs need careful planning and should involve as many faculty members as possible. The goals are to communicate (1) valuable information about the department and how to succeed in it; (2) department policy, established by the chair or dean, that hostile, sexist, racist, unethical, or destructively competitive behaviors will not be tolerated (to be effective this policy must be backed up by actual behavior, not just hope); (3) that students are active participants in shaping their own education, and should think about their intellectual objectives and future employment; and (4) that obtaining information, advice, and mentoring from faculty requires action on both sides—faculty have to be accessible but students must ask for what they may need and are not receiving. Examples of such first-year orientation meetings are found in the neurosurgery department at the University of Iowa and in the astronomy department at UC Berkeley.

Meetings of the various groups of graduate students should be repeated during the first and subsequent years so that discussion of the topics moves along as the students do, and the students move in a cohort. The second meeting could discuss examinations, if one is administered in that particular department in the first year. The third could focus on developing a research topic, and so on, with meetings preceding each major event. Social events should also be part of the program so that students at different levels and students and faculty have a chance to get acquainted (Katz and Hartnett 1976; Nettles 1990).

A complement to a general orientation is an initial individual advising and evaluation session with an engaged faculty adviser. The incoming student and the adviser should discuss the student's background, scientific interests, and possible employment goals, and explore how the student may best navigate the program, covering classes, faculty members, and departments. The student thus receives a clear idea of what is expected of her and what she needs to do to progress. Although she will find other academic advisers, her progress should be monitored by this professor, who should remain available as a general program mentor. In the department of biomathematics at UCLA, this program's success is reflected in the fact that students finish within five years and generally go directly into employment, since a postdoctoral position is considered superfluous (personal communication from Professor Carol Newton).

Mentoring is a significant element in the success of women and persons of color in STEM—not to change the student, but to provide information and guidance as necessary—and is something which departments should develop (Frierson 1998; COSEPUP 1995). Mentors are usually made, not born, so it would be valuable to have ongoing programs on how to be a mentor for both students and faculty. Each party in the relationship needs to understand its parameters and how to make it successful. Students in turn can mentor undergraduate students as well as younger graduate students. There is a fine line, too, between providing advice and guidance and not allowing the mentee sufficient latitude. While many respondents in the study complained about the laissez-faire approach of some faculty, some used it to develop independently and more or less favored being left alone. It is also debatable whether mentors are or should be role models. Having a white man as a mentor may foster excellence in science, but not provide insight on either dealing with the sexist or racist elements of the environment (Turner and Myers 2000) or balancing child raising with academic life (Zelditich 1997). That is why women and minority group members want more faculty like them, even though matching students and mentors by social identity categories stereotypes such faculty and such individuals may not always be good role models. This is a significant issue, since graduate students often do not find the lives of their advisers worthy of aspiration (MacLachlan 2000). Consequently only a small percentage of Ph.D.s actively seek faculty positions at research universities.[2] Also, students need to be aware of potential "negative" mentoring—another reason for training students in mentoring—as this can undermine both student confidence and career aspirations (Wilson 1997).

First-year students would profit greatly from being exposed to the inner workings of subfields in their disciplines by having faculty offer seminars or talks about their own research areas, the significant problems within them, and how the problems are being addressed. Particularly important for the student is a discussion about what one needs to know to actively participate in the field and how one acquires this knowledge. Often students only discover in their third year or even later that they need to take a vital class or master some research technique. Acquiring such information early makes the path clearer and is likely to shorten the student's time to degree as a result.

Although many STEM students seek work after the Ph.D. outside of the academy, the usual assumption is that students will follow the path of the adviser and become faculty. The irony of this is that faculty advisers often do not prepare their students for work in the academy, and neither does the department. Being effective in an academic environment requires several distinct skills, few of which are currently deliberately taught in graduate school. The most obvious is teaching. How much is really taught about instruction in teaching assistant orientations or assigned pedagogy courses is an open question, since their content varies greatly, depending on the faculty in charge of such courses. The faculty themselves may or may not have ever received formal teacher training. Since many introductory and even upper-division courses in science and mathematics are taught by graduate students, often bad teaching (and attitudes) drives women and minorities, as well as majority men, out of science majors (Seymour and Hewitt 1997). Teaching can be highly discriminatory (Sandler, Silverberg, and Hall 1996) and thus it is critical that graduate students receive solid training to fulfill their student teaching obligations effectively and to maintain an inclusive environment in the classroom—to prepare them to be successful teachers after the Ph.D.

More than just relying on the current, rather haphazard structure, departments should offer seminars in teaching methods in which every student would have to participate. Several examples now exist, which can be accessed through the *Preparing Future Faculty* program, the Council of Graduate Schools (CGS), or the American Association of Colleges and Universities (AACU). Organization would undoubtedly vary, but students should have at least a semester-long course on teaching taught by a combination of professional pedagogy teachers and department faculty before being placed in the classroom. This implies that all students in a department, whether on an external fellowship or not, would be required to teach as a prerequisite for graduation. The logic behind this is that even if a student does not ultimately go into teaching, she will still be in a professional environment in which she needs to communicate her ideas clearly, listen effectively to what is being communicated to her, be able to work cooperatively with others, and be able to involve and motivate others (Richlin 1993). Effective teaching is based on these skills, and these also are the areas of expertise that industry employers would like to see in Ph.D.s (COSEPUP 1995; Poock 2001). For those who do become faculty, the knowledge of how to teach effectively in their field makes the transition to academic employment much easier.

Career development for graduate students, whether female or minority, requires much more than occasional workshops. To be an effective scientist one needs to learn many different skills. In an ideal world, the principal investigators (PIs) would make sure that, in the years a student works with them, these skills are imparted through both their active teaching and the student's careful observation. Unfortunately, in the judgment of the study participants, almost every PI fell short in some area, although very rarely in all areas. These shortcomings included all the things related to effective lab management, as well as grant writing, article writing and submission, and even effective training in the field. Advisers also often failed to support students in job hunting, networking, and developing specific job-related skills. Many students requested classes or more organized training in all of the skills they did not obtain. Consequently, in the ongoing orientations, all of these missing pieces should be discussed, with extra sessions taught by department or other experts in grant writing, publishing, and laboratory management. Fischer and Zigmond (1998, 39–40) noted the need for "explicit instruction in survival skills" like these and indicated that the University of Pittsburgh provides a program of instruction as well as training in honorary authorship, avoiding plagiarism, oral presentations, and much more, in eight seven-hour workshops, one each month. In all of this instruction, attention needs to be paid to ways in which women and minority group members can be poorly served and how to overcome these impediments.

At least a year before graduation the department should offer focused job search workshops which deal with the purpose of postdoctoral positions and how to go about selecting the one most advantageous to career development; how to obtain an academic job, beginning with the kind of position the student might be interested in; how to prepare job search materials and organize the search effectively, so as to get the most help from advisers; how to give a job seminar and prepare a teaching demonstration; how to be effective in an on-campus interview; how to evaluate an institution's suitability; and how to negotiate a job offer. Similar workshops should be offered for other forms of employment.

While the recommendations made in this chapter do not provide a detailed blueprint of what a department ought to do, they do provide a guide. How these recommendations would be actually implemented depends greatly on the configuration of each specific department. The overriding goal of all these activities is to make progress through graduate school intellectually informed and to provide substantial information about skills and future employment. However, the implementation of even some of these activities and programs would serve to make departments more collegial and collaborative for graduate students and faculty (Etzkowitz, Kemelgor, and Uzzi 2000).

The various remedies suggested here will not necessarily fit all fields and departments. The idea is to provide a broad picture of departmental activities that could be adapted for local circumstances, emphasizing critical elements which should always be present. Implicit in this discussion is the need for graduate education to transform itself from within to make it more successful for more graduate students. Departments and their faculty have to be the loci of this change, not external administrative entities such as graduate divisions, although they, and the higher administration, need to strongly encourage such efforts. Federal funding agencies can also help by requiring specific supportive and professional development activities for graduate students as part of the grant activity and requiring their assessment. Likewise, various professional organizations can make valuable contributions (Applegate 2002). These are forms of acknowledgment that the "standard model" of graduate education, still bearing traces of its nineteenth-century Germanic origins as a program of study for a male social elite, must change to accommodate the current, more democratic ideology of access based on merit. Women and minority group members must be full participants in every form of education in the country in order to acquire the tools for full participation in the professions. The current system of graduate education in science and engineering is still far from this goal.

Notes

My thanks to the Spencer Foundation, the UC Industry-University Cooperative Research Program, the UC Berkeley Graduate Division, and the Center for Studies in

Higher Education for their financial support of this project. My opinions, findings, and conclusions are my sole responsibility.

1. See similar findings in LaPidus (1998) and Nettles and Millet (1999).
2. See Ivie, Stowe, and Czujko (2001), who report that of the 150 Black academic physicists, two-thirds teach at a historically Black college or university.

References

Applegate, James L. 2002. *Engaged Graduate Education: Seeing with New Eyes.* Washington, D.C.: Association of American Colleges and Universities.

Austin, Ann E. 2002. "Preparing the Next Generation of Faculty: Graduate School as Socialization to the Academic Career." *Journal of Higher Education* 73(1): 94–122.

Baird, Leonard L. 1993. "Studying Graduate Student Retention and Degree Attainment: Resources for Researchers." *New Directions for Institutional Research* 80: 81–90.

Bowen, William G., and Neil L. Rudenstine. 1992. *In Pursuit of the Ph.D.* Princeton, N.J.: Princeton University Press.

COSEPUP (Committee on Science, Engineering, and Public Policy). 1995. *Reshaping the Graduate Education of Scientists and Engineers.* Washington, D.C.: National Academy Press.

Etzkowitz, Henry, Carol Kemelgor, and Brian Uzzi. 2000. *Athena Unbound: The Advancement of Women in Science and Technology.* New York: Cambridge University Press.

Fischer, Beth A., and Michael J. Zigmond. 1998. "Survival Skills for Graduate School and Beyond." In *The Experience of Being in Graduate School: An Exploration,* ed. Melissa S. Anderson, 29–40. San Francisco: Jossey-Bass.

Frierson, Henry T., Jr., ed. 1998. *Examining Protégé-Mentor Experiences.* Stamford, Conn.: JAI.

Golde, Chris M. 1998. "Beginning Graduate School: Explaining First-Year Doctoral Attrition." In *The Experience of Being in Graduate School: An Exploration,* ed. Melissa S. Anderson, 55–64. San Francisco: Jossey-Bass.

Golde, Chris M., and Timothy M. Dore. 2001. *At Cross Purposes: What the Experiences of Today's Doctoral Students Reveal about Doctoral Education.* Report prepared for the Pew Charitable Trusts, Philadelphia, Penn. http://www.phd-survey.org. Accessed February 16, 2005.

Hill, Susan. 2003. *Science and Engineering Doctorate Awards: 2002.* Arlington, Va.: National Science Foundation, Division of Science Resource Statistics.

Hoffer, T. B., et al. 2003. *Doctorate Recipients from United States Universities: Summary Report 2001.* Chicago: National Opinion Research Center.

Hollenshead, Carol S., et al. 1996. "The Graduate Experience in the Sciences and Engineering: Rethinking a Gendered Institution." In *The Equity Equation: Fostering the Advancement of Women in the Sciences, Mathematics, and Engineering,* ed. Cinda-Sue Davis et al., 122–62. San Francisco: Jossey-Bass.

Hollis, Ernst V. 1950. "Graduate School." In *Encyclopedia of Educational Research,* ed. W. S. Monroe, 510–19. New York: Macmillan.

Ibarra, Robert A. 2001. *Beyond Affirmative Action: Reframing the Context of Higher Education.* Madison: University of Wisconsin Press.

Ivie, Rachel, Katie Stowe, and Roman Czujko. 2001. *2000 Physics Academic Workforce Report.* College Park, Md.: American Institute of Physics.

Katz, Joseph, and Rodney T. Hartnett. 1976. *Scholars in the Making: The Development of Graduate and Professional Students.* Cambridge, Mass.: Ballinger.

LaPidus, Jules B. 1998. "If We Want Things to Stay as They Are, Things Will Have to Change." In *The Experience of Being in Graduate School: An Exploration,* ed. Melissa S. Anderson, 95–102. San Francisco: Jossey-Bass.

Long, J. Scott, ed. 2001. *From Scarcity to Visibility: Gender Differences in the Careers of Doctoral Scientists and Engineers.* Washington, D.C.: National Academy Press.

MacLachlan, Anne J. 1996. *The Graduate Experience of Women and Minorities at UC Berkeley, 1980–1989.* Berkeley, Calif.: Center for Studies in Higher Education.

———. 2000. *The Impact of the Novartis Agreement on the Graduate Students of Plant Microbial Biology.* Report to the Vice Chancellor. Berkeley, Calif.: Center for Studies in Higher Education.

Mason, Mary Ann. 2002. "Do Babies Matter? The Effect of Family Formation on the Lifelong Careers of Academic Men and Women." *Academe* 88(6): 21–27.

NAGPS (National Association of Graduate-Professional Students). 2000. *The 2000 National Doctoral Program Survey.* http://survey.nagps.org. Accessed February 16, 2005.

Nelson, Donna J. 2004. "Congressional Report on Diversity among Science University Faculty." http://www.now.org/issues/diverse/diversity-report.pdf. Accessed June 12, 2003.

Nerad, Maresi, and Deborah S. Miller. 1996. "Increasing Student Retention in Graduate and Professional Programs." In *Assessing Graduate and Professional Education: Current Realities, Future Prospects,* ed. Jennifer Grant Haworth, 61–76. San Francisco: Jossey-Bass.

Nettles, Michael T. 1990. "Success in Doctoral Programs: Experiences of Minority and White Students." *American Journal of Education* 27(3): 494–522.

Nettles, Michael T., and Catherine M. Millet. 1999. "The Human Capital Liabilities of Underrepresented Minorities in Pursuit of Science, Mathematics, and Engineering Doctoral Degrees." *Making Strides* 1(2): 11–14.

NSB (National Science Board). 2003. *The Science and Engineering Workforce: Realizing America's Potential.* Arlington, Va.: National Science Foundation (NSB 03-69).

NSF (National Science Foundation). Division of Science Resources Statistics. 2003. *Women, Minorities, and Persons with Disabilities in Science and Engineering: 2002.* Arlington, Va.: National Science Foundation (NSF 03-312). http://www.nsf.gov/sbe/srs/nsf03312. Accessed February 16, 2005.

Poock, M. C. 2001. "A Model for Integrating Professional Development in Graduate Education." *College Student Journal* 35(3): 345–52.

Pruitt-Logan, Anne S., Jerry G. Gaff, and Joyce E. Jentoft. 2002. *Preparing Future Faculty in the Sciences and Mathematics: A Guide for Change.* Washington, D.C.: Council of Graduate Schools, Association of American Colleges and Universities.

Rapoport, Alan I. 1998. *Summary of Workshop on Graduate Student Attrition.* Arlington, Va.: Division of Science Resources Studies, Directorate for Social, Behavioral, and Economic Sciences, National Science Foundation.

Richlin, Laurie, ed. 1993. *Preparing Faculty for the New Conceptions of Scholarship.* New Directions in Teaching and Learning 54. San Francisco: Jossey-Bass.

Sandler, Bernice R., Lisa A. Silverberg, and Roberta Hall. 1996. *The Chilly Classroom Climate: A Guide to Improve the Education of Women.* Washington, D.C.: National Association for Women in Education.

Seymour, Elaine, and Nancy M. Hewitt. 1997. *Talking about Leaving: Why Undergraduates Leave the Sciences.* Boulder, Colo.: Westview.

Solorzano, Daniel G. 1998. "Critical Race Theory, Race and Gender Microaggressions, and the Experience of Chicana and Chicano Scholars." *International Journal of Qualitative Studies in Education* 11(1): 121–36.

Turner, Caroline S. V. 2002. "Women of Color in Academe: Living with Multiple Marginality." *Journal of Higher Education* 73(1): 74–93.

Turner, Caroline S. V., and Samuel L. Myers. 2000. *Faculty of Color in Academe: Bittersweet Success.* Boston, Mass.: Allyn and Bacon.

Wilson, Reginald. 1997. "Negative Mentoring: An Examination of the Phenomenon as It Affects Minority Students." In *Diversity in Higher Education,* ed. Henry Frierson, 177–86. Greenwich, Conn.: JAI.

Zelditch, Morris. 1997. "Mentor Roles in Graduate Studies." In *Diversity in Higher Education,* ed. Henry Frierson, 23–37. Greenwich, Conn.: JAI.

Websites of Organizations Mentioned

American Association for the Advancement of Science, Minority Scientists Network. http://nextwave.sciencemag.org/miscinet/

American Chemical Society, Women Chemists Committee. http://membership. acs.org/w/wcc

Association for Women in Science. http://www.awis.org/

Diversity information, Electrical Engineering and Computer Science Department. University of California, Berkeley. http://www.eecs.berkeley.edu/Programs/ grad/Diversity/diversity.html

Mathematicians of the African Diaspora. http://www.math.buffalo.edu/mad/ index.html

National Association of Graduate-Professional Students. http://www.nagps2.org

National Association of Mathematicians. http://www.math.buffalo.edu/mad/NAM/ index.html

National Science Foundation, ADVANCE Institutional Transformation Grants. http://www.nsf.gov/funding/pgm_summ.jsp?pims_id=5383

National Science Foundation, Alliances for Graduate Education and the Professoriate. http://www.nsf.gov/funding/pgm_summ.jsp?pims_id=5474

National Society of Black Physicists. http://nsbp.org

Preparing Future Faculty program. http://www.preparing-faculty.org

Re-envisioning the Ph.D. project. http://www.grad.washington.edu/envision/

Society for Advancement of Chicanos and Native Americans in Science. http://www.sacnas.org/

13 How Can Women and Students of Color Come to Belong in Graduate Mathematics?

Abbe H. Herzig

The decreasing proportions of women and students of some racial and ethnic groups in most science, technology, engineering, and mathematics (STEM) disciplines as they progress to higher educational and professional levels has been the subject of extensive research in recent years. Despite this body of research, little work has been done to examine the *processes* by which diverse students choose and leave STEM disciplines, and even fewer specific guidelines have been advanced for what can be done to achieve a more equitable post-secondary educational environment. In this chapter, I attempt to do just that, by building a framework for understanding the retention of women and students of color in graduate STEM study with the hope that this framework can move faculty, administrators, and other interested parties closer to understanding the actions they can take to recruit and retain a more diverse group of scientists, mathematicians, and engineers.

The analysis in this chapter is focused on mathematics, and is supported by data from interviews conducted in 1999–2000 with twenty-three continuing and attrited graduate students and twenty-one faculty in one highly ranked mathematics department at a Research I university. Faculty participants represented most major research areas in the department, were all full professors, had been in the department for most of their careers, and included some who teach first-year graduate courses. Interviews explored their "mathematical autobiographies," experiences as mathematicians, and roles as professors and advisers. Graduate student participants had completed between one and more than six years of study at the time of their interviews. Some had left or were about to leave the doctoral program, some had previously considered leaving the program but had decided to persist, and some were planning to or about to complete their Ph.D.s. Interviews with graduate students explored their "mathematical autobiographies," experiences in undergraduate and graduate school, and decisions about continuing or leaving.[1]

Previous works propose that building students' sense of *belonging* within mathematics is a critical feature of an equitable K–12 education (Allexsaht-

Snider and Hart 2001; National Council of Teachers of Mathematics 2000; Ladson-Billings 1997; Tate 1995). Martha Allexsaht-Snider and Laurie Hart (2001) argue that when structural aspects of schools, beliefs about students and the learning of mathematics, and classroom processes are aligned in a way that facilitates diverse students' sense of belonging in mathematics, then we are more likely to achieve the goal of "mathematics for all" so often cited in reform and policy documents. This chapter tries to extend these ideas to the post-secondary level, and in particular to postgraduate mathematics, arguing that building a sense of belonging is a critical feature of an equitable mathematics education at these advanced levels as well.

While all graduate students face some obstacles to developing a sense of belonging in mathematics, women and students of color face additional obstacles. The premise here is that in order to recruit and retain more women and students of color in mathematics (and, by analogy, into other STEM disciplines), more attention needs to be paid to eliminating these obstacles and to building more paths by which students can come to feel that they belong in mathematics.

Belonging in Graduate Mathematics

In describing aspects of classroom processes and teaching practices that are critical for equitable education in K–12 mathematics, Allexsaht-Snider and Hart (2001, 97) define *belonging* as "the extent to which each student *senses* that she or he belongs as 'an important and active participant in all aspects of the learning process' (Ames 1992, p. 263) in mathematics. This sense of belonging can be fostered by many different aspects of classroom processes."

As in K–12 and at the undergraduate level, a sense of belonging at the graduate level can and must be fostered by aspects of classroom processes, but also by other important components of the educational experience in and out of class. Jean Lave and Etienne Wenger (1991) describe the work of students as being peripheral to some "community of practice," and argue that student learning necessarily takes place through participation in that community of practice. The nature of that participation determines students' opportunities to learn, and learning is intertwined with, and inseparable from, the social functioning of the people and the activities in which they engage (Boaler 2002; Wenger 1998). This is consistent with Vincent Tinto's (1993) argument that doctoral student persistence is a function of both social and academic integration within the communities of the local department or program. Tinto argues that "social membership within one's program becomes part and parcel of academic membership, and social interaction with one's peers and faculty becomes closely linked not only to one's intellectual development, but also to the development of important skills required for doctoral completion" (1993, 232). That is, doctoral students need to do more than just learn mathematics; they need to learn to participate in social and cultural practices.

I extend Allexsaht-Snider and Hart's (2001) work to define belonging for a graduate student as her[2] sense that she is an important and active participant in both the academic and social communities of her department and program. Important *academic* influences on belonging include faculty beliefs about teaching and learning, such as beliefs about who can do mathematics, and both faculty and student beliefs about mathematics and mathematics epistemology. Important *social* influences on belonging include students' relationships with faculty both in and out of class, encompassing teaching, mentoring and advising, and informal interactions and relationships, as well as students' relationships with other students.

Graduate students represent a different population than K–12 or undergraduate mathematics students. While fostering the success of K–12 students includes helping them form an identity that includes mathematics (Martin 2000; Boaler 2002), doctoral students already have developed identities as mathematics students, as evidenced by their enrollment in a graduate mathematics program. In this sense, graduate students arrive in graduate school already with some degree of belonging in mathematics, which has led them to pursue an advanced degree. Rather than *building* these students' belonging within the discipline, the challenge is to *cultivate* and *develop* the belonging they already have when they begin graduate school; if graduate school acts to distance students from mathematics or to decrease their belonging with the discipline (Herzig 2004), then something is wrong. While agency on the part of individual students is clearly important, we need to move beyond developing strategies to "fix" women and students of color (Campbell 1995), to building an educational context that is accessible for all students. Consequently, the focus of this chapter is the *retention* of graduate students, rather than their *persistence,* as the former implies action on the part of the faculty, while the latter implies action on the part of students. It discusses those institutional or structural aspects of the educational environment that are most clearly influenced by faculty, rather than the behavior of individual graduate students.

Coming to Belong in Mathematics

The idea of belonging is related to several other theoretical perspectives in education: integration (Lovitts 2001; Tinto 1993), enculturation and identity (Martin 2000), and participation in a community of practice (Lave and Wenger 1991; Wenger 1998). It is the last of these that will be most helpful here, as it describes a *process* by which students move to fuller participation. Developing a sense of belonging can be conceptualized as a process by which students initially participate in "peripheral" ways in the activity of the community of practice. Through that participation, their skills and knowledge increase, and they move toward greater participation (Lave and Wenger 1991) and an enhanced sense of belonging.

Peter Wiles (1999) identifies two distinct communities of practice in which mathematics graduate students participate: the course-taking community and the research community. For their first several years in graduate school, most of students' activity centers on coursework and preparation for qualifying examinations (and, for some, working as teaching assistants) (National Research Council 1992). Their activity at the course-taking stage is far from the activity of practicing mathematicians; although the work of mathematicians involves doing research, often collaboratively, graduate students' early experiences have little to do with research and are often individual. Once students pass their qualifying exams and courses, they begin to have access to the research community. This is one critical difference between mathematics and many other STEM disciplines, since early, authentic research experiences remain rare in most mathematics doctoral programs. The isolation of doctoral students from authentic mathematical practice represents one significant obstacle they have to overcome in order to develop a sense of belonging in mathematics.

Across a variety of disciplines, students' integration in their programs has been identified as an important predictor of their progress toward the doctoral degree; conversely, students who are not well integrated have been more likely to leave graduate school without the degree (Lovitts 2001; Girves and Wemmerus 1988; Tinto 1993; Nerad and Cerny 1993; Herzig 2002). But students are members of many communities of practice, such as those of work, family, and home community. To some extent, retention depends on whether they are able to successfully negotiate the competing demands of these different communities, or, alternatively, the way that these other communities interfere with or enhance students' participation and belonging in their programs (Tinto 1993). This model of interference with participation may help explain the myriad of personal, institutional, and disciplinary factors that have been previously implicated in doctoral student attrition (Herzig 2002). Designing flexible programs and services to allow students to balance these competing demands is one way in which policymakers can have a significant impact on recruiting and retaining women and students of color. For example, African American women were particularly successful in one doctoral program that made a commitment to "accommodating the busy professional and personal lives of the women, many of whom are working mothers" (Manzo 1994); students and graduates of the program reported that such flexibility was critical in their persistence in the program. Other campus resources, such as affordable housing, child-care, and convenient transportation, have also been associated with lower attrition and shorter time to degree (Nerad and Cerny 1993).

Combining these ideas of participation and inclusiveness generates a model in which experiences that enhance participation—both socially and academically —lead to increased belonging, which increases the likelihood of retention. Experiences that inhibit social and academic participation lead to decreased belonging, which decreases the likelihood of retention. If graduate education is conceived and structured in a way that facilitates students' academic and social participation and cultivates their sense of belonging, then more students are likely to succeed. As will become clear in the subsequent sections, graduate education in mathematics is not structured this way; instead, it comprises a series of explicit obstacles to participation and belonging. Many of these obstacles are artificial, in the sense that they are not related in meaningful ways to students' potential for participating effectively in mathematical activity, and, unfortunately, many of these barriers are more formidable for women and students of color.

Academic Influences and Obstacles to Belonging in Graduate Mathematics

Beliefs about Teaching and Learning Mathematics

Two interleaved assumptions are often made about mathematics: that mathematics is a very difficult field of study, and that only some people have the talent required to be successful at mathematics (Love 2002). Most of the mathematicians I interviewed, like the one quoted here, expressed the belief that mathematics is harder than other fields:

> In mathematics you really have to do something which may in fact be impossible and you don't know in advance whether it is or isn't impossible. . . . In that sense it's hard. You have to come up with really new ideas. . . . I don't really think that in most areas to write a Ph.D. thesis you need a really new idea. You have to do some new work that hasn't been done before, but I don't think you really need a new idea. That, I think, is harder.

It may be this belief in the difficulty of mathematics that encourages a belief that it takes a special kind of talent to succeed in mathematics (Love 2002; Allexsaht-Snider and Hart 2001; Herzig 2002). In contrast, it is important to note that research has consistently reported that students who leave STEM do not perform less well on standard measures of achievement; there is no evidence that women and students of color leave STEM because of any difficulty they encounter in performing at the same level as those who persist (Bair and Haworth 1999; Zwick 1991; Bowen and Rudenstine 1992; Lovitts 2001).

Unfortunately, making ability a salient feature of an educational environment interferes with students' motivation to learn, their use of effective learning strategies, and their engagement with the content of the curriculum (Ames 1992). Having extensively reviewed the literature on students' motivation to learn, Carole Ames concludes that "[a] sense of 'I belong here' . . . is an identification with the purposes of schooling, and *this self-perception ought to be more easily attained under those conditions in which the focus is not on the adequacy of one's abilities*" (1992, 263; italics added). Furthermore, when the focus of instruction is on correctness, absence of errors, and success relative to public norms—as it often is in advanced mathematics—students are oriented *away* from learning and toward producing those things that they believe are expected of them. When tasks are meaningful, support the development and use of effective learning strategies, and give students opportunities to develop responsibility and independence, then students become actively engaged, expend sustained effort, focus on learning (instead of performing), and use effective learning strategies (Ames 1992). William Tate (1995) and Gloria Ladson-Billings

(1997) argue much the same point in the case of culturally relevant pedagogy for African American students, in which tasks that are meaningful and related to the lives of students engage them in high-level mathematical problem solving and reasoning. Tate goes on to argue that when African American students are forced to reason about mathematics in "standardized," school-accepted ways, surrounded by white middle-class students, they are distanced from mathematics rather than being engaged with it.

Not only does a belief that talent is an important predictor of success interfere with student belonging, it also removes the responsibility for instruction from the teachers. This was precisely the reaction in the mathematics doctoral program I studied (Herzig 2002), where faculty belief in the importance of talent led them to virtually ignore doctoral students in their first several years of the program, describing instruction as providing an opportunity for students to discover whether (or prove that) they possessed that talent. Some of these faculty felt that they were doing students a favor by helping them avoid wasting time if they were unlikely to make it. In this way, the focus of instruction moved away from fostering the learning and development of mathematicians-in-training to "weeding out" students who were "wasting their time."

The question of who survives this harsh weeding out is still an open one: are the survivors really those who are best able to make contributions to mathematical thought, or are they the ones who possess the cultural knowledge that allows them to fit the social norm of a mathematics graduate student? According to Tinto,

> Attainment of [doctoral] candidacy can be viewed as a form of social initiation into a group whose members have a vested interest in maintaining the norms of that group. . . . [I]t can also be argued that successful completion of a doctoral degree calls for the successful performance of a social role called "graduate student." (1993, 255)

Part of that social role is talking like a full participant (Lave and Wenger 1991). A student needs to master the various forms of discourses that appropriately demonstrate her ability to "think mathematically" and allow her to be accepted as a legitimate member of the mathematics community (Gerholm 1990). The forms of communication used by some women and students of color may make it particularly difficult for their mathematical communications to be accepted (Rosser 1995; Orr 1997).

It therefore seems that students who survive in mathematics do so not simply because they master the intellectual content of mathematics, but because they are able to adopt other aspects of the sociocultural practices of mathematicians. The emphasis on talent may lead to a self-selection of those students who are best able to adopt those practices. As one prominent mathematician noted,

> For people who view the profession as a kind of priesthood, it is appealing to reduce numbers by keeping out all but the most worthy. However, there might be several negative consequences to such an approach. First, there would be the terrible human waste of labeling a large group of our most talented people as fail-

ures and choking them out. . . . Second, while Darwinian selection appeals to many mathematicians as a fair way to choose who succeeds, the playing field is often not as level as many would like to believe. In many cases, it's as though someone taught some of the animals how to use weapons and then accepted the outcome of which animals survived as having been dictated by nature. (Douglas 1997, 43)

Students who leave mathematics may be those who are unable or unwilling—or unaccepted as they try—to assume the social and cultural roles considered appropriate for mathematicians. Those responsible for graduate mathematics education need to attend to mechanisms that can build a more flexible, accepting educational context in which a more diverse range of students *can* participate and come to belong in mathematics.

Beliefs about Mathematics

Leone Burton (1995, 220–21) proposed a definition of knowing in mathematics based on five categories: its person- and cultural/social-relatedness; the aesthetics of mathematical thinking it invokes; its nurturing of intuition and insight; its recognition and celebration of different approaches, particularly in styles of thinking; and the globality of its applications. Interviews with practicing mathematicians confirm that these features are important to their work (Burton 1999; Henrion 1997; Herzig 2002), and yet these features play a minor role in post-secondary mathematics instruction.

Social Relatedness and Globality of Applications

Current research-based notions of what it means to understand mathematics indicate not that understanding is something someone either has or does not have, but rather that understanding grows and deepens as the learner develops a growing web of connections among related ideas. By helping learners to connect mathematics to things they already know, both in and out of mathematics, teachers foster the growth of students' understanding (Hiebert et al. 1997). The mathematicians interviewed by Burton (1999) said that they knew an idea worked when it could be verified by connections with other ideas, either in other areas of mathematics or in "real-world" applications. These connections can and should be built into the instruction of graduate students, to foster students' understanding of important mathematical ideas and to help them develop a sense of belonging within the discipline.

Mathematics is often taught in highly abstracted ways, with little or no explicit connection to other mathematical ideas, ideas outside of mathematics, or the mathematical "big picture." Betty Johnston (1995) argues that abstraction in mathematics is a masculine artifact, a way of denying the social nature of mathematics, and is a consequence of modern industrial society's tendency to separate things into manageable pieces, alienated from their context. Some au-

thors have reported that women are more interested in relationships and interaction among ideas than are men, who may be more content to examine information out of context (Belenky et al. 1986; Rosser 1995). Although graduate
students have described being attracted to mathematics because of its abstract
nature (Becker 1984), some women have reported disillusionment with how far
removed their mathematical studies were from meaningful or relevant questions (Stage and Maple 1996; Herzig 2004).

Postgraduate instructors in STEM need to help students learn by developing
connections among important ideas, both within and among disciplines. Questions such as "Why is this an interesting result?" "How does this help us?" "How
does this relate to ideas in other areas of mathematics? to applications outside
of mathematics?" are integral parts of mathematical thinking and reasoning,
and should become integral parts of mathematics instruction at all levels. This
approach would both enhance the development of all students' understanding
and encourage more women and students of color to choose and stay in mathematics.

The Aesthetics of Mathematical Knowing

Mathematicians often talk about the affective nature of their work. The
mathematicians I interviewed described their work as exciting and said they
appreciated the aesthetics of mathematics, freely using words like "beauty,"
"pleasure," "delightful," and "pretty." However, like Burton, I found that "My
participants, as teachers, are not exploiting their experiences. Nor are most of
them giving learners a sense of the fun, excitement, challenge which holds them
in the discipline" (1999, 139). Burton's conversations with mathematicians indicated "how little they attempted to convey the struggle and the pleasure which
they had described to me of doing mathematics" (1999, 140). Similarly, most
of the mathematicians I interviewed described their goals for teaching graduate courses in terms of communicating the main ideas of the subject, rather
than sharing their excitement about or vision of mathematics. The students
described wanting to know more about how their professors thought about
mathematics, how they approached solving problems, and how their work fit
into the broader mathematical landscape, like this female graduate student:

> I've never had a conversation of that kind of any depth with any of the
> professors at this place so I really have no idea of what anyone here feels. . . .
> I perceive they're just spewing material at me. . . . I can't perceive the feel-
> ings behind it.

Sharing the enthusiasm, excitement, and joy which makes mathematics meaningful to mathematicians can only help students develop connections of their
own with mathematics; leaving these aesthetics out of the classroom makes
mathematics appear to be cold, abstract, and sterile, which at best fails to invite

students into mathematics, and at worst establishes a serious obstacle to their sense of belonging in the classroom.

Nurturing Intuition and Insight

While mathematicians rely on intuition, insight, creativity, and trial and error, most instruction presents mathematics as a sequence of deductive proofs of clearly stated theorems. As one female graduate student explained,

> He'd come in and he would race through the stuff on the board and we would furiously copy down what he was doing and it seemed like just streams and streams of words, signifying nothing. . . . No indication of what on earth was going on, not much in terms of why we were doing what we were doing or where this was going. Just, "here's a lemma," "here's the proof," "here's another lemma," "here's the proof," "here's a theorem." Very little motivation, and I think I didn't see the whole big picture.

This mode of presentation skips other important stages and methods of reasoning, often without regard for the intuitive or other processes that gave rise to the result or that give it meaning. Several graduate students described ways they had reached out to faculty to try to discuss the deeper mathematical ideas and connections behind the results presented in class; almost unanimously, they reported having had these efforts rebuffed.

This type of presentation, although somewhat standard, has several important drawbacks. First, it obscures from students' view how mathematicians think about problems, and fails to mentor them in mathematical thinking. Second, it makes it difficult for students to discern the meaning of results, the important ideas, and their connections with other ideas. Third, as discussed earlier, women have reported leaving STEM when connections and the meaning of results are not apparent (Rosser 1995; Sonnert and Holton 1995; Stage and Maple 1996); this pedagogical approach would then be an impediment to retaining women. And finally, by teaching mathematics so differently from how they actually do mathematics, mathematicians are denying students the opportunity to see how *people* do mathematics, and by removing this personal component from instruction, they actually distance students from the field rather than enhancing their belonging.

Social Influences and Obstacles to Belonging in Graduate Mathematics

> I had one female professor in my time here and it was so much easier to get interested because I could imagine myself being her. . . . The math department seemed so much like an old white guys' club, and I didn't really see that I had a place in the old white guys' club.

Faculty-Student Relationships, in and out of Class

Encouragement and moral support from mentors play important roles in students' decisions to enroll and persist in graduate studies in mathematics (Manzo 1994; Hollenshead, Younce, and Wenzel 1994; Stage and Maple 1996; National Research Council 1992; Cooper 2000; Carlson 1999; Herzig 2002). Faculty members help socialize students to understand the norms and practices of the department and discipline (Cooper 2000; Etzkowitz, Kemelgor, and Uzzi 2000; Gerholm 1990). Unfortunately, such encouragement and mentoring can be the exception, rather than the rule, with graduate students complaining that they have few substantive relationships with faculty (Herzig 2002).

Women students in STEM receive less mentoring from male faculty than do men students (Hollenshead, Younce, and Wenzel 1994; Berg and Ferber 1983; Sonnert and Holton 1995; Etzkowitz, Kemelgor, and Uzzi 2000).[3] Further, it has been suggested that faculty tend to mentor students of the same sex category as themselves (Reskin, Koretz, and Francis 1996; Berg and Ferber 1983); unfortunately, in the fall of 2002, only 13 percent of full-time doctoral mathematics faculty were women while 31 percent of full-time mathematics graduate students were women (Kirkman, Maxwell, and Priestly 2003). Women in STEM have faced discrimination in finding mentors, ranging from professors who would not take on women students or did not seem to tap into their professional networks as vigorously for their women students as they did for men (Sonnert and Holton 1995), to blatantly sexist behavior from faculty, including unwanted sexual advances, tolerance of public sexist comments, and public statements that women are not as smart, dedicated, or talented as men (Committee on the Participation of Women 2003). Women have also been excluded from the informal social networks of their laboratories or departments, treated as "invisible" or otherwise had their contributions marginalized, and been stereotyped as less capable and competitive than men (Becker 1990; Etzkowitz, Kemelgor, and Uzzi 2000; Sonnert and Holton 1995; Stage and Maple 1996). As a result, they may not be taken seriously by faculty, as this mathematics professor explained:

> I think women can be as good as men in mathematics. I frequently see them less motivated to work quite as hard. . . . maybe it's an incorrect perception but I tend in general to see the women slightly less motivated than men and I don't know why that is.

Women's socialization, starting from a young age, leads them to look for interaction, attention, and reinforcement, rather than to be autonomous and independent learners (Fennema and Peterson 1985; Etzkowitz, Kemelgor, and Uzzi 2000). This socialization can work against them in the eyes of their advisers, especially in a disciplinary culture like that found in mathematics, where work is expected to be individualistic and independent. Claudia Henrion (1997) argues that the image of a mathematician as a loner serves as a filter to keep certain types of people out of mathematics. The ways in which females (and people of other groups) interact may be different from those expected by male faculty; those behaviors may be misinterpreted as inferior, rather than different, as reported in one study:

> In [one science] department a female academic model based on interpersonal relationships, affiliation and nurturance had become accepted as legitimate and had even become the departmental norm. This was in strong contrast to another research site where the expression by women of a need for these characteristics in the laboratory environment was derided as a desire for dependence and emotionality by the adherents of the patriarchal system that was in place. (Etzkowitz et al. 1992, 174)

The perceptions of women as dependent may result both from women's socialization to build more interactive relationships and from the ways that they need to work harder than men to develop relationships in the male-dominated culture of STEM.

The importance of relationships with faculty, the tendency of faculty to mentor students of the same sex category, the negative perceptions faculty may have of women students, and the small numbers of female faculty in many STEM disciplines combine to pose serious obstacles for women doctoral students. The number of faculty of color in most STEM disciplines is so low that similar studies have not been conducted on these groups; however, in a survey of Latino/a, African American, and Native American graduate students in mathematics, few students cited graduate faculty as having been helpful to them (Alexander and Hawkins 1997). It seems likely that students of color face a similar—and perhaps more extreme—series of obstacles. Unfortunately, these obstacles can provide powerful evidence to students and women of color that they do *not* belong in the culture of academic STEM.

Students' Relationships with Other Students

Mathematicians and mathematics students have commonly been stereotyped as lacking in social skills (Campbell 1995). Nel Noddings argues that

> There seems to be something about the subject or the way it is taught that attracts a significant number of young people with underdeveloped social skills. . . . If this impression of students who excel at math is inaccurate, researchers ought to produce evidence to dispel the notion, and teachers should help students to reject it. If it is true, math researchers and teachers should work even harder to make the "math crowd" more socially adept. Because that group so often tends to be exclusive, girls and minority youngsters may wonder whether they could ever be a part of it. But when the group is examined from a social perspective, many talented young people may question whether they *want* to be a part of it. (1996, 611)

Since mathematics students are commonly perceived as being white, male, childless, without interests outside of mathematics, and socially inept, it may be that members of various groups recognize tangible ways in which they do *not* fit in with this group.

Women and students of color have reported feeling isolated in mathematics, computer science, and physics (Becker 1990; Hollenshead, Younce, and Wenzel 1994; Alexander and Hawkins 1997). Like this woman, each of the six women mathematics graduate students I interviewed described ways in which she felt she did not fit in:

> I sometimes walk into a room, look around, realize I'm the only woman in the room, *again*, and it has an effect. . . . It makes me feel like on some level most of the people I interact with are missing one particular thing in common with me and I find that discouraging. . . . There are still some times when it feels uncomfortable that there aren't more people like me.

Students in several programs have reported the importance of having a "critical mass" of women or students of color (Cooper 2000; Manzo 1994). Mutual support among graduate students is important, as they study and complete assignments together, share information, and provide one another with important moral support (Stage and Maple 1996; Golde 1996). This can be particularly important for students from underrepresented groups. Duane Cooper (2000) reported that African American doctoral students in the mathematics doctoral program at the University of Maryland highly valued the academic and social support of other Black students.

Summary: What Should Be Changed?

Traditional approaches to equity issues in education have called for interventions aimed at arming the underserved or underrepresented students with "survival skills" intended to help them develop behaviors and attitudes toward learning that are more like those of students of the dominant group. This approach has significant ethical and political implications, in that it poses white, European males as the standard-bearers for appropriate educational behavior. It locates gender issues in mathematics as a "girl problem" rather than a problem with teaching, curriculum, the structure of education, or other social forces that influence learning (Campbell 1995). A different path to change may be more appropriate: transforming the educational context to be more reflective of the more diverse audience it attempts to serve.

All students—male and female, of all classes, ethnicities, and races—need to participate in mathematical practice and feel that they belong in mathematics in order to succeed. In some sense, graduate education in mathematics consists of a series of obstacles to participation and belonging, with an institutionalized belief that "talented" students will overcome those obstacles to prove that they have "what it takes" to succeed (Herzig 2002). In addition to these obstacles faced by all students, women and students of color face *additional* obstacles, which may influence their low rates of retention. These obstacles spanned all four of the dimensions discussed: faculty beliefs about teaching and learning, faculty beliefs about mathematics, students' relationships with faculty, and students' relationships with other students.

If mathematics faculty and policymakers are serious about recruiting and retaining more women and students of color in mathematics, then they need to consider what they can do to create an educational experience that is built around helping *all* students participate in the discipline and develop a sense of belonging. In particular, STEM faculty should consider the following recommendations:

- Give all students early and frequent opportunities to participate in authentic practices of the academic communities to which they are apprentices, including research, conferences, writing, and other professional activities. This participation is critical to their learning to become scientists, engineers, and mathematicians.
- Implement flexible supports to help students balance the competing demands of school, family, and other commitments in their lives. For example, the demands of parenting present a more serious obstacle for women graduate students than for men (Sonnert and Holton 1995; Nerad and Cerny 1993; Lovitts 2001). Similarly, students may have family, community, or cultural conflicts with the traditional organization of graduate programs. Programs that accommodate the real lives of diverse students will be more accessible to them.
- Focus instructional experiences less on the identification of "talented students" and more on facilitating *learning*. "Talented" may in fact be a misnomer, as it may refer to students who are predisposed to think or act in ways that are already common in the culture of the discipline. Keep an open mind about *who* can succeed in STEM and *what* that success looks like. Given appropriate educational experiences, it may be that a much broader range of students can make valuable contributions to STEM disciplines.
- Construct course curricula in ways that help students explore connections among ideas, both in other areas of the discipline and with other disciplines, on an ongoing basis. While students can and should infer some of these connections themselves, pedagogy can lead them in appropriate directions and help them learn with deeper *understanding*. Pedagogy focused on connections and understanding helps all students learn, and will be more accessible to a broader range of students.
- Share with students the enthusiasm, excitement, and vision that professionals find in the discipline. *Mentor* students in how scientists and mathematicians actually engage in their work, including the roles of insight, intuition, and creativity in thinking about problems. Look to develop the aesthetic, emotional, and intuitive aspects of the thought and work of STEM, so that the diverse range of intellectual activity used by STEM professionals is valued and accepted at the graduate level as well. These aspects of scientific and mathematical thought will help *all* students engage with their studies.
- Facilitate explicit opportunities for graduate students to develop sustained and meaningful *positive relationships* with other students and program faculty. Make a particular effort to ensure that women and Latino/as, African Americans, and Native Americans have meaningful, positive, and caring mentoring relationships, since those relationships are particularly scarce for those students in STEM. Recognize that all students are individuals, and that while some may not communicate, interact, think, or otherwise behave in ways that are expected of scientists and mathematicians, this diversity of behavior is in fact a rich resource for the further development of mathematical and scientific thought, rather than being

an impediment to it. Build a flexible, accepting educational context in which a diverse range of students *can* participate and come to belong in science, technology, engineering, and mathematics.

Notes

I would like to express my appreciation to Laurie Hart, whose thoughtful comments on an earlier draft of this paper were instrumental in developing and communicating the ideas of belonging presented here. This project would not have been possible without the graduate students and faculty who gave generously of their time to discuss their experiences in mathematics with me.

1. For more details on these interviews, see Herzig 2002, where I report on a subset of these data.
2. I use the feminine pronoun throughout. While it can be argued that this is just as sexist a convention as the more customary use of the masculine pronoun, I expect that in using female pronouns there is little risk that anyone will overlook the fact that some STEM students and faculty are indeed male (after Maddy 1990).
3. The small numbers of women STEM faculty have made students' relationships with female faculty difficult to analyze.

References

Alexander, John W., and William A. Hawkins. 1997. *Survey of Minority Graduate Students in U.S. Mathematical Sciences Departments.* Washington, D.C.: Mathematical Association of America and National Association of Mathematicians.

Allexsaht-Snider, Martha, and Laurie E. Hart. 2001. "'Mathematics for All': How Do We Get There?" *Theory into Practice* 40(2): 93–101.

Ames, Carole. 1992. "Classrooms: Goals, Structures, and Student Motivation." *Journal of Educational Psychology* 84(3): 261–71.

Bair, Carolyn Richert, and Jennifer Grant Haworth. 1999. "Doctoral Student Attrition and Persistence: A Meta-synthesis of Research." Paper presented at the annual meeting of the Association for the Study of Higher Education (ASHE), November 20, San Antonio, Texas.

Becker, Joanne Rossi. 1984. "In Pursuit of Graduate Education in Mathematics: Factors That Influence Women and Men." *Journal of Educational Equity and Leadership* 4(1): 39–53.

———. 1990. "Graduate Education in the Mathematical Sciences: Factors Influencing Women and Men." In *Gender and Mathematics: An International Perspective,* ed. Leone Burton, 119–30. Boston, Mass.: Cassell.

Belenky, Mary Field, Blythe McVicker Clinchy, Nancy Rule Goldberger, and Jill Mattuck Tarule. 1986. *Women's Ways of Knowing: The Development of Self, Voice, and Mind.* New York: Basic Books.

Berg, Helen M., and Marianne A. Ferber. 1983. "Men and Women Graduate Students: Who Succeeds and Why?" *Journal of Higher Education* 54(6): 629–48.

Boaler, Jo. 2002. "The Development of Disciplinary Relationships: Knowledge, Practice, and Identity in Mathematics Classrooms." *For the Learning of Mathematics* 22(1): 42–47.

Bowen, William G., and Neil L. Rudenstine. 1992. *In Pursuit of the Ph.D.* Princeton, N.J.: Princeton University Press.

Burton, Leone. 1995. "Moving towards a Feminist Epistemology of Mathematics." In *Equity in Mathematics Education: Influences of Feminism and Culture,* ed. Pat Rogers and Gabriele Kaiser, 209–25. London: Falmer.

———. 1999. "The Practices of Mathematicians: What Do They Tell Us about Coming to Know Mathematics?" *Educational Studies in Mathematics* 37: 121–43.

Campbell, Patricia B. 1995. "Redefining the 'Girl Problem in Mathematics.'" In *New Directions for Equity in Mathematics Education,* ed. Walter G. Secada, Elizabeth Fennema, and Lisa Byrd Adajian, 225–41. New York: Cambridge University Press.

Carlson, Marilyn P. 1999. "The Mathematical Behavior of Six Successful Mathematics Graduate Students: Influences Leading to Mathematical Success." *Educational Studies in Mathematics* 40: 237–58.

Committee on the Participation of Women. 2003. "Improving the Persistence of Women in Graduate Mathematics." Symposium presented at the joint meeting of the American Mathematical Society and the Mathematical Association of America, Baltimore, Maryland.

Cooper, Duane A. 2000. "Changing the Faces of Mathematics Ph.D.'s: What We Are Learning at the University of Maryland." In *Perspectives on African Americans,* ed. Marilyn E. Strutchens, Martin L. Johnson, and William F. Tate, 179–92. Changing the Faces of Mathematics. Reston, Va.: National Council of Teachers of Mathematics.

Douglas, Ronald G. 1997. "Educating Mathematical Sciences Graduate Students." In *Preserving Strength while Meeting Challenges: Summary Report of a Workshop on Actions for the Mathematical Sciences,* 41–44. Washington, D.C.: National Academy Press.

Etzkowitz, Henry, Carol Kemelgor, Michael Neuschatz, and Brian Uzzi. 1992. "Athena Unbound: Barriers to Women in Academic Science and Engineering." *Science and Public Policy* 19(3): 157–79.

Etzkowitz, Henry, Carol Kemelgor, and Brian Uzzi. 2000. *Athena Unbound: The Advancement of Women in Science and Technology.* Cambridge: Cambridge University Press.

Fennema, Elizabeth, and Penelope Peterson. 1985. "Autonomous Learning Behavior: A Possible Explanation of Gender-Related Differences in Mathematics." In *Gender Influences in Classroom Interaction,* ed. Louise Cherry Wilkinson and Cora B. Marrett. Orlando, Fla.: Academic Press.

Gerholm, Tomas. 1990. "On Tacit Knowledge in Academia." *European Journal of Education* 25(3): 263–71.

Girves, Jean E., and Virginia Wemmerus. 1988. "Developing Models of Graduate Student Degree Progress." *Journal of Higher Education* 59(2): 163–89.

Golde, Christine M. 1996. "How Departmental Contextual Factors Shape Doctoral Student Attrition." Ph.D. diss., Stanford University.

Henrion, Claudia. 1997. *Women in Mathematics: The Addition of Difference.* Bloomington: Indiana University Press.

Herzig, Abbe H. 2002. "Where Have All the Students Gone? Participation of Doctoral Students in Authentic Mathematical Activity as a Necessary Condition for Persistence toward the Ph.D." *Educational Studies in Mathematics* 50(2): 177–212.

———. 2004. "'Slaughtering This Beautiful Math': Graduate Women Choosing and Leaving Mathematics." *Gender and Education* 16(3): 379–96.

Hiebert, James, et al. 1997. *Making Sense: Teaching and Learning Mathematics with Understanding.* Portsmouth, N.H.: Heinemann.

Hollenshead, Carol, Patricia Soellner Younce, and Stacy A. Wenzel. 1994. "Women Graduate Students in Mathematics and Physics: Reflections on Success." *Journal of Women and Minorities in Science and Engineering* 1: 63–88.

Johnston, Betty. 1995. "Mathematics: An Abstracted Discourse." In *Equity in Mathematics Education: Influences of Feminism and Culture,* ed. Pat Rogers and Gabriele Kaiser, 226–34. London: Falmer.

Kirkman, Ellen E., James W. Maxwell, and Kinda Remick Priestly. 2003. "2002 Annual Survey of the Mathematical Sciences (Third Report)." *Notices of the American Mathematical Society* 50(8): 925–35.

Ladson-Billings, Gloria. 1997. "It Doesn't Add Up: African American Students' Mathematics Achievement." *Journal for Research in Mathematics Education* 28(6): 697–708.

Lave, Jean, and Etienne Wenger. 1991. *Situated Learning: Legitimate Peripheral Participation.* Cambridge: Cambridge University Press.

Love, Nancy. 2002. *Using Data/Getting Results: A Practical Guide for School Improvement in Mathematics and Science.* Norwood, Mass.: Christopher Gordon.

Lovitts, Barbara E. 2001. *Leaving the Ivory Tower: The Causes and Consequences of Departure from Doctoral Study.* Lanham, Md.: Rowman & Littlefield.

Maddy, Penelope. 1990. *Realism in Mathematics.* Oxford: Clarendon.

Manzo, Kathleen Kennedy. 1994. "American University: Success Is in the Numbers, African American Women Excel in Math Ph.D. Program." *Black Issues in Higher Education* 11: 40–43.

Martin, Danny Bernard. 2000. *Mathematics Success and Failure among African American Youth: The Roles of Sociohistorical Context, Community Forces, School Influence, and Individual Agency.* Mahwah, N.J.: Erlbaum.

National Council of Teachers of Mathematics. 2000. *Principles and Standards for School Mathematics.* Reston, Va.: National Council of Teachers of Mathematics.

National Research Council. 1992. *Educating Mathematical Scientists: Doctoral Study and the Postdoctoral Experience in the United States.* Washington, D.C.: National Academy Press.

Nerad, Maresi, and Joseph Cerny. 1993. "From Facts to Action: Expanding the Graduate Division's Educational Role." *New Directions for Institutional Research* 80: 27–39.

Noddings, Nel. 1996. "Equity and Mathematics: Not a Simple Issue." *Journal for Research in Mathematics Education* 27(5): 609–15.

Orr, Eleanor Wilson. 1997. *Twice as Less: Black English and the Performance of Black Students in Mathematics and Science.* New York: W. W. Norton.

Reskin, Barbara F., Jane F. Koretz, and Lesley Lee Francis. 1996. "Women in Science." *Academe* 82(3): 57–65.

Rosser, Sue V. 1995. "Reaching the Majority: Retaining Women in the Pipeline." In *Teaching the Majority: Breaking the Gender Barrier in Science, Mathematics, and Engineering,* ed. Sue V. Rosser, 1–21. New York: Teachers College Press.

Sonnert, Gerhard, and Gerald Holton. 1995. *Who Succeeds in Science? The Gender Dimension.* New Brunswick, N.J.: Rutgers University Press.

Stage, Frances K., and Sue A. Maple. 1996. "Incompatible Goals: Narratives of Graduate Women in the Mathematics Pipeline." *American Educational Research Journal* 33(1): 23–51.

Tate, William F. 1995. "Returning to the Root: A Culturally Relevant Approach to Mathematics Pedagogy." *Theory into Practice* 34(3): 166–73.

Tinto, Vincent. 1993. *Leaving College: Rethinking the Causes and Cures of Student Attrition.* 2nd ed. Chicago: University of Chicago Press.

Wenger, Etienne. 1998. *Communities of Practice: Learning, Meaning, and Identity.* Cambridge: Cambridge University Press.

Wiles, Peter. 1999. "Graduate Students as Legitimate Peripheral Participants in a Mathematical Community." Madison, Wisc. Unpublished paper.

Zwick, R. 1991. *Differences in Graduate School Attainment Patterns across Academic Programs and Demographic Groups.* Princeton, N.J.: Educational Testing Service.

14 Designing Gender-Sensitive Computer Games to Close the Gender Gap in Technology

Anna M. Martinson

Media technologies, including those used primarily for leisure, can reinforce societal divisions (Bryce 2001; Winner 1980). Investigation into the relationship between gender and technology includes considering the symbolic role technology plays in gender stereotypes, as well as the demographic characteristics of people being educated in technical fields, assembling hardware components, writing software programs, and using specific technologies (e.g., office technology, home computers, and communication technologies). Interest in the representation of gender in the design of software has included attention to educational software and computer games (Agosto 2000; Cassell and Jenkins 1998; Huff and Cooper 1987; Provenzo 1991). This chapter focuses on gender and the sexual politics of computer games, specifically their implications for the recruitment and retention of women in computer science.

Gender differences in computer interest and skills "begin in the early grades, are reinforced by peers and the home environment, and continue into adulthood" (Upitis 1998, 295). While women make up 56.8 percent of the U.S. workforce, they make up only 8.5 percent of the country's engineers (Goodman Research Group 2002). In 1990, 70 percent of computer specialists were men (Cassell and Jenkins 1998). More than a decade later, in an article titled "Keeping Women in the Tech Industry," BBC News (2000) reported that only 18 percent of Dell's U.S. directors were women. Alarmingly, the percentage of women graduating with degrees in computer science has been *declining* since the 1980s, further exacerbating the gender gap in computing (Camp 1997; Camp, Miller, and Davies 2000).

Recognizing the need to get more women involved in technology fields, women in Science, Technology, Engineering, and Math (STEM) and related programs (such as Women in Science and Engineering [WISE] and Women in Technology [WIT]) have been developing initiatives to attract and retain women in degree programs. Recommendations often emphasize the need to change the culture of computer science by revising admission procedures, revising the curriculum—especially contextualizing computer science—paying more attention to good teaching, hiring more female faculty, increasing the mentoring and encouragement of female students, and fostering institutional and community support (Cohoon 2001; Cuny and Aspray 2001; Margolis and Fisher 2002).

The importance of attracting girls to STEM fields during their elementary and high school years is also suggested (e.g., Sanders 1994), as is changing the home environment (Facer et al. 2001; Honey et al. 1991; Shashaani 1994; Young 2000). Anne-Marie Huurre, executive producer of Women Wise, a software company, proposes that emphasis be placed on bringing "technology into the home and leisure markets in an entertaining, nonintimidating and enjoyable way" (WomenGamers.com 2001). Justine Cassell, an associate professor at MIT's Media Laboratory, agrees, claiming that there is a "vicious circle of fewer girls being interested in computer games, leading to fewer women working in computer science and technology fields. Which leads to fewer women developing computer games and fewer girls being interested in technology" (MIT News 1997, 1).

Conventional wisdom suggests that one of the best ways to learn about computers is to tinker with them: to play with the various software commands and options, or even to dismantle and attempt to reassemble the hardware components. Gender differences in childhood tinkering have been identified: boys are more likely to tinker and continue on into computer science and engineering programs, while girls are less likely to tinker and also less likely to enroll in engineering courses (Kahle n.d.; Margolis, Fisher, and Miller 1998; Martinson 1998).

Another image of computer play is that of "surfing the Web," or following links somewhat at random. Gender differences in patterns of exploring the Web have also been found. Men are more likely to start at one site and then go off from there, whereas women are more likely to start at a site, follow a link or two, and then return to the original site (Passig and Levin 1999). Web sites without "back" buttons are less usable for people who prefer the latter style of browsing the Web. One of the implications for design would be that both "home" and "back" buttons should always be included.

The best-known form of computer play, however, is engagement with computer games. Computer-based games are already part of the daily routine of many U.S. households (Griffiths 1997) and constitute many children's first encounter with computers (Miller, Chaika, and Groppe 1996). The computer game industry in the U.S. was a $6 billion industry in 2001 (in comparison, Hollywood films accounted for $8.1 billion in the same year) (Glaubke et al. 2001; Mirapaul 2002). Allucquère Rosanne Stone argues that "it is entirely possible that computer-based games will turn out to be the major unacknowledged source of socialization and education in industrialized countries" (Stone 1995, 27).

This chapter explores the relationship between gender and computer games and the potential for computer games to motivate and sustain women's interest in computing. First, computer games are discussed as a microcosm of the masculine culture of computing, focusing on violence and sexist representations in the games. Second, different versions of gender equality in the design of computer games are explored. The role of design and attempts to design female-friendly computer games are also reviewed. Third, the need for research into the role of computer games in the leisure pursuits of adult women is considered. The chapter concludes with a brief exploration of the potential for courses on game design to contextualize computer science education and provide female computer science students with opportunities to design computer games in their own image.

The Male Culture of Computer Games

Children at all grade levels continue to perceive computers as a male domain (Collis 1985; Krendl, Broihier, and Fleetwood 1989; Young 2000). By third grade, if girls have continually lacked access to computers, their lack of access often transmutes into lack of interest (Cassell and Jenkins 1998). Additionally, computer gaming is widely perceived as a male pastime (Benston 1989; Cassell and Jenkins 1998; Kiesler, Sproull, and Eccles 1985; Pereira 1994; Shneiderman 1998; Spilker and Sorensen 2000), a view also held by executives of game companies, game retailers, and parents (de Jean et al. 1999). Further, boys begin playing computer games earlier and play more regularly than do girls (Griffiths 1997; Martinson 2003). "Seventy-five percent of video games are bought for boys" (Pereira 1994, A5).

While some women enjoy the current gaming options, there seems to be a potential for a strong female market if there were more games that appealed to girls and women. Games targeted to adolescent males leave their female peers with limited gaming choices (AAUW 2000; Kiesler, Sproull, and Eccles 1985). Women make up between 5 and 43 percent of computer game players, with higher numbers reported in marketing reports and lower numbers in academic research (de Castell and Bryson 1998; Greenspan 2002; Smith 2001). Acquiring an accurate count of women gamers may be difficult, since many female gamers take pains to protect their privacy in order to avoid online harassment (Banshee 2000). Anecdotal evidence suggests that at least a few of the serious female gamers are initially interested in games because they are romantically interested in a male gamer (GameDiva 2002; Yates and Littleton 1999).

The early exposure to computers boys receive by playing computer games may partially explain subsequent findings that boys are more confident and comfortable in using computers in other contexts (Kiesler, Sproull, and Eccles 1985; Margolis and Fisher 2002; Morahan-Martin 1997; Pereira 1994). Gender differences in enjoyment of electronic games have been related to the gender gap in computing, for example, via the relationship between educational and recreational activities (Upitis 1998). Playing computer games can improve a number of skills important in life, including reaction time, eye-hand coordination, visual attention, and iconic recognition, as well as visual-spatial and problem-solving skills (Gailey 1993; Glaubke et al. 2001). Marcia C. Linn (1985,

238) contends that "once enrolled in computer courses, females perform as well as or better than males and appear to be equally interested in classroom activities. What has not been assessed is whether females maintain this interest as much as males do when they leave the classroom. In particular, if the main out-of-school computer activity is the arcade game, females are unlikely to maintain their interest in computers."

High levels of violence in current games (Glaubke et al. 2001; Provenzo 1991), the topics and themes of games (Miller, Chaika, and Groppe 1996; Subrahmanyam and Greenfield 1998), and the sexist representations of women in games (Glaubke et al. 2001; Provenzo 1991) have all been proposed as reasons why girls display less interest in computer games than boys. Katharine Anderson-Davila, who writes online under the pseudonym "Banshee" and is a consulting editor for WomenGamers.com (a Web site aimed at women players of computer games), uses a restaurant analogy when she argues that "[f]emale gamers are getting leftovers—someone else has ordered from the menu, and if we like it, great, but if we don't, too bad—it's that or don't eat" (Banshee 2000).

Nevertheless, there is evidence of a potential market for female-friendly computer games. Melissa Chaika (1997) points to an industry report that 85 percent of girls said they would be interested in games if the games were designed with them in mind. Design features which might appeal to girls include open-ended exploration (e.g., not built around completion or winning), virtual reality options (e.g., cross-cultural communication with an online pen pal), and educational (e.g., career exploration), entertainment (e.g., soap operas or game shows), or travel themes (Miller, Chaika, and Groppe 1996).

Violence

A substantial proportion of computer games employ metaphors (referring to war, sports, and street fighting) that traditionally are gendered male and contribute to cultural associations of technology with masculinity (Edwards 1990; Glaubke et al. 2001; Wilder, Mackie, and Cooper 1985). Not only are many video games violent, but also many of these games feature violence aimed at females (Pereira 1994). Characters in video games tend to reflect "exaggerated gender stereotypes of macho, dominant males and submissive or sexual females; both of which may explain why they appeal more to boys than girls" (Christine Gailey, quoted in Morahan-Martin and Schumacher 1998, 32).

Sexist Representations

Images of women are rarely featured on the covers of games (Chaika 1997; Glaubke et al. 2001; Provenzo 1991); and when they are featured, women are frequently depicted in an overtly sexual manner and usually as victims needing rescue (Chaika 1997; Provenzo 1991). Anderson-Davila points out that "female gamers are insulted by how women are depicted on the boxes on the shelves—it's either *Barbie Digital Makeover* or *Tomb Raider 14,* if women are present at all. A digital female body of normal proportions wearing practical clothing is a rare commodity indeed, much less a minority digital woman" (Banshee 2000).

As well as their representation in cover art, the characterization of females within games is also problematic. In a study by Children Now, 64 percent of characters in top-selling video games were male human characters, while only 17 percent were female human characters (the remaining 19 percent were non-human characters). Similarly, 73 percent of the player-controlled characters were male while only 12 percent were female. Even in games rated "E" (for everyone), 85 percent of the characters were male and 15 percent were female. Among games with female characters, roughly 50 percent featured females as props or bystanders rather than directly engaged in game action. Both male and female characters behaved in gender-stereotypical ways: males were more likely to be physically aggressive, while females were more likely to be verbally aggressive or to scream (Glaubke et al. 2001). Further, characters tend to reflect racial, as well as gender, stereotypes: " 'tough' women are rarely blond; 'princesses' are rarely anything else" (Gailey 1993, 88).

The perception that video and computer games are the province of males (adolescent or otherwise) is occasionally used to justify the continuing use of sexist representations in game design. In response to a question on the Internet message board for the game *Icewind Dale* about why female characters have low-cut clothes, Chris Parker, producer of the game, explained in 1999 that "we don't want to disappoint the gamers by stepping outside of these established expectations." On the same message board, the game site's Webmaster posted this announcement on June 8, 2000: "Jason Manley [concept artist for the game *Icewind Dale*] has done it again with the creation of 2 new portraits for your adoring pleasure in the Downloads & Archives section. The portraits are of 2 female characters from IWD and can be best described as seductive, playful, and down right hot. But dont take my word for it check out out {*sic*] for yourself and dont forget to download the High Resolution versions so you can check out every detail of these 2 hot babes" (Banshee 2000). Such comments assume that all gamers are male and that male gamers enjoy sexist representations of women. While the barrier of sexist representations is not insurmountable, female gamers do take notice and object even as they continue to play (Yates and Littleton 1999).

Some researchers argue that "it is overly deterministic to assume that there is a causal relationship between female representation in a text, and the nature of consumption of that text by female gamers" (Bryce and Rutter 2002, 248). Nevertheless, it seems clear that sexist representations can constitute a barrier to full female participation in the gaming community. Lara Croft, the female protagonist of *Tomb Raider,* has been touted by some as a main character that girls can identify with. However, the popularity of game patches which render the already scantily clad Croft completely nude (Schleiner 2001) leads one to question whether females are really the target audience for *Tomb Raider.* At the very least, girls and women have to reinterpret the representations of female characters (or lack thereof).

In a media environment saturated with sexist representations (in books, movies, television, radio, and other media) many women develop the ability to enjoy activities despite sexist representations (e.g., Radway 1987). One strategy they use is to imagine themselves as managers in charge of creating game scenarios rather than identifying with stereotyped female game characters. Yasmin Kafai found that "irrespective of the considerable gender stereotyping found in many video games (for example, portraying women as victims or prizes), girls seem to resolve the dilemma by redefining their positions in casting themselves in managerial roles" (Kafai 1998, 92). Some might consider disassociation with stereotyped characters a problematic response, since the underlying sexism is not addressed.

Computer Games and Gender Equality

Prior to the development of analyses of sexism in computer games, the prevailing attitude was that the main barrier was in the girls themselves; other interests kept them away from computers. In response to the common refrain that "girls just don't like computer games," Megan Gaiser, president and CEO of Her Interactive (developer of the award-winning *Nancy Drew* CD-ROM interactive games), draws a parallel with the history of girls' involvement in the physical play of sports: "As more choices appear that appeal to a broader spectrum of women and girls, we believe the girls' game market will grow in the same way that girls' college sports did after Title IX. There was a time when the standard belief was that 'girls don't like sports.' As soon as girls' sports received equal emphasis and funding, boom—it turns out that girls LOVE sports, and female sports has become a huge industry" (GameDiva 2001).

While many people are interested in equality with respect to technology, the meaning of "equality" frequently varies. Two understandings of equality with respect to computer games are common: the "separate but equal" approach and the nongendered equal opportunity approach. The "separate but equal" approach finds equality in contexts where games for girls are as readily available as games for boys, that is, roughly equal numbers of games are specifically targeted at girls (such games are sometimes called "pink software") and boys. In this version of equality the content of the games matters less than the fact that girls are playing with technology. Jan Russo, a software reviewer for SuperKids (a Web site featuring reviews of educational software), remarked, "When our reviewers' eyes light up at the sight of familiar logos and feminine packaging, I find it impossible to deny their validity. No matter what we might want to believe, the world of computers is still a man's world, and our girls will need all the help they can get to enter and succeed in it" (Russo 1997, 1). The "separate but equal" approach attempts to work within the status quo, and some maintain such efforts may be an important step toward eventual transformation. Others contend that a "separate but equal" approach will only reinforce stereotypes about gender and technology.

A second understanding of equality argues for nongendered equal opportunity and access. This involves the creation of new games for both boys and girls that avoid reinforcing gender stereotypes. In this vision of equality, what is problematic is the status quo in which the Mattel Hot Wheels computer for boys comes bundled with extra mathematics, geographic, and logic software, while the Mattel Barbie computer does not. Rather, girls should be given the same exposure to math, geography, and logic. At most, the same games might be packaged in both pink and blue purely for marketing purposes, although most likely games would not be marketed specifically to either girls or boys.

A third, less common approach looks beyond the design of any specific game to consider the ways that cultural stereotypes about gender and computers play a role in the process of software development. This approach includes an analysis of software designers, as well as of the advertising, marketing, and distribution of computer games (Chaika 1997; McDonough 1999; WomenGamers.com 2001). For example, Chaika (1997) found that games for girls were featured in the educational sections of toy stores, while games for boys were prominent in the entertainment sections. Huurre reported the following experience when she requested reviews of her company's female-oriented software: "Three major, mainstream women's magazines told me that although they reviewed movies, books, audio CDs etc., 'women DON'T use computers or play games so don't send us any press release or information for our readers.' A major young women's magazine said teens would rather 'PLAY with boys and makeup, not computers'" (WomenGamers.com 2001).

A focus on the social context of computer game development also reveals that few women are involved in design, not only of computer games, but also of software in general. Many WISE programs operate with this understanding and work to transform the social context of computing so that more females are encouraged to pursue their interest in computers. While it is too simplistic to argue that women computer scientists will design software that other women will like, people concerned with diversity at the point of design argue not only that is it important to design games for girls, but also that it is important to have more women involved in software development. Pamela Mendels (2000) quotes Pamela S. Haag, director of research for the American Association of University Women Educational Foundation, on the importance of design. Haag states, "We want a diversity of perspectives when we are designing new technologies, and if girls are not at the table when the technology is being designed and created, this technology is less likely to speak to all of us" (Mendels 2000, B10).

Designers and Design Bias

The Gender and Learning Technologies group at the University of Michigan suggests that rather than focus on "trying to change girls' attitudes and behavior toward technology . . . a different approach to the problem [is to focus] on how the design of educational technology affects the gap" (*Black Issues in Higher Education* 2001, 32). Charles Huff and Joel Cooper's (1987) research documents another troubling aspect of the design of computer games: the apparently unconscious tendency to use a male-gendered style when designing ostensibly neutral software (see also McDonough 1999). Huff and Cooper's (1987) study of designers of educational software found that designers unintentionally introduced gender bias into their designs. When asked to design for boys, designers made educational software that resembled video games. When they were asked to design for girls, the educational software they produced utilized more conversational dialogue. Finally, when they were asked to design for students, the resulting software resembled the games designed for boys.

Issues of gender and computer games extend beyond the themes of gender representation in a specific game. In some games, the analysis of bias also needs to "account for the virtual presence of the designer (including the designer's conception of the user) in technological artifacts" (McDonough 1999, 857). Not only computer games, but also the designers of the games, are embedded in social (and, by implication, gendered) relations. Designers' expectations of users influence subsequent designs (Miller, Chaika, and Groppe 1996).

Designers determine the background theme of the virtual environments, the avatars available to players, whether players can remain anonymous, and whether they can communicate privately with others in the game. Each of these design choices has the potential to be more appealing to one social group and less appealing to others. For example, designers predominantly choose science fiction and fantasy themes for virtual reality environments—genres associated with middle-class white male identity (McDonough 1999). Jerome McDonough argues that deploying these themes in a design establishes middle-class white male interests and values as normative in the virtual environment. The inclusion of stereotyped nonwhite avatars—such as a black Rastafarian or a Japanese geisha—is more likely to appeal to a white audience in search of "exotic" experiences than to a truly multicultural audience (McDonough 1999; Nakamura 1995). In his study of the design of virtual environments, McDonough found that

> Virtual environment designers as a group are predominantly young, white, middle-class men; this predominance is particularly marked among the software engineers who appear to wield the greatest control over the design process. The designers' own conceptions of what constitutes an enjoyable environment and experience online inform the products they create, resulting in virtual environments designed by a white, middle-class culture and for a white, middle-class culture. Users' enactments of identities in these environments thus take place in worlds where the structures of dominance in the larger world have already been reproduced, and white, masculine identity normalized. (1999, 867)

Furthermore, the designers of the graphical environment software conceived of their users as being young, middle-class, computer-literate (albeit without programming expertise) white males, living in the United States (McDonough 1999). The parallel, in terms of demographic characteristics, between designers and the presumed user group is striking.

The tendency for a designer to design for him- or herself is understandable (McDonough 1999; Wright 2000). However, the consequences for usability (broadly defined) for different user groups are worth attending to. Research findings that women are less interested in, or make less use of, certain technologies may reflect a complexity of social factors that become manifest in the design process. In addition to attempting to spur female interest in computing, we must also try to create female-friendly designs (Martinson, Schwartz, and Vaughan 2002). McDonough characterizes the problem by explaining that "designers' intellectual construction of their product's users, and the encoding of that construction within the technologies designers develop, therefore can constitute a form of social power which designers wield over users. If users are to employ the technology . . . they must conform themselves to the designers' understanding of their world" (McDonough 1999, 857). McDonough stresses that bias is incorporated into software despite the best intentions of designers. He and others argue that technological design is also social design (Bijker and Law 1992; McDonough 1999; Wajcman 1991; Winner 1980).

Attempts to Design Female-Friendly Games

Concerns about gender bias in game software have spurred a number of attempts to design computer software with special appeal for girls, such as Mattel's "Barbie Fashion Designer" and games designed by GirlGames, Girl Tech, and Her Interactive. Designing games for girls has largely been a hit-or-miss proposition (Subrahmanyam and Greenfield 1998). Standard claims have been that girls object to the levels of violence in many computer games, and have trouble identifying with the male heroes and highly sexualized female figures common in some genres. Under the weight of this argument, developers have neglected to explore other design elements that might appeal to girls. Kaveri Subrahmanyam and Patricia Greenfield's 1998 analysis indicates that while gender differences in game preferences exist, the differences are more complex than the stereotype that boys like violence and girls do not.

Various female-friendly design guidelines have been developed (Agosto 2000). Using Subrahmanyam and Greenfield's work as a base, Christina Glaubke and colleagues (2001) identified features of a computer game that are likely to appeal to girls: a creative component, puzzle elements, cooperative play, available solicited help, female player-controlled characters, a realistic setting, positive unsolicited feedback, a slow or variable pace, predictable rules, a clear explanation of those rules, and an absence of violence (especially the absence of killing and evil characters). Notably absent from this list are ways to counteract the oversexualization of female characters. Kallah, a contributor to GameGirlz (a Web site devoted to supporting women's enjoyment of computer games), identifies six criteria for a positive female character: she should be a character first and a woman second; not in the game just to be rescued; smart and savvy; an adult, not a child; of human proportions and appropriately dressed; and not in love with the hero (Kallah n.d.).

Females and males also differ in the type of system feedback they enjoy. Girls prefer supportive responses that encourage learning from mistakes, while boys prefer more aggressive feedback (Bertozzi 2001; Miller, Chaika, and Groppe 1996). Color schemes may also play a role: the dark colors frequently used in game environments do not appeal to everyone, and some women might prefer game environments with brighter colors (McDonough 1999; WomenGamers.com 2001). Identification of these features from studies of girls (Inkpen et al. 1994) raises questions of what games might look like if they were designed by females for girls and women.

Computer Games, Leisure, and Adult Women

While the issue of computer games for girls has received considerable research attention (*Black Issues in Higher Education* 2001; Cassell and Jenkins 1998; de Jean et al. 1999; Griffiths 1997; Huff 1997; Kafai 1998; Provenzo 1991), few studies to date have examined the use of computing technology in the leisure pursuits of adult women. Some argue that to understand the leisure experience of any group, one needs to understand the context of that leisure experience (see, e.g., Steeves, Becker, and Choi 1988). In addition to gender, age and degree of prior experience with computers make a difference in enjoyment of computer gaming (Miller, Chaika, and Groppe 1996). Figures from the Interactive Digital Software Association (IDSA) indicate that 61 percent of computer game players are eighteen or older (IDSA 2001).

Any examination of the leisure practices of adult women must begin with the recognition that women's leisure is not monolithic, but rather is shaped by the opportunities and constraints faced by women (Henderson et al. 1989). The nature of leisure tends to differ for males and females; in part, males simply have more time for pursuing leisure activities (Henderson et al. 1989). Men also receive more cultural support for their leisure activities (Chaika 1997; Henderson et al. 1989). Elena Bertozzi (2001) argues that "women's play is not usually defined as such." Leisure time for women is often fragmented and usually occurs throughout the day in tandem with work (Deem 1986; Henderson et al. 1989), and women tend to make less clear distinctions between work and leisure than men (Henderson et al. 1989).

To understand the issues involved in designing entertainment technology for women, Anna Martinson, Nancy Schwartz, and Misha Vaughan (2002) studied the leisure practices of single, college-educated women. These women's leisure activities were multidimensional and fit into well-established routines such as reading before going to bed, reading an online humor magazine during a lunch break, or playing a quick game of solitaire between phone calls.

Designers may also want to consider that popular leisure technology products for females—especially adult women—may not resemble the current slate of computer games (Martinson, Schwartz, and Vaughan 2002). Ben Shneiderman believes that "accommodating a broader spectrum of usage situations forces researchers to consider a wider range of designs and often leads to innovations that benefit all users" (Shneiderman 2000, 87).

The capabilities and offerings of the Internet, such as e-mail, newsgroups, and chat programs, are transforming the nature of leisure (Bryce 2001). Just as women enjoy talking on the phone, more recent research suggests that in the digital realm, e-mail and chat rooms constitute leisure environments for women (Boneva, Kraut, and Frohlich 2001; Howard, Rainie, and Jones 2001; Martinson, Schwartz, and Vaughan 2002). In fact, female computer and video game players make up nearly half (43 percent) of all gamers (IDSA 2001, 7). Signs of new kinds of leisure software are evident, exemplified by the introduction of yoga on CD-ROMs, and virtual pets (Bloch and Lemish 1999; Rood 1998). Huurre's company, Women Wise, founded to incorporate technology into adult women's leisure time, is developing software programs with nontraditional themes: "Upcoming Women Wise titles will feature topical themes including: health and wellness, nutrition, fun titles on international spas and retreats, aromatherapy, Feung Shui, in addition to other nonviolent entertainment properties. All titles will incorporate the latest research into interface design, story/character creation and visual treatments that appeal specifically to women" (WomenGamers.com 2001, 2).

Despite the well-documented male bias in the design, marketing, and use of computer games, there is ample evidence that girls and women are able to successfully negotiate these barriers. At the same time, there is a clear need for computer games designed with the female market in mind—either by designing leisure technologies specifically for females, or by avoiding sexist representations and including a greater diversity of characters and themes.

To determine the extent to which the trends identified thus far can be generalized, larger-scale studies need to be conducted. Continued research into how the leisure practices of women in specific contexts differ (such as stay-at-home mothers or graduate students in computer science) is also needed.

Another avenue for exploration is the academic study of computer game design. Academic study of computer games is a nascent specialty, and some institutions are beginning to offer courses and degrees in computer game design (e.g., the Art Institute of California, DigiPen Institute of Technology, Indiana University's Masters in Immersive Mediated Environments, the University of Washington's Certificate Program in Game Development). The study of computer game design may also make it easier to integrate an understanding of social issues, including gender issues, with the knowledge of design skills (Schirra 2001). Ideally, professors would avoid specifying in advance what such technologies would look like—computer games, yoga CD-ROMs, chat rooms, or perhaps something new. Courses on designing leisure technologies for girls and women might afford increased insight into, as well as provide a convenient springboard for discussion of, issues of gender and computer science.

There are encouraging signs—the International Game Developer's Association (IGDA) has established a special interest group on Women in Game Development with an accompanying discussion list (see their Web site at http://www.igda.org/women/), and one of its founders has published a book on gender and game design (Graner Ray 2004).

It is likely that the gender gap in computer games persists through the combined effects of gender asymmetry at each point in the cycle of technological development and use. To be sure, cultural associations of masculinity and femininity with specific roles (e.g., programmer) and objects (e.g., telephones) change over time. At the same time, technological artifacts and practices still carry gender-specific associations. Fully understanding the cultural importance of gender is ultimately necessary in order to explain the gender gap in information technology. Conversely, understanding gender and information technology may provide insight into the broader patterns of gender in society.

References

AAUW (American Association of University Women). 2000. *Tech-Savvy: Educating Girls in the New Computer Age.* Washington, D.C.: American Association of University Women Educational Foundation.

Agosto, Denise. 2000. "Evaluating Electronic Information Resources for Young Women: General Research Concepts." http://girlstech.douglass.rutgers.edu/PDF/completereport.pdf. Accessed February 16, 2005.

Banshee [Katharine Anderson-Davila]. 2000. "What's in a Name? Who's in the Game?" http://www.womengamers.com/articles/namegame1.html. Accessed February 16, 2005.

BBC News. 2000. "Keeping Women in the Tech Industry." August 9.

Benston, Margaret Lowe. 1989. "Feminism and System Design: Questions of Control." In *The Effects of Feminist Approaches on Research Methodologies,* ed. Winnie Tomm, 205–23. Waterloo, Ontario: Wilfrid Laurier University Press.

Bertozzi, Elena. 2001. "Women and Play: Will She or Won't She?" Unpublished manuscript.

Bijker, Wiebe E., and John Law, eds. 1992. *Shaping Technology/Building Society: Studies in Sociotechnical Design.* Cambridge, Mass.: MIT Press.

Black Issues in Higher Education. 2001. "Michigan Research Group Seeks to Boost Female Use of Software and Computers." *Black Issues in Higher Education* 17(26): 32.

Bloch, Linda-Renee, and Dafna Lemish. "Disposable Love: The Rise and Fall of a Virtual Pet." *New Media and Society* 1(3): 283–303.

Boneva, Bonka, Robert Kraut, and David Frohlich. 2001. "Using E-Mail for Personal Relationships: The Difference Gender Makes." *American Behavioral Scientist* 45(3): 530–49.

Bryce, Jo. 2001. "The Technological Transformation of Leisure." *Social Science Computer Review* 19(1): 7–16.

Bryce, Jo, and Jason Rutter. 2002. "Killing Like a Girl: Gendered Gaming and Girl

Gamers' Visibility." In *Computer Games and Digital Cultures—Conference Proceedings,* ed. Frans Mäyrä, 243–55. Tampere, Finland: University of Tampere Press.

Camp, Tracy. 1997. "The Incredible Shrinking Pipeline." *Communications of the ACM* 40(10): 103–10.

Camp, Tracy, Keith Miller, and Vanessa Davies. 2000. "The Incredible Shrinking Pipeline Unlikely to Reverse." http://www.mines.edu/fs_home/tcamp/new-study/new-study.html. Accessed February 16, 2005.

Cassell, Justine, and Henry Jenkins, eds. 1998. *From Barbie to Mortal Kombat: Gender and Computer Games.* Cambridge, Mass.: MIT Press.

Chaika, Melissa. 1997. "Computer Game Marketing Bias." *ACM Crossroads* 3(2). http://www.acm.org/crossroads/xrds3-2/girlgame.html. Accessed February 16, 2005.

Cohoon, J. McGrath. 2001. "Toward Improving Female Retention in the Computer Science Major." *Communications of the ACM* 44(5): 108–14.

Collis, Betty. 1985. "Psychosocial Implications of Sex Differences in Attitudes towards Computers: Results of a Survey." *International Journal of Women's Studies* 8(3): 207–13.

Cuny, Jan, and William Aspray. 2001. *Recruitment and Retention of Women Graduate Students in Computer Science and Engineering.* Washington, D.C.: Computing Research Association.

de Castell, Suzanne, and Mary Bryson. 1998. "Re-tooling Play: Dystopia, Dysphoria, and Difference." In *From Barbie to Mortal Kombat: Gender and Computer Games,* ed. Justine Cassell and Henry Jenkins, 232–61. Cambridge, Mass.: MIT Press.

de Jean, Jillian, Rena Upitis, Corina Koch, and Jonathan Young. 1999. "The Story of Phoenix Quest: How Girls Respond to a Prototype Language and Mathematics Computer Game." *Gender and Education* 11(2): 207–23.

Edwards, Paul N. 1990. "The Army and the Microworld: Computers and the Politics of Gender Identity." *Signs: Journal of Women in Culture and Society* 16(1): 102–27.

Facer, Keri, Rosamund Sutherland, Ruth Furlong, and John Furlong. 2001. "What's the Point of Using Computers? The Development of Young People's Computer Expertise in the Home." *New Media and Society* 3(2): 81–97.

Gailey, Christine Ward. 1993. "Mediated Messages: Gender, Class, and Cosmos in Home Video Games." *Journal of Popular Culture* 27(1): 81–97.

GameDiva. 2001. "Chicks Who Make Games—Megan Gaiser." http://us.thrustmaster. com/news/read.php3?newsid=335&skin=GameDiva. Accessed July 20, 2002.

———. 2002. "Chicks Who Make Games—Stevie Case." http://us.thrustmaster.com/ news/read.php3?newsid=369&skin=GameDiva. Accessed June 22, 2002.

Glaubke, Christina R., Patti Miller, McRae A. Parker, and Eileen Espejo. 2001. *Fair Play? Violence, Gender, and Race in Video Games.* Oakland, Calif.: Children Now.

Goodman Research Group. 2002. Final report of The Women's Experiences in College Engineering (WECE) Project. http://www.grginc.com/WECE_FINAL_ REPORT.pdf. Accessed February 16, 2005.

Graner Ray, Sheri. 2004. *Gender-Inclusive Game Design: Expanding the Market.* Hingham, Mass.: Charles River Media.

Greenspan, Robyn. 2002. "Games People Play." http://cyberatlas.internet.com/big_

picture/applications/print/0;s1;s3;s01_1152221,00.html. Accessed February 16, 2005.

Griffiths, Mark D. 1997. "Computer Game Playing in Early Adolescence." *Youth and Society* 29(2): 223–37.

Henderson, Karla A., M. Deborah Bialeschki, Susan M. Shaw, and Valeria J. Freysinger. 1989. *A Leisure of One's Own: A Feminist Perspective on Women's Leisure.* College Park, Penn.: Venture.

Honey, Margaret, Babette Moeller, Cornelia Brunner, Dorothy Bennett, Peggy Clements, and Jan Hawkins. 1991. "Girls and Design: Exploring the Question of Technological Imagination." *Transformations* 2(2): 77–90.

Howard, Philip E. N., Lee Rainie, and Steve Jones. 2001. "Days and Nights on the Internet: The Impact of a Diffusing Technology." *American Behavioral Scientist* 45(3): 383–404.

Huff, Charles W. 1997. "The Internet Is a Fine Place for Women." *Computers and Society* 27(4): 27.

Huff, Charles W., and Joel Cooper. 1987. "Sex Bias in Educational Software: The Effect of Designers and the Software they Design." *Journal of Applied Social Psychology* 17(6): 519–32.

IDSA (Interactive Digital Software Association). 2001. *State of the Industry Report, 2000–2001.* Washington, D.C.

Inkpen, Kori, Maria Klawe, Joan Lawry, Kamran Sedighian, Steve Leroux, and David Hsu. 1994. "'We Have Never-Forgetful Flowers in Our Garden': Girls' Response to Electronic Games." *Journal of Computers in Mathematics and Science Teaching* 13(4): 383–403.

Kafai, Yasmin B. 1998. "Video Game Design by Girls and Boys: Variability and Consistency of Gender Differences." In *From Barbie to Mortal Kombat: Gender and Computer Games,* ed. Justine Cassell and Henry Jenkins, 90–114. Cambridge, Mass.: MIT Press.

Kahle, Jane Butler. n.d. "Gender Equity in Science Classrooms." http://www.te-mat. org/kahle/. Accessed February 10, 2005.

Kallah. n.d. "What Makes a Good Female Character?" http://www.gamegirlz.com/ editor/goodfemchar.shtml. Accessed February 16, 2005.

Kiesler, Sara, Lee Sproull, and Jacquelynne S. Eccles. 1985. "Pool Halls, Chips, and War Games: Women in the Culture of Computing." *Psychology of Women Quarterly* 9: 451–62.

Krendl, Kathy A., Mary C. Broihier, and Cynthia Fleetwood. 1989. "Children and Computers: Do Sex-Related Differences Persist?" *Journal of Communication* 39(3): 85–93.

Linn, Marcia C. 1985. "Fostering Equitable Consequences from Computer Learning Environments." *Sex Roles* 13(3–4): 229–40.

Margolis, Jane, and Allan Fisher. 2002. *Unlocking the Clubhouse: Women in Computing.* Cambridge, Mass.: MIT Press.

Margolis, Jane, Allan Fisher, and Faye Miller. 1998. "Computing for a Purpose: Gender and Attachment to Computer Science." Working paper of the Carnegie Mellon project on Gender and Computer Science: Women in Computer Sciences: Closing the Gender Gap in Higher Education. http://www-2.cs.cmu.edu/ ~gendergap/purpose.html. Accessed August 14, 2002.

Martinson, Anna M. 1998. "Perceptions of Gender and Technology." Paper presented at the annual meeting of the American Psychological Association, San Francisco, August 14–18.

——. 2003. "The Use, Enjoyment, and Gendering of Computer Games." Paper presented at the annual meeting of the Southeastern Women's Studies Association, Blacksburg, Virginia, March 20–22.

Martinson, Anna M., Nancy Schwartz, and Misha W. Vaughan. 2002. "Women's Experiences of Technology and Leisure: Implications for Design." *New Media and Society* 4(1): 25–45.

McDonough, Jerome. 1999. "Designer Selves: Construction of Technologically Mediated Identity within Graphical, Multiuser Virtual Environments." *Journal of the American Society for Information Science* 50(10): 855–69.

Mendels, Pamela. 2000. "Changing Girls' Attitudes about Computers." *New York Times,* April 12: B10.

Miller, Leslie, Melissa Chaika, and Laura Groppe. 1996. "Girls' Preferences in Software Design: Insights from a Focus Group." *Interpersonal Computing and Technology* 4(2): 27–36.

Mirapaul, Matthew. 2002. "Shoot Away. You Can't Keep the Digital Avenger Down." *New York Times,* January 7: 2. http://www.nytimes.com/2002/01/07/arts/design/07ARTS.html?ex=1064980800&en=ffa55a09694eeb75&ei=5070. Accessed April 26, 2002.

MIT News. 1997. "MIT Conference to Explore Gender, Computer Games." http://w3.mit.edu/newsoffice/nr/1997/43472.html. Accessed February 16, 2005.

Morahan-Martin, Janet. 1997. "Gender Differences in Internet Use." Paper presented at the annual meeting of the American Psychological Association, Chicago, August.

Morahan-Martin, Janet, and Phyllis Schumacher. 1998. "Are Internet and Computer Experiences and Attitudes Related? Gender Differences." Paper presented at the annual meeting of the American Psychological Association, San Francisco, August 14–18.

Nakamura, Lisa. 1995. "Race in/for Cyberspace: Identity Tourism and Racial Passing on the Internet." *Works and Days 25/26* 13(1–2): 181–93.

Passig, David, and Haya Levin. 1999. "Gender Interest Differences with Multimedia Learning Interfaces." *Computers in Human Behavior* 15: 173–83.

Pereira, Joseph. 1994. "Computers, the Gender Divide: A Tool for Women, a Toy for Men; Video Games Help Boys Get a Head Start." *Wall Street Journal,* March 1: A5.

Provenzo, Eugene F. 1991. *Video Kids: Making Sense of Nintendo.* Cambridge, Mass.: Harvard University Press.

Radway, Janice A. 1987. *Reading the Romance: Women, Patriarchy, and Popular Literature.* London: Verso.

Rood, J. 1998. "Exploring Yoga in 3D." Seeking the Light series. Colorado Springs, Colo. http://www.explore-it.com/yoga.html. Accessed August 4, 2001.

Russo, Jan. 1997. "Software for Girls: A Mother's Perspective." http://www.superkids.com/aweb/pages/features/girls/jrc1.shtml. Accessed February 16, 2005.

Sanders, Jo. 1994. *Lifting the Barriers: 600 Tested Strategies That Really Work to Increase Girls' Participation in Science, Mathematics, and Computers.* Seattle, Wash.: Washington Research Institute.

Schirra, Jörg R. J. 2001. "'Computer Game Design': How to Motivate Engineering Students to Integrate Technology with Reflection." In *Proceedings of the Fourth Annual Conference of the UNESCO International Centre for Engineering Education*, ed. Z. J. Pudlowski, 165–69. UNESCO International Centre for Engineering Education.

Schleiner, Anne-Marie. 2001. "Does Lara Croft Wear Fake Polygons?" *Leonardo* 34(3): 221–26.

Shashaani, Lily. 1994. "Socioeconomic Status, Parents' Sex-Role Stereotypes, and the Gender Gap in Computing." *Journal of Research on Computing in Education* 26(4): 433–51.

Shneiderman, Ben. 1998. *Designing the User Interface: Strategies for Effective Human-Computer Interaction.* Reading, Mass.: Addison Wesley Longman.

———. 2000. "Universal Usability." *Communications of the ACM* 43(5): 84–91.

Smith, Jonas H. 2001. "What Women Want (and It Ain't Counter Strike)." http://www.game-research.com/art_what_women_want.asp. Accessed February 16, 2005.

Spilker, Hendrik, and Knut H. Sorensen. 2000. "A ROM of One's Own or a Home for Sharing?" *New Media and Society* 2(3): 268–85.

Steeves, H. Leslie, S. L. Becker, and H. C. Choi. 1988. "The Context of Employed Women's Media Use." *Women's Studies in Communication* 11(1): 21–46.

Stone, Allucquère Rosanne. 1995. *The War of Desire and Technology at the Close of the Mechanical Age.* Cambridge, Mass.: MIT Press.

Subrahmanyam, Kaveri, and Patricia M. Greenfield. 1998. "Computer Games for Girls: What Makes Them Play?" In *From Barbie to Mortal Kombat: Gender and Computer Games,* ed. Justine Cassell and Henry Jenkins, 46–71. Cambridge, Mass.: MIT Press.

Upitis, Rena. 1998. "From Hackers to Luddites, Game Players to Game Creators: Profiles of Adolescent Students Using Technology." *Journal of Curriculum Studies* 30(3): 293–318.

Wajcman, Judy. 1991. *Feminism Confronts Technology.* Cambridge: Polity.

Wilder, Gita, Diane Mackie, and Joel Cooper. 1985. "Gender and Computers: Two Surveys of Computer-Related Attitudes." *Sex Roles* 13(3–4): 215–28.

Winner, Langdon. 1980. "Do Artifacts Have Politics?" *Daedalus* 109(1): 121–36.

WomenGamers.com. 2001. "For the Love of a Game." http://www.womengamers.com/interviews/lotus.php. Accessed February 10, 2005.

Wright, Kathryn. 2000. "GDC 2000: Race and Gender in Games." http://www.womengamers.com/doctork/racegender.php. Accessed February 8, 2005.

Yates, Simeon J., and Karen Littleton. 1999. "Understanding Computer Game Cultures: A Situated Approach." *Information, Communication, and Society* 2(4): 566–83.

Young, Betty J. 2000. "Gender Differences in Student Attitudes towards Computers." *Journal of Research on Computing in Education* 33(2): 204–16.

15 Making Sense of Retention: An Examination of Undergraduate Women's Participation in Physics Courses

Heidi Fencl and Karen R. Scheel

Retention of women in science, technology, engineering, and mathematics (STEM) fields has been studied from multiple perspectives by researchers in many disciplines. Social science research illuminates numerous factors associated with low retention rates among women in STEM. Among these are classroom climate, teaching quality, students' reasons for entering a STEM major, and psychological factors such as self-efficacy (the situation-specific belief that an individual can succeed at a given task) and confidence (Seymour 1995; Hackett et al. 1992). The study reported in this chapter adds to the understanding of retention by examining effects of teaching strategies on the retention of women, and men, in introductory physics courses. Results suggest that student-active pedagogies, originally proposed to increase students' understanding of physics concepts, also improve retention of both women and men students and may help to narrow the retention gap between them. Findings also indicate future directions for exploring the mediating effects of student self-efficacy on the relationship between teaching styles and retention. The study has implications for more systemic approaches to improved teaching in science courses at the undergraduate level.

Given the complexity of the issue, it is no surprise that retention of women in STEM fields is under examination from multiple perspectives and in multiple academic fields. For clarity in pulling together several bodies of literature with relevance to the current study, the following two sections highlight[1] work in two key areas: examination of reasons why women leave STEM fields; and psychological factors that show promise for understanding and predicting retention and related behaviors.

Reasons Women Leave Scientific Study

Studies do not support the suggestion that women leave the sciences at a greater rate than do men because of lower GPAs, less natural aptitude, or poorer preparation (Seymour 1992b, 1995). On the other hand, gender differences have been found in factors such as classroom climate experiences and loss of self-esteem (e.g., Crawford and MacLeod 1990; Sandler, Silverberg, and Hall 1996; Seymour 1995); these factors, therefore, merit further scrutiny.

Mary Crawford and Margo MacLeod (1990), in an examination of students' assessment of climate at one small state university and one liberal arts institution, found significant gender differences in students' levels of personalized interaction (how well the student is known and valued by the instructor as an individual in the course), with women reporting lower levels than men. Personalized interaction also varied with class size and academic area, with science as the worst area. Crawford and MacLeod also reported that levels of student assertiveness in the classroom differed at the liberal arts institution (but not at the state university): male students there were more assertive than female students.

Crawford and MacLeod did not find main effects for student gender for participatory climate (a variable measuring the classroom climate as a whole based on frequency of students' active participation), but female and male students gave different reasons for not participating. Women more frequently cited reasons such as not having formulated their ideas well enough, not knowing enough about the subject, and not wanting to appear unintelligent than did men, while men were more likely to cite not having done the reading. Although Crawford and MacLeod's findings are consistent with those of other studies, in their study the link between classroom climate and retention is not a direct one. Elaine Seymour (1992a, 1992b, 1995) studied the relationship between retention and factors including climate and course considerations. Participants from a spectrum of campuses, ranging from a private college with a strong teaching emphasis to a large public research institution, were surveyed or interviewed. Both students who stayed in STEM majors and students who switched to other fields were included in the sample group. Not surprisingly, Seymour found that both women and men identified many of the same concerns in their classes and their majors, with additional climate variables cited by women. The most frequently cited concern of both sexes, mentioned by 85 percent of the women and 80 percent of the men, was poor teaching (Seymour 1995). Additional climate variables which women students identified as contributing to their stress (rudeness, failure to treat women as serious students or relate to them as colleagues, unpleasant treatment of women who got good grades, and sexually suggestive remarks and jokes) were largely attributed to their fellow students rather than faculty members, although faculty and TA complicity, by ignoring these behaviors, was also noted (Seymour 1995).

As indicated above, these student concerns do not show a direct link to retention. Seymour found poor teaching to be the most common student concern, but it was *not* the most important factor in changing majors; women and men who changed majors ranked it only fourth and third, respectively, among their reasons for switching. Behavior of other students did not directly make the top eighteen reasons for changing majors, with the possible exception of "morale undermined by competitive culture," identified as a contributing factor by more of the switching men (26 percent) than of the switching women (4 percent) (Seymour 1995).

Student concerns also do not always show a consistent picture across studies. Although Crawford and MacLeod (1990) found class size to have a large effect on climate variables for both individual students and the class as a whole, participants in Seymour's study did not cite class size, poor facilities, poor TAs, or language barriers as concerning factors (Seymour 1992a). This does not necessarily mean that the results are contradictory. For example, class size might be indirectly included as a factor contributing to other concerns. (One might ask if larger classes contribute to poor teaching or to worse behavior of male students toward their female counterparts.)

A further complication to understanding retention is that both students who remained in STEM majors and those who changed to nontechnical fields identified many of the same concerns about their science programs. Interestingly, those who did and those who did not change majors did not differ in ability (Seymour 1992a).

In spite of the complexity of the issue, Seymour (1995) was able to draw some interesting generalizations related to retention. Consistent with attribution theory and trends for girls even in grade school, women participants experienced a drop in self-esteem regardless of their actual performance, while men's self-esteem rose over the same period. In probing more deeply into reasons for switching majors, Seymour found that women's reasons for leaving STEM study indeed had to do with low self-esteem (cited by 77.9 percent of the women compared to 42.9 percent of the men) and psychological alienation, while men were more likely than women to change majors for job-related reasons. However, Seymour also concluded that the largest difference between women and men was not in their reasons to leave STEM fields but rather in their reasons for entering those fields. Women were much more likely than men to study science because they had been encouraged by others to do so, and, with the exception of African American women, "performing for others" was key to understanding their subsequent behaviors (Seymour 1995).

The Effect of Self-Efficacy and Pedagogy on Retention

A key point in these climate and retention findings is that given the same set of circumstances, one individual may choose to remain in a STEM major while another individual of equal ability may choose to switch. Understanding retention, therefore, also requires examining individual behaviors and the interaction between the individual and the environment.

Studies in educational and vocational psychology demonstrate a relationship between an individual attribute, self-efficacy, and retention of students (e.g., Betz and Voyten 1997; Brown et al. 1996; Hackett et al. 1992; Hackett and Lent 1992; Lapan, Shaughnessy, and Boggs 1994; Lent, Brown, and Larkin 1987; Nauta, Epperson, and Kahn 1998; Schaefers, Epperson, and Nauta 1997; Siegel,

Galassi, and Ware 1985). Self-efficacy is an individual's situation-specific belief that she or he can succeed at a given task (Bandura 1977, 1986; Betz and Hackett 1981). Yet to be thoroughly explored, however, is whether teaching approaches (one factor in the educational environment) might affect retention through a mediating variable such as self-efficacy, and whether such links operate differently for women and men.

Gail Hackett and colleagues (1992) studied the relationship of academic achievement to undergraduate engineering students' self-efficacy, interest, outcome expectations (what outcomes participants expected for successful completion of their major), stress, strain, coping, high school GPA, and SAT scores. They found that academic milestone self-efficacy (an individual's confidence in her/his ability to complete educational requirements for a variety of STEM occupations) was the strongest predictor of achievement and performance, as measured by cumulative college GPA.

Similar results directly tied to retention come from Robert Lent, Steven Brown, and Kevin Larkin's (1986; 1987) work, which studied students enrolled in a career/educational planning course for undergraduates considering STEM majors. Participants exhibited high ability as determined by their past achievement. They completed two self-efficacy measures rating their perceptions of their ability to complete the educational requirements and academic milestones required for STEM majors. No significant gender effects were found for either measure. For both measures, high self-efficacy respondents were significantly more successful and persistent than those in the lowest self-efficacy quartile (Lent, Brown, and Larkin 1986). Furthermore, self-efficacy gave a unique variance to grades and persistence beyond math ability (math PSAT score), past achievement (high school rank), and expressed interest.

Lent, Brown, and Larkin (1987) also compared the relative contribution of self-efficacy to that of two variables emphasized by other vocational choice theories: interest congruence (agreement between a person's interests and those of practitioners in her/his field) and consequence thinking (anticipation of major consequences in decision making). Of these three variables, they found self-efficacy to be the most useful predictor of both success (grades) and persistence as measured over a one-year follow-up period.

The picture that begins to emerge from educational/vocational studies suggests that self-efficacy is an important part of individual retention decisions. It might help to explain, for example, why some women persist in the same chilly climate which causes other women to change majors. And if, as Seymour (1995) suggests, women are more likely than men to enter STEM study because of the influence of others rather than for intrinsic reasons, development of self-efficacy suggests an important intervention toward their retention.

Self-efficacy is understood as a dynamic attribute, influenced by a person's experiences. For example, Hackett et al. (1992) found that faculty encouragement was positively related to self-efficacy. It is therefore not unreasonable to hypothesize that teaching methods themselves might influence retention through their effect on self-efficacy. An increasing body of research has focused on science teaching (methods and curriculum) and student learning, and has shown that a variety of student-active pedagogies (teaching approaches in which students are actively, rather than passively, engaged with the material) improve student performance. An obvious question to ask is if these pedagogical approaches also improve students' self-efficacy and ultimately retention. Before exploring such potential links between teaching and retention, a brief review of pedagogies suggested by science education research is in order.

Teaching by inquiry is one pedagogy that has found good success (e.g., Shaffer and McDermott 1992; Thacker et al. 1994). For example, Beth Thacker and colleagues compared student performance on two electric circuit problems for students in an inquiry-based physics course and three traditionally taught (largely lecture-based) courses. At the time of the test, inquiry students had experienced circuits conceptually but had not yet studied their mathematical formulation. All students were given the same problems: a two-part analytic problem requiring some math for the solution, and a conceptual problem which could be solved by either conceptual or analytic means. In spite of having less math background than the majority of the students in the traditional courses, the inquiry students outscored all groups on the conceptual circuit problem, and all but the honors physics students on the analytic problem (Thacker et al. 1994).

Cooperative learning strategies such as group problem solving are also very successful. Successes include higher scores on traditional exam problems (Gautreau and Novesky 1997; Heller, Keith, and Anderson 1992) and greater ability to solve difficult problems (Heller, Keith, and Anderson 1992). Tutorials are a variation on cooperative learning designed to take small groups of students through a process of developing concepts and reasoning skills and applying them to the real world (McDermott 2001). McDermott found that students using physics tutorials, like those learning through inquiry, do much better than traditionally taught students on conceptual problems, and somewhat better than traditionally taught students on quantitative problems, even though they spent less time solving quantitative problems.

Science Teaching/Retention Study

Taken together, retention research and science education research pose an interesting, and as yet unexplored, pair of questions: Do pedagogies used in the classroom and laboratory affect retention? If so, is self-efficacy an important mediating factor in that relationship? Exploration of these questions is particularly timely. Interventions to encourage women to succeed in STEM have thus far focused largely on providing women with skills to succeed in the climate of science, but it is increasingly recognized that the organization and culture of science, too, share responsibility for retention. Given that students spend such a large amount of time in classes, it seems reasonable that their routine experiences could be at least as important for retention as outside experiences.

The study presented in this chapter explores the hypothesis that teaching methods affect the retention of students (especially women) in introductory classes. Steps to explore the role of self-efficacy as a possible mediating factor are described in the chapter's last section.

Nine introductory physics sections with a total enrollment of 611 students were included in the study. The sections came from three midwestern institutions: a small private liberal arts college, a small state university, and a state research university. Instructors voluntarily allowed researchers access to their classes. Students were asked to complete a survey instrument during a course meeting. Three hundred twenty-one students (196 men and 125 women) completed the survey questionnaires.

The questionnaire included three sections: demographic information such as sex, age, ethnicity, and math background; descriptive information about pedagogies used in class;[2] and students' plans for future science study, as well as their enjoyment of aspects of the class and how helpful they were. Drop-rate data for the courses were obtained from the course instructors or from the university registrar when not provided by the instructor. Both those actual drop rates (percent of students who attended the first day but subsequently dropped the course) and "desire-to-drop" rates (percent of students answering "yes" to the question, "If this course were not required for your major/career plans, would you have preferred to drop it?") were used as retention measures. The desire-to-drop variable was included because both majors and nonmajors can be found in any science class. A nonmajor may decide to stick out a required course in order to complete a different major, but that student might, with the same set of experiences, drop the course and change majors if it were in her/his major field. It is possible that actual drop rates in the course, then, do not tell a complete story of course retention decisions.

Each section's questionnaires were identified as coming from a calculus- or algebra-based physics course, and as from a course with large or small enrollment. Courses were labeled traditional, nontraditional, or mixed according to student ratings of the teaching methods used; a course was labeled nontraditional if, for instance, question and answer, conceptual laboratory activities, and/or audiovisual presentations were the largest elements of the course, in terms of the time devoted to each. Traditional courses were defined as those in which lecture accounted for at least 50 percent of total course time (approximately 54 percent of time was spent in lecture in traditional sections) and the time devoted to quantitative lab activities was greater than that spent on conceptual lab activities. Mixed courses were those that fell between the two extremes. In practice, there was no gray area in assigning definitions, as the courses very clearly divided into those that relied almost exclusively on lecture and quantitative lab (traditional), those that used virtually no lecture (nontraditional), and those that used a variety of traditional and nontraditional techniques (mixed). Questionnaires were combined according to class type for analysis. Of the 321 participants, 186 (128 men and 58 women) were in traditional classes, 88 (46 men and 42 women) were in mixed classes, and 47 (22 men and 25 women) were in nontraditional classes.

Unfortunately, all large classes were traditional (although not all traditional

Table 15.1. Percent of Students Who Dropped Introductory
Physics, by Course Type

Course Type	Traditional	Mixed	Nontraditional
All Students	12	3.0	1.9
Men	12	1.9	3.9
Women	11	4.2	0.0

Table 15.2. Percent of Students Completing a Survey
Who Would Have Preferred to Drop Introductory Physics,
by Course Type

Course Type	Traditional	Mixed	Nontraditional
All Students	54	42	70
Men	42	35	59
Women	79	50	80

classes were large), so effects of class size could not always be controlled by this study. Variables not addressed were sex of instructor,[3] ethnicity of students,[4] and percent of women in the classes.[5] Both instructor and student sex present possible confounding variables to be addressed in future work. Ethnic diversity among participants was low for all sample groups, and so the conclusions stated here are related to teaching approaches for European American students. Additional work with broader demographics is needed in order to formulate more general results for a diverse population of students.

Analysis of Data

Table 15.1 shows the actual drop rates, and table 15.2 shows the desire-to-drop rates, broken down according to sex and course type. Even the pure numbers show cause for concern, as the lowest proportion of students desiring to drop (men in mixed pedagogy courses) is 35 percent, and the lowest percentage of women indicating a desire to drop is 50 percent (also in mixed pedagogy courses). One cautionary note should be added about both the pure numbers and the subsequent analyses. The bulk of enrolled students who did not complete questionnaires came from traditional courses with large enrollments, and were not in class the day the survey was administered. This introduces a possible bias that should be considered in evaluating the results, as the two other pedagogical categories did not experience as high a noncompletion rate.[6]

The data in tables 15.1 and 15.2 show differences in retention variables between students in courses of different pedagogical types and between women and men in the same course type. Because information for these variables was collected as frequency data, chi-square (χ^2) analyses were performed for drop rates and for desire-to-drop rates in order to test whether the differences were significant. These analyses test the significance of the relationship between two variables when data are expressed in terms of frequency of joint occurrence. In other words, the analytic approach used in the study assessed whether course type and retention variables, and whether gender and retention variables, are significantly related. Results are summarized in table 15.3. Throughout the discussion, "expected values" means the drop or desire-to-drop rates that would be expected for the population if the variables were unrelated (for example, if students' desire to drop the course were independent of gender). "Observed values" means the values actually obtained in the study. Alpha (α), or the level beyond which results would be considered statistically significant, was set in advance at .05. Significance, then, indicates that the discrepancy between expected and observed values is great enough that results of that magnitude would occur by chance (rather than because of real relationships between the variables) just 5 percent of the time.

One category of analyses examined retention variables (actual drop rates and desire-to-drop rates) in courses of different pedagogical types. Actual drop rates were found to be significantly related to the type of course. Students in traditional courses were more likely to drop than students in nontraditional courses and in mixed courses: $\chi^2 = 4.89$ and $\chi^2 = 7.60$, respectively. Numbers of expected drops were too small to perform meaningful statistical comparisons between nontraditional and mixed courses. Because all large enrollment courses were traditional, an additional test was done to compare drop rates between large-enrollment, calculus-based, traditional courses and small-enrollment, calculus-based, traditional courses. Differences based on course size were not found to be significant, although numbers were small enough that the expected number of drops (4.94) was marginal for this analysis.

Desire-to-drop rates showed significant results between all course types. For traditional vs. nontraditional, traditional vs. mixed, and mixed vs. nontraditional courses, the χ^2 values were 4.98, 4.50, and 15.36, respectively. An additional χ^2 test rejects the null hypothesis that a student's desire to drop is independent of the teaching methods used in class: $\chi^2 = 9.85$, p<0.05.

These two sets of results (differences in drop and desire-to-drop rates for different course types) address the primary point of the study, indicating that alternative teaching methods originally introduced to improve student understanding also improve student retention in introductory physics classes. However, additional χ^2 analyses were performed to determine if gender is significantly related to retention, both within and across course types. Expected drop numbers were too small to compare actual drop rates of men and women in nontraditional and mixed courses; no significant gender difference was found in the percentage of students dropping traditional courses, $\chi^2 = 0.34$.

Additional χ^2 tests were performed to test the null hypotheses that a stu-

Table 15.3. χ^2 Results for Retention Analyses

Null Hypothesis	χ^2
There are no differences in drop rates between students in courses of different pedagogical types.	
traditional vs. nontraditional courses	4.89*
traditional vs. mixed courses	7.60*
mixed vs. nontraditional courses	
There are no differences in desire-to-drop rates between students in courses of different pedagogical types.	
traditional vs. nontraditional courses	4.98*
traditional vs. mixed courses	4.50*
mixed vs. nontraditional courses	15.36*
There are no differences in drop rates between men and women in courses of the same type.	
traditional	0.34
nontraditional	
mixed	
There are no differences in desire-to-drop rates between men and women in courses of the same type.	
traditional	13.8*
nontraditional	4.5*
mixed	4.3*
all courses combined	22.75*

* indicates rejection of the null hypothesis ($\alpha = .05$). No values are listed in cases where numbers are too small to allow meaningful inferences.

dent's desire to drop introductory physics is independent of her/his gender for 1) all courses combined, 2) traditional courses, 3) nontraditional courses, and 4) mixed courses. All null hypotheses were rejected, with χ^2 values of 22.75 (for all courses taken together), 13.8 (for traditional courses), 4.5 (for nontraditional courses), and 4.3 (for mixed courses). In all cases, women were more likely to prefer to drop than men. Because retention, especially of women, is the underlying interest area for this study, phi coefficients (a measure of the strength of the relationship between gender and desire to drop in each of the course type categories) were also computed for these results. Scores may range from 0 to 1, with larger values representing stronger relationships. The phi coefficients were .27, .31, and .22 for traditional, nontraditional, and mixed courses respectively. In other words, while women were more likely to prefer to drop in all cases, this effect was weakest in the mixed courses.

Discussion of Results

Drop rates for students in traditionally taught introductory physics courses were found to be significantly greater than drop rates for students in either nontraditional or mixed pedagogy courses. Rates at which students desired to drop the class were also higher in traditional courses than in mixed pedagogy courses, but were higher for students in nontraditional courses than in traditional courses.[7] In all cases, women desired to drop the courses at higher rates than did men. Taken together, these results suggest that teaching approaches used in introductory physics courses do affect retention of students taking those classes, and to some degree do so differently by gender. Classes which use a mixture of lecture and student-active pedagogies such as cooperative learning and inquiry or conceptual laboratory activities fared the best overall for the retention variables.

Several points about these results deserve further discussion. Most immediately, the results suggest that science educators can make a substantial difference in retention (almost 30 percentage points as measured by desire-to-drop rates for both women and men across course types in this study) with the pedagogies they bring into their classrooms. It is also clear, however, that pedagogical choice is not the only factor that determines course retention. Even in the best case observed in this study, 50 percent of the women and 35 percent of the men would drop their introductory physics course if given the choice. The fact that other factors beyond pedagogy are important is not unexpected, given results of other studies (e.g., Seymour 1995) which suggest that climate, career decisions, and a number of other factors unrelated to teaching are all part of students' retention decisions, and that although science students cite poor teaching as their top concern, they do not cite it as their leading reason for switching majors. Additional insight comes from the few participants in this study who included comments on their survey, most of whom said that they wanted to drop the course because it was difficult.

Another point worthy of discussion is the success of mixed pedagogy courses over courses of other types. There are a number of plausible explanations. It is possible that students in mixed courses are more likely to connect with a way of teaching that makes sense to them. It is also possible that including variety in the course helps to engage students more actively. There may also be a selection effect among instructors, so that those who include a mix of pedagogies may also be more in touch with, for example, climate, their students, or educational research. And, finally, it is possible that a key teaching approach was present significantly in the mixed courses in this sample but not in the others. In that case, the apparent retention-related performance of mixed courses could actually be an effect of a single approach. The design of the next stage of this project will help to unfold some of these possibilities by evaluating effects of each pedagogy individually, including questions about student-student and instructor-student climate, and matching professors' backgrounds where possible.

What should be made of the fact that students in the nontraditional courses showed a lower actual drop rate, but a higher desire-to-drop rate, than students in the traditional courses? As mentioned earlier, traditional courses showed the highest non-response rate of the three course categories. It is plausible that the students not in class when the survey was administered are more likely to desire to drop than those attending. In that case, the desire-to-drop rates for the traditional courses would be artificially low. Even if that were found to be true, however, there remains a substantial difference in desire to drop between students in the nontraditional and the mixed courses. The key difference between these course types is the absence or presence of lectures. Interestingly, when asked to rate the helpfulness and usefulness of the pedagogies employed in their course, study participants rated lectures lowest of all pedagogies in both categories. This suggests that future work should explore the role that lecture plays in the comfort level of students.

Although desire-to-drop rates were originally introduced to explore retention of students with majors other than physics, the gap between drop rates and desire-to-drop rates was present in calculus-based courses (largely physical science and engineering majors) as well as in algebra-based courses (largely biological science majors).[8] The persistence of the gap highlights again the complexity of retention. Students make retention decisions based on a spectrum of experiences and criteria. Experience in a single course is part of that spectrum, but it is not the entire basis for a decision. Retention in a course (as examined in this study) and retention in a major are related but are not the same. The difference between drop and desire-to-drop rates suggests that, to the degree that pedagogies affect retention decisions, departmentwide curricular reform has an important role to play.

Finally, the large differences in desire-to-drop rates between women and men, and between women in traditional courses and those in mixed courses, suggest that future work should continue to examine gender effects. Retention trends (for example, lower drop and desire-to-drop rates for students in mixed courses compared to traditional courses) are the same for men as for women, but the pattern is stronger for women. Science educators have long argued that educational reform is not a zero-sum game, that making courses more woman-friendly does not make them less man-friendly. This study supports that claim. The list of factors contributing to lower retention of female science students than of male science students is long (Crawford and MacLeod 1990; Sandler, Silverberg, and Hall 1996; Seymour 1992a, 1992b, 1995) and it should be expected that a solution to the retention gap must therefore also be multifaceted. By illustrating that the gender gap in desire-to-drop rates can be narrowed by including a mix of pedagogies in a course,[9] this study shows that pedagogical approach is one of the important factors in the unequal retention of students.

Results of this project, then, support the hypothesis that teaching methods affect retention of students, especially women, in introductory classes. The complexity of the results suggests that further study, including exploration of the role self-efficacy might play in mediating the relationship between course pedagogy and retention, is in order. In particular, given that self-efficacy has shown such promise in predicting retention (e.g., Lent, Brown, and Larkin 1987), it is possible that the increased retention in mixed pedagogy courses comes about as particular teaching approaches, or inclusion of multiple teaching approaches, helps students to build physics self-efficacy.

Follow-up Study on Teaching Styles, Self-efficacy, and Retention

In preparation for pursuing the question of retention and self-efficacy of physics students, a survey instrument was developed and given to students in seven introductory chemistry courses on two-year campuses of a large state university system (Scheel et al. 2002). Data on instructor background were collected on a brief instructor survey. The student survey, completed by 155 students, contained a demographic section, a section about the frequency with which fourteen teaching strategies were used in class, and a course-specific sources-of-self-efficacy scale. The scale was modeled after existing scales and probed the four sources of efficacy (performance accomplishment, vicarious learning, social persuasion, and emotional arousal) as postulated by Albert Bandura (1977). Collection of sources-of-efficacy information, rather than just an academic milestones self-efficacy score, is a key component of the examination of any self-efficacy-mediated link between pedagogies and retention. For example, it could be hypothesized that collaborative learning might build self-efficacy, and thus increase retention, by contributing to a student's sense of achievement, by fostering beneficial social interactions within the group, or by providing a climate which reduces negative or stressful emotional reactions to the course. Analysis of the effects of specific pedagogies on sources-of-efficacy scores will help to distinguish between such possibilities and thus to better understand the influence of teaching approaches on self-efficacy.

Analysis of data from the chemistry sample (Scheel et al. 2002) supports the preliminary conclusions that 1) pedagogies used in undergraduate introductory chemistry courses appear to be related to development of students' chemistry self-efficacy;[10] 2) in agreement with the results presented here, a mix of traditional and nontraditional pedagogies appears to be beneficial; 3) collaborative learning appears as the single most self-efficacy-enhancing teaching strategy; and 4) teaching strategies appear to have similar impacts on the development of self-efficacy for men and women, but the use of discussion and cooperative learning may be especially important for female chemistry students, as these two strategies were related to less negative emotional arousal for women. Emotional arousal was the only source of self-efficacy in which women and men differed significantly. Although the above are preliminary conclusions, the consistency between these results and the retention study, combined with analysis supporting the reliability and validity of the developed survey instrument, sets the stage for a detailed examination of the mediating effects of self-efficacy on the link between pedagogies and retention.

Following the chemistry pilot of the self-efficacy/retention instrument, the survey was modified to include a section of climate questions relating to the course. The next steps are to administer the tested and revised survey to a sample of introductory physics students and to examine relationships between teaching strategies, self-efficacy, climate, and retention. The analysis will examine the mediating role that self-efficacy might play between pedagogical approaches and retention, as well as possible links between pedagogical approach and course climate.

Retention of women in technical fields is clearly not a straightforward issue, and no single answer exists for increasing women's representation. Bringing together relevant research from sociology, psychology, and physics education, however, provides a richer understanding of women's persistence in STEM fields than is given by a single perspective. It also suggests a broader array of approaches to increasing retention than comes from any single field.

Outside the classroom, interventions to improve the retention of women in science have long included, for example, assertiveness training and other climate-related education, and mentoring. Some authors (e.g., Betz and Schifano 2000) have explored training interventions for women intended to build self-efficacy. This study suggests that additionally addressing retention by including a mix of pedagogies within a student's normal course of science classes also shows promise for improving retention of physics students and for narrowing the retention gap between women and men.

The indication that retention can be affected through pedagogical choices made daily in the classroom is particularly noteworthy for two reasons. First, it addresses the problem that attempts to increase diversity through interventions aimed at changing underrepresented populations are inherently flawed—because they put increased pressure on the underrepresented population, and because they reduce the benefits that a more diverse set of practitioners can bring to the field as those practitioners are trained to, in some way, be more like the majority. Second, physics education researchers and women's studies scholars have shown that through increased use of student-active pedagogies there is a common ground in which the goals of both improving the teaching of physics content and of changing the gendered climate in the science classroom can be addressed. By showing a link between these same teaching strategies and actual retention, this study adds improved retention, especially but not exclusively of women, to the list of benefits which come from widespread use of student-active pedagogies in physics courses. The complexity of retention decisions and the unacceptably large desire-to-drop rates among science students reinforce the need to address pedagogical change systemically throughout science courses in higher education.

The immediate next step of the ongoing project, then, is to examine the effect of pedagogies on retention more carefully by addressing the role that different teaching approaches might play in the development of student self-efficacy and by analyzing the relative contribution of that effect to overall self-efficacy and retention decisions.

Notes

1. It is not the goal of this essay to provide an exhaustive literature review of each of the disciplines contributing to retention research. Interested readers are encouraged to contact the authors for a more complete bibliography.

2. Students were asked about the amount of class time devoted to lecture, group learning, question and answer, student problem demonstration, quantitative laboratory activities, conceptual laboratory activities, audiovisual presentations, and other activities.

3. In order to preserve the anonymity of instructors, courses were not identified by instructor. Therefore there was no way to compare results for the (one) female instructor to those for the male instructors with the same course type (mixed courses) under this study design.

4. Data regarding ethnicity of students were collected, but the percent of students of non-European descent was too small to make statistical comparisons.

5. Approximately 50 percent of the students in both the nontraditional and mixed courses were women, but only about 18 percent of the students in the traditional courses were women. No subgroups of classes involving traditionally taught and other courses had the same gender makeup, so that examination of the effect of other women students in the class was not possible with this sample.

6. One plausible argument is that actual desire-to-drop rates for traditional courses are higher than those found by the study, which would be true if students not attending class are, on average, less happy with the course than those in attendance. If this is indeed the case, results of this study would be strengthened rather than weakened. However, further data collection is clearly required before such an assumption can be made.

7. Note that this result is sensitive to any sampling bias due to nonattendance by students in large-enrollment, traditional courses.

8. Forty-five percent of participants in the mixed algebra-based courses, 25 percent of participants in the mixed calculus-based courses, 58 percent of participants in the traditional algebra-based courses, and 47 percent of participants in the traditional calculus-based courses indicated a desire to drop the course.

9. Women's desire-to-drop rate decreased from 1.9 times that of men (traditional courses) to 1.4 times that of men (mixed courses).

10. Effect sizes, however, are small, suggesting that other important variables also influence chemistry self-efficacy.

References

Bandura, Albert. 1977. "Self-Efficacy: Toward a Unifying Theory of Behavior Change." *Psychological Review* 84: 191–215.

———. 1986. *Social Foundation of Thought and Action: Social Cognitive Theory.* Englewood Cliffs, N.J.: Prentice-Hall.

Betz, Nancy E., and Gail Hackett. 1981. "The Relationship of Career-Related Self-Efficacy Expectations to Perceived Career Options in College Women and Men." *Journal of Counseling Psychology* 28: 399–410.

Betz, Nancy E., and Ross S. Schifano. 2000. "Evaluation of an Intervention to Increase Realistic Self-Efficacy and Interests in College Women." *Journal of Vocational Behavior* 56: 35–52.

Betz, Nancy E., and Karla Klein Voyten. 1997. "Efficacy and Outcome Expectations Influence Career Exploration and Decidedness." *Career Development Quarterly* 46: 179–89.

Brown, Steven D., Robert W. Lent, Nancy E. Ryan, and Eileen B. McPartland. 1996. "Self-Efficacy as an Intervening Mechanism between Research Training Environments and Scholarly Productivity: A Theoretical and Methodological Extension." *Counseling Psychologist* 24: 535–44.

Crawford, Mary, and Margo MacLeod. 1990. "Gender in the College Classroom: An Assessment of the 'Chilly Climate' for Women." *Sex Roles* 23(3–4): 101–22.

Gautreau, Ronald, and Lisa Novesky. 1997. "Concepts First—A Small Group Approach to Physics Learning." *American Journal of Physics* 65: 418–28.

Hackett, Gail, Nancy E. Betts, J. Manuel Casas, and Indra A. Rocha-Singh. 1992. "Gender, Ethnicity, and Social Cognitive Factors Predicting the Academic Achievement of Students in Engineering." *Journal of Counseling Psychology* 39: 527–38.

Hackett, Gail, and Robert W. Lent, 1992. "Theoretical Advances and Current Inquiry in Career Psychology." In *Handbook of Counseling Psychology,* 2nd ed., ed. Steven D. Brown and Robert W. Lent, 419–51. New York: Wiley.

Heller, Patricia, Ronald Keith, and Scott Anderson. 1992. "Teaching Problem Solving through Cooperative Grouping. Part 1: Group vs. Individual Problem Solving." *American Journal of Physics* 60: 627–36.

Lapan, Richard T., Peter Shaughnessy, and Kathleen Boggs. 1994. "Efficacy Expectations and Vocational Interests as Mediators between Sex and Choice of Math/Science College Majors: A Longitudinal Study." *Journal of Vocational Behavior* 49: 277–91.

Lent, Robert W., Steven D. Brown, and Kevin C. Larkin. 1986. "Self-Efficacy in the Prediction of Academic Performance and Perceived Career Options." *Journal of Counseling Psychology* 33: 265–69.

———. 1987. "Comparison of Three Theoretically Derived Variables in Predicting Career and Academic Behavior: Self-Efficacy, Interest Congruence, and Consequence Thinking." *Journal of Counseling Psychology* 34: 293–98.

McDermott, Lillian. 2001. "Oersted Medal Lecture 2001: Physics Education Research—The Key to Student Learning." *American Journal of Physics* 60: 1127–37.

Nauta, Margaret, Douglas L. Epperson, and Jeffrey H. Kahn. 1998. "A Multiple-Groups Analysis of Predictors of Higher Level Career Aspirations among Women in Mathematics, Science, and Engineering Majors." *Journal of Counseling Psychology* 45: 483–96.

Sandler, Bernice R., Lisa A. Silverberg, and Roberta M. Hall. 1996. *The Chilly Classroom Climate: A Guide to Improve the Education of Women.* Washington, D.C.: National Association for Women in Education.

Schaefers, Kathleen G., Douglas L. Epperson, and Margaret M. Nauta. 1997. "Women's Career Majors?" *Journal of Counseling Psychology* 44: 173–83.

Scheel, Karen R., Heidi S. Fencl, Mahnaz N. Mousaui, and Karen E. Reighard. 2002. "Teaching Strategies as Sources of Self-Efficacy in Introductory Chemistry." Paper presented at the annual meeting of the American Psychological Association, Chicago, August.

Seymour, Elaine. 1992a. "Undergraduate Problems with Teaching and Advising in SME Majors—Explaining Gender Differences in Attrition Rates." *Journal of College Science Teaching* 21(5): 284–92.

———. 1992b. "'The Problem Iceberg' in Science, Mathematics, and Engineering Education: Student Explanations for High Attrition Rates." *Journal of College Science Teaching* 21(4): 230–38.

———. 1995. "The Loss of Women from Science, Mathematics, and Engineering Undergraduate Majors: An Explanatory Account." *Science Education* 79: 437–73.

Shaffer, Peter S., and Lillian C. McDermott. 1992. "Research as a Guide for Curriculum Development: An Example from Introductory Electricity. Part II: Design of Instructional Strategies." *American Journal of Physics* 60: 1003–13.

Siegel, Ross G., John P. Galassi, and William B. Ware. 1985. "A Comparison of Two Models for Predicting Mathematics Performance: Social Learning versus Math Aptitude-Anxiety." *Journal of Counseling Psychology* 32: 531–38.

Thacker, Beth, Eunsook Kim, Kelvin Trefz, and Suzanne Lea. 1994. "Comparing Problem Solving Performance of Physics Students in Inquiry-Based and Traditional Introductory Physics Courses." *American Journal of Physics* 62: 627–33.

16 Creating Academic Career Opportunities for Women in Science: Lessons from Liberal Arts Colleges

Neal B. Abraham

Recruiting and retaining women in science (both as undergraduates and as faculty members) is a challenge for liberal arts colleges and research universities alike. As pools of job applicants shrink, new strategies for attracting prospective science faculty and nurturing new faculty members are also being developed and tested out of a mixture of necessity and desperation. This critical review draws on experiences at two liberal arts institutions—Bryn Mawr College and DePauw University—exploring the evolution of special programs designed to attract and retain undergraduate majors, to attract women and minority group members to faculty positions in fields in which they are underrepresented, and to ensure their retention, engagement, and success. At all career stages, from the first introductory undergraduate course to tenure, promotion, and beyond, special programs that have been successful for women and minorities prove to be beneficial for all.

The insights reported here on the growth and nurturing of women scientists are drawn from varied insider perspectives as a faculty member, department chair, and dean, and from completing sixteen program reviews at other institutions. Most notably, as a result of the lingering effects of gender inequity from twenty and more years ago, strategies to support and encourage women in the sciences, as both students and faculty members, have been more intentional and visible at DePauw than at Bryn Mawr.

Dimensions of the Problem

It was only a little more than a decade ago that the number of applicants for each faculty position in math and science was in the hundreds. This was a period of high production of new Ph.D.s, combined with a slight economic downturn which generated a glut of experienced applicants in holding patterns in postdoctoral positions (*Physics Today* 1998, 49), exacerbated by the shutdown of the Superconducting Supercollider project (Seife 2003, 36). Search committees struggled to select finalists, often enjoying the luxury of considering marginally relevant extra criteria. Hiring departments and their search committees could be very (perhaps even excessively) specific in their job definitions and in their selection of candidates; faculty members could be relatively ruthless in their limited support of new junior colleagues and relatively critical in their analyses of the performance of candidates for tenure.

In recent years, the numbers of science graduate students have declined while the opportunities for industrial employment in many subfields of science have increased. These factors have reduced the numbers of serious applicants for many academic positions in the sciences to ten or twenty.[1] Though the percentages of women graduates and job applicants in these fields have grown, the actual number has decreased in some cases and remained steady in others. This seeming statistical oddity is explained by the decline of the total numbers. The combination of these factors has placed extraordinary pressure on hiring committees. Even so, we are well advised to observe the studies which demonstrate that academia has continued to underutilize women and minority candidates earning Ph.D.s, attracting less than a proportionate share to faculty positions (Nelson 2002).

The exception to this pattern have been the liberal arts colleges. Remarkably, while research universities report little or no progress in expanding their complements of women faculty members in science, math, and engineering fields, a recent survey of 136 four-year colleges (Doyle 2000) found that they have expanded the percentage of women in their science, technology, engineering, and mathematics (STEM) faculty from 21 percent in the 1980s to 40 percent in the 1990s. The message for those interested in substantial improvement in recruitment, mentoring, and retention of women STEM faculty members in research universities is that many of the solutions may be found in the colleges. However, if some of the strategies which have worked in the colleges are successfully adopted in research universities, we will then merely heighten competition for a limited resource. Hence, alongside the focus on hiring, institutions in all sectors of higher education must continue to focus on the recruitment and retention of women majors and graduate students if they wish to find more women among their job candidates in the future.

Perhaps it is no coincidence that the same predominantly undergraduate institutions, particularly liberal arts colleges, have a long track record of producing a disproportionate number of the undergraduate STEM majors who go on to complete Ph.D.s in these fields (PKAL 1991). Such colleges often graduate women and members of underrepresented minority groups with degrees in mathematics and physical sciences in percentages of their graduating classes that are ten or twenty times the national average. In particular, the women's colleges and the historically Black colleges and universities (HBCUs) have been unusually successful in the proportion of their students who earn bachelor's degrees and go on to earn Ph.D.s in STEM fields. But the data show that even leaving aside the extraordinary record of women's colleges and HBCUs, liberal arts colleges have been more successful in recruiting and retaining women and minorities as undergraduate STEM majors. This sector thus bears examination of the sort completed under Project Kaleidoscope (PKAL 1991).[2] Those seeking enhancements for undergraduates and graduate students at research universities would be well advised to form partnerships and collaborations with colleagues at liberal arts colleges, who may be excellent sources of both students and programmatic ideas.

In short, there are institutions that have been very successful on two levels. It is likely that visits to successful sites, or the development of consulting relationships with representatives of schools which have had success, will enhance the planning and program development activities at institutions wishing to make improvements.

Results at DePauw and Bryn Mawr

Both Bryn Mawr College and DePauw University have been successful in recruiting and retaining women students and faculty in STEM fields. At Bryn Mawr, a women's college devoted to advancing women in the professions, we replaced a modestly successful record with an extraordinary one; by the mid-1980s 5 percent of our graduates (who are all women) had majored in physics and 10 percent had majored in mathematics (Wright 2002; Abraham 1999). These rates are twenty or more times higher than the national average. There also were almost equally strong records for women graduating with degrees in chemistry (averaging six times the national average). These successes have been sustained for twenty years.

Because advancement of women has always been a strong and explicit institutional goal at Bryn Mawr, the college has instituted few programs specifically devoted to advancing women in science. For decades, roughly 40 percent of the faculty members in STEM at Bryn Mawr were women and nearly 30 percent of all bachelor's graduates completed majors in the sciences. Institutional culture and commitment were so strong that, contrary to what many say is necessary for a climate to be supportive of women in coeducational environments, some of the successes at Bryn Mawr were created when there were few women faculty members in the math and physics departments. Neither were there special mentoring programs or a "women in science" program. Rather, success came from building communities of students at all levels with faculty members, through the creation of gathering places for evening study, through collaborative research projects, and through effective interaction and engagement in teaching and learning. Much of this was developed spontaneously, perhaps as a consequence of a critical mass of teachers and students from which both good ideas and volunteers willing to try them out could emerge. I also attribute it to an evolved pedagogical approach which focused on providing learning and challenging experiences for all rather than filtering and testing vigorously to find the elite few who could survive.

By contrast, at DePauw one finds evidence of the residual impact of the persistent and pervasive sexism of the past. Across all departments, women constitute barely 25 percent of the full professors on the faculty, while among assistant and associate professors they are fully half. The steady march of time is advancing women hired in periods of gender equity. They are being tenured, promoted, and retained at least as often as men in these fields, so that overall parity is growing. But we are just reaching the maturity of an eighteen-year-long effort of gender-sensitive affirmative action, started with the arrival of DePauw's current president, Dr. Robert G. Bottoms. Many of the more senior women were trailblazers, and they carry memories of encounters with the institutionalized sexism that was prevalent at the time—doubts they would stay, doubts about their research, and doubts about their expertise, expressed especially by students in their classes. At DePauw, not surprisingly, there are support programs for women in science, for both students and faculty members, which were constructed ten or more years ago when a small group of embattled women in the sciences and mathematics were struggling for survival while hoping they were the vanguard of systemic change. They received key support from strong senior women in other disciplines who exerted pressure on administrators and faculty colleagues alike. Indeed, the strong and persistent women-in-science programs began when there was but one woman on the faculty of each of the math and science departments. And as a caution to those who hope to make quick changes to redress decades of gender imbalances, I note that even with a successful eighteen-year record of gender-balanced hiring at DePauw, we still find institutional barriers which must be overcome.

Support for Undergraduate Majors

The successes at Bryn Mawr and DePauw for undergraduate majors indicate that women are fully capable of succeeding as STEM majors and going on to graduate schools and science careers. What contributes to higher rates of success in these and a number of other liberal arts colleges? Among the answers are small classes, supportive advisers, and research opportunities (Abraham 1999). A further answer is that departments must recognize that their introductory courses, those which draw students who have normal—rather than extraordinary—preparation and science interest in high school, can be sources of additional majors. There is a temptation to draw as majors only those who do exceptionally well in introductory courses, who are often those with AP or honors course credits from high school. It is not hard when students arrive with widely different credentials and experiences to seek the majors from among the most obviously qualified, those who by prior experience are most likely to excel in introductory courses. However, those departments which recruit, mentor, and encourage students from all levels of the introductory courses they offer will find they have larger numbers of successful majors to show for that effort, especially women and minority group members, since these groups are less likely to have taken those AP and honors courses in science in the first place.

Early in the college careers of many students, a little extra encouragement may have a significant impact on their self-confidence and self-esteem. An important feature of DePauw's Women in Science program is that it puts prospective women science majors into early contact with each other and with supportive and enthusiastic upperclass women students and women faculty members. A salient feature of the Bryn Mawr physical science and math departments is that faculty members personally approach individual students, encouraging them to consider doing further work in their subjects. While many faculty members agree that one might reasonably encourage all students with grades of B or better, often we would hear from women earning grades of B+ or A– that they doubted that they had the talent to succeed. Specific outreach by faculty members to promising individual students, encouraging them to take additional courses or to declare a major, significantly enhanced the number of majors and the number of enrollments in intermediate-level courses.

Another question to ask thoughtfully is whether the early measures of successful performance in introductory courses accurately reflect how students will do later in the major field. While most advanced scientific work is related to research, experimenting with the unknown, imagination, teamwork, and problem solving, many schools continue to use multiple-choice questions that test the students' knowledge of facts at the introductory level. This choice, often compelled by the large numbers of students and the corresponding difficulty of prompt grading, may discourage those with the imagination, creativity, and teamwork skills needed to succeed at more advanced levels. The lesson learned from many different schools is that scientists and science students can be measured and mentored by a variety of teaching and testing strategies. We should remember the importance of using a diverse set of measures of successful performance, if we wish to gather the largest possible number of science majors and successful scientists.

The lessons learned from Project Kaleidoscope (PKAL 1991) and in studies of success in undergraduate science education are instructive. This project began as a National Science Foundation–sponsored study of science education, one of four in different sectors of higher education (undergraduate colleges, research universities, comprehensive universities, and two-year colleges). The group that chose the title "Project Kaleidoscope" was to study science education in the undergraduate colleges, and the title reflected our perception that there were many ways to success. After completing our report, and with the assistance of other funding agencies, we broadened our mandate to the study of what works more generally in science education in colleges and universities. The data confirmed our impressions that the liberal arts colleges were unusually productive of scientists, by all national measures, and while we had not specifically set out to study efficacy for educating women and minorities, we also found that women's colleges were generally unusually productive of women scientists and that HBCUs were the most productive of Black science graduates who have become Ph.D. scientists. Analysis of the data strongly suggested that what works for women and for minorities also works for majority men.

Successful and productive programs emphasize such things as

- connected learning;
- lean, laboratory-rich curricula;
- investigative and research experiences at all levels of the curriculum;
- facilities and spaces which invite students to study in a personally and ergonomi-
  cally encouraging environment (including an "after hours" space where majors
  can gather in study groups);
- learning as a social and cooperative activity;
- the diversity of possible careers following scientific training;
- a variety of ways to assess progress, strength, and success;
- faculty who personally engage with, encourage, and support students;
- active recruitment by faculty members of students to the major;
- accessible curricula that invite students to continue and that facilitate completion
  of a major regardless of a student's level of expertise upon inception—that is,
  curricula with many different entry points.

At Bryn Mawr, the most important features of physics programs which enhanced the numbers of women were a broad and challenging "conceptual physics" course for all students, even those with enough calculus to start with a standard introductory course; curricula which made it possible for a student to take
an introductory course as a sophomore and still complete the major on time;
and a gathering place (a "majors' office" in the physics wing) where physics majors could gather to study and where they could store materials between meetings. Each of these features strikes down a barrier to conventional educational
strategies in physics. Rather than starting with a broad calculus-based survey
course (which privileges those who took the same course in high school and
those with advanced mathematics preparation), we started with broad concepts, modern physics, and the intersection of physics and society, emphasizing
analysis and discussion to achieve conceptual understanding. While some might
say this put those with advanced high school training at a disadvantage by not
giving them a fast track to achieve ready success, we found that this approach
empowered all students, provided motivation for the rigorous study of physical
laws and mathematics that was to come, and meanwhile created a learning environment that strengthened the eventual accomplishments of all students who
continued. The sophisticated understanding and perspective we were fostering
in the conceptual physics course was greater than we had previously seen among
senior physics majors prepared in a more conventional curriculum.

The rigid sequential nature of the conventional physics curriculum had also
been a barrier to access. We at Bryn Mawr had been as guilty as any physics
department in enshrining this impediment, as a student who started introductory physics in the sophomore year could not complete a major in four years.
Students who tried to double up (taking two years' worth of courses in one year)
often struggled. However, this became vastly easier when students had a solid
conceptual foundation. Indeed, the success of the conceptual physics course in
preparing students for any more specialized course challenges the assumption
that it is essential to have a sequential curriculum that revisits topics two and
three times to help students "get it."

The enhanced sociability of the learning environment also proved empowering. We added lockers and computers to the majors' office, and we gave students keys so that they had access to classrooms and seminar rooms and to the kitchenette. Long into the night, students looking for physics majors, students considering physics as a major, and the majors themselves could be found working together or working alone near others. The social environment promoted hard work, and a little fun occasionally helped to break the tension.

It is also noteworthy that much can be learned by faculty members from listening to students before deciding to be critical or encouraging. Faculty members are often survivors of a "sink or swim" environment, and they often imagine that students experience courses just as they did when they were students. In contrast, by listening to students comment on how they had done on tests, or how they assessed their scores on those tests, I learned that some students count only the errors they make, and others focus on every problem they could not easily solve. The latter group of students, whose objective performance measures were so high that I thought the graded work gave obvious encouragement, were concluding that they could not achieve the requisite level of mastery. Listening to each student, and adding additional encouragement to continue where appropriate, became one of my most important recruiting tasks as a faculty member in entry-level courses; giving high grades to those who earned them was not enough.

Encouragingly, the PKAL study did not find that women had special deficits or needs which had to be met with special approaches; rather we found that the characteristics of successful programs for women were successful in recruiting more white men as well. The point is, more staid and conventional instructional programs in the sciences have not only excluded most women and minorities, they have excluded most men as well. What works to bring women and minorities into science works as good science education for all students.

The institutions we studied often were not only flexible, but innovative. Some sponsored learning clinics, often run by students. Many emphasized teamwork in homework and in-class assignments. Some offered research on unconventional topics and in interdisciplinary fields. Others promoted internships in the summer, in an intersession of some kind, or for semester-long projects. Many emphasized creating good teaching opportunities for their more advanced majors.

Remarkably, at a time when there is an excess of rhetoric that college faculty members should teach, and not be drawn away from teaching to research, the successful programs reject this dualistic and divisive rhetoric and the paradigms and the categories it suggests. At the successful institutions, the faculty members are engaged with their students in research, which invigorates both the students and the faculty members. Engagement with the unknown and engagement with each other are synergistic.

In PKAL, we took a hard look at the numbers and percentages of graduates who were science majors and the origins of those who earned Ph.D.s in all types of institutions, because we wanted to identify clear hallmarks of successful programs.[3] Even when we correct for the more limited set of possible majors at liberal arts colleges, where typically one does not find engineering and other specialized programs, liberal arts college graduates are disproportionately likely to end up in the professoriate and as teachers, and they are disproportionately likely to seek careers as teacher-scholars on college and university faculties. Sometimes this productivity is very puzzling. Often liberal arts colleges can offer only a limited number of courses in such fields, whereas research universities offer many more. Yet the success rate persists. And it seems that this is because science at these institutions more nearly resembles the careers that students might pursue—with emphases on reasoning rather than rote recall, on clear speaking and writing rather than multiple-choice exams, and on cooperation rather than competition.

It is worth noting that not every program in women's colleges and liberal arts colleges is exemplary (PKAL 2:107; Doyle 1992). There are coeducational colleges and research universities, as well as women's colleges, with good productivity, and there are other schools of each type with less successful records, with the reasons hiding just below the surface in the stories their students and faculty members can tell. But the best of the women's colleges and the best of the liberal arts colleges match or exceed the best of the research universities in terms of the proportion of women among their undergraduate science majors.

Graduate Education

Surviving the Initial Year

The first year of graduate school holds the key for many students as they make a transition from a broad and diverse experience as undergraduates to the disciplinary focus and intensity of a graduate program. Many liberal arts college graduates find this transition particularly stressful, as they often encounter graduate curricula designed to serve entering graduate students who have completed more courses than are required in a standard liberal arts college's undergraduate major. The liberal arts science graduates must compete with some U.S. undergraduates who began their college-level work with advanced placement, and hence finish more comprehensive majors, and with those who did their undergraduate work at schools whose programs offer graduate courses for their more advanced undergraduates. The liberal arts graduate must also compete with many international graduate students who have already completed master's degrees or equivalently advanced coursework. While there is no disputing that these advanced students should be welcomed to graduate programs, it is the responsibility of the graduate department to ensure that a reasonably well-prepared university or liberal arts college graduate can succeed, through proper mentoring and placement.

Access to and success in graduate education for liberal arts science majors should be a priority for everyone, because a disproportionate number of women and minority graduates with undergraduate science majors come from the undergraduate college environment. Investment in their success is of particular value to graduate programs because liberal arts college graduates are likely to be more experienced in problem solving, laboratory work, and research, while graduates of larger universities are often prepared and selected through multiple-choice exams and quizzes in the early years and theoretical courses in the later years. The laboratory and research experience of liberal arts college graduates often significantly exceeds that of their fellow graduate students and prepares them for the rigors of the research endeavors leading to advanced degrees. Graduate schools and departmental graduate programs must recognize and support these students.

Pathways to Success in Graduate Programs

Once the graduate students are incorporated into the life of the graduate departments, there are many successful strategies for encouraging them to continue. First, it is important for graduate departments to keep a wide range of career options open to the students. Far too many graduate research advisers speak negatively, or not at all, of careers in industry or of opportunities for teaching and scholarship in the undergraduate colleges. Similarly, one might reexamine the great honor that graduate advisers and students attach to the "research fellowship" which allows students to "avoid having to teach." Though research progress is important and research focus may lead to more theoretical or experimental results sooner, and thus to better job offers, graduate schools which are dismissive of the value and quality of good teaching opportunities for their graduate students may be alienating many of their students and many of their potential faculty recruits. It may have to be the responsibility of deans and department chairs to ensure that their graduate departments, particularly the graduate advisers but also a broad range of graduate faculty members, remain informed about the wide range of career choices and of the benefits of a mix of teaching and research activities during the graduate school years.

*Partnerships between Colleges and Universities
to Aid Graduate Student Persistence*

Liberal arts colleges, hoping to recruit talented and well-trained scientists to their faculties, can implement another set of strategies that encourage persistence in graduate work. Here are some examples.

Graduate internships. In fields, such as computer science, in which DePauw has trouble finding qualified candidates for faculty positions, we recruit promising students with bachelor's or master's degrees and offer them graduate internships, under which we cover their tuition and pay them a salary (one more attractive than most graduate fellowships) in return for their teaching at DePauw, with careful mentoring from experienced faculty colleagues, while they are working on their Ph.D.s. This program also helps to recruit identified can-

didates from underrepresented groups, including women and members of ethnic minorities.

Links to Preparing Future Faculty (PFF) programs. We search out those universities which have strong programs designed to prepare future faculty (*Liberal
Education* 2002) and offer to host interns, who shadow established college faculty members several times a week, sit in on classes, and discuss course design
and professional activities. In my own experience, there has been no greater
challenge to my pedagogy than that which has come from undergraduate or
graduate teaching interns who watch and query what I do, and offer suggestions
for alternative strategies.

When appropriate, we hire PFF students recommended by their universities
for teaching internships or for predoctoral or postdoctoral fellowships. They
gain more experience with the liberal arts college sector of higher education,
and both they and we have opportunities to consider future employment relationships. If these interns are women or members of underrepresented minorities, they may add to the cadre of mentors and role models for our undergraduate majors.

Contacts with graduate departments and deans. We develop close connections
with graduate departments and with graduate school deans. Faculty leaders in
many of our departments not only mentor PFF graduate students, but often
visit the graduate counterpart departments to participate in research colloquia
or to speak on teaching careers. Our faculty members also often visit research
universities to follow up on the progress of our undergraduate majors pursuing
graduate degrees. My advice is that graduate departments should make similar,
reciprocal visits more systematically. Such regular contacts encourage undergraduate majors, expand the appreciation of the graduate admission committees for the credentials of undergraduate college graduates, and increase the
number of these applicants who are highly qualified and likely to succeed.

Predoctoral dissertation/teaching fellowships. We look for opportunities to recruit promising graduate students in their "dissertation year" to teach at De-
Pauw while completing their theses. We direct a number of these fellowships to
recruiting women and members of underrepresented minority groups; having
them on campus working with students and faculty colleagues benefits both
them and us.

Postdoctoral research/teaching fellowships. On occasion, there are advantages
to recruiting a recent Ph.D. for a teaching/research postdoctoral fellowship
rather than holding all openings on the faculty for full-time term or tenure-
track positions. Explorations with strong recruits are often rewarding, leading
to longer-term appointments when appropriate.[4]

Attracting and Retaining New Faculty

Liberal arts colleges, far more than most other schools, place a premium
on the growth and development of their new faculty colleagues. Advertising this
in interviews and briefing materials eases some of the candidates' concern about

getting tenure and helps us recruit those candidates interested in developing their skills in teaching as well as in research.

Unlike many research universities, liberal arts colleges do not place the full weight and responsibility for professional growth and development on the grant-writing and grant-getting skills of the candidate. In addition to providing well-equipped laboratory spaces for faculty research and student-faculty collaborative research, we provide adequate start-up funding for initial equipment and supplies. And in addition to promising appropriate matching funds for grant applications, liberal arts colleges also provide internal grants programs and funding entitlements. But beyond competing with research universities to offer attractive "research support," liberal arts colleges also provide attractive support which indicates an interest in faculty development in the craft of teaching, a priority many candidates find attractive.

Our faculty professional development programs at DePauw (like those at many other liberal arts colleges) offer such features as

- workshops in the craft of teaching;
- a diverse set of valued options for teaching, including not only introductory and advanced courses in the major field, but also first-year seminars and experimental interdepartmental topics courses;
- opportunities for interdisciplinary and team teaching;
- opportunities to develop new pedagogies, ranging from improved lecture and demonstration techniques to more interactive pedagogies and active learning strategies;
- a reduced teaching load in the first year with the time reassigned to course development or research projects;
- a pretenure one-semester leave at full pay for professional growth and development;
- options for reassigning some teaching duties to a curriculum development or scholarly project approved under a competitive peer-review process;
- seed money and grants of time for proposal writing or for the support of work for which grant proposals are pending;
- sabbatical leave support for salary, and for travel and relocation expenses;
- an internal grants program for projects ranging from travel to equipment purchases to workshop attendance;
- a professional conference fund with both unrestricted components and supplements for presenting papers or organizing sessions;
- competitive awards for summer stipends for student-faculty collaborative research, independent curriculum development, and scholarly or creative projects;
- institutional support to supplement external awards such as major fellowships and recognitions, so faculty members can take advantage of those awards without salary penalties;
- at the senior level, recognition via fixed-term (2-5 years) grants of time and salary supplements for sustained excellence in teaching, service, and scholarship.

Here are two examples I share with candidates I interview for faculty positions. First, I mention the opportunity a faculty member has to develop a first-year seminar on a topic of special interest. I also mention the more "active-learning courses" in our January term. Then I ask the candidate to propose courses or topics that might be appropriate for the audience of mainly first-year students. Most candidates are thrilled to have the opportunity to think about ways to bring their interests and experiences into courses for first-year students. Many are also puzzled about how such discussion-based courses can have depth if the students do not yet have expertise to share. And some worry that they have not yet mastered such a wide range of topics themselves. In response to these concerns I can offer the services of our faculty-led workshops on pedagogical styles for teaching first-year students in seminar-sized classes on interdisciplinary topics. Even when working with senior faculty members with well-established teaching styles, we have found it important to support risk-taking pedagogy, to prevent the intimidation or the situation from overwhelming the excitement.

The second example arises in our discussions about general education and how courses must serve the nonscience major as well as the science major. Heads nod, but there is a sudden interest when I explain that introductory math and science courses are not "intrinsically certified" as the way for students to acquire their required "competency" in quantitative reasoning. I explain that not only do we certify course content and pedagogy as suitably challenging of quantitative reasoning skills, but we certify the instructors only after they have taken a faculty-led workshop on student fears and the appropriate pedagogies for such a course. Here, again, the workshops help new faculty members to navigate the difficulties and challenges of helping all students succeed, in all sorts of science and quantitative courses.

The way in which an institution addresses other, more whole-life-centered issues also can be effectively advertised and addressed in the recruiting and interviewing process. These include spousal career development and employment search support, job sharing options, and tenure clock adjustments for medical and parental leaves. Recruiters and interviewers can also provide in-depth information about local schools and cultural opportunities, including those offered by the hiring institution through cultural affairs or performing arts programs. If subsidized rental housing or mortgage subsidies are available, they should be noted, and candidates should be told of safe and supportive residential communities.

Surprisingly, candidates are often quite interested in faculty development programs that are available only to more senior faculty members. New faculty members are not only impressed by the range and duration of the college's commitment to their continued growth and development, but are also persuaded to think with us about their own goals in that regard. Also, just as studies show that students learn from peer mentoring, it is well documented that among the most successful faculty development programs are those in which faculty members lead for each other (*Diversity Digest* 2002). The importance of demonstrating that there is a collaborative and cooperative community of teacher-scholars cannot be overestimated, as such a community attracts many prospective faculty members to the profession and to particular job opportunities.

Much of this may seem exhausting or overzealous to those who are focused on recruiting the best and brightest person to do ground-breaking research. We find, on the contrary, that offering a range of information and opportunities, and making a clear commitment to flexibility in growth and development, attracts far more candidates and leads to more sophisticated discussions about career futures (not to mention less stress for everyone as those careers unfold).

Faculty Development Programs

When recruited candidates arrive to start work, it is important to immediately set a tone and climate that ensure overall success. Most importantly, the details must be tailored to the individual needs of the new faculty members and to the tenure and promotion criteria of the department and the institution. The following are strategies for welcoming and involving new faculty members, with a heavy emphasis on facilitation by faculty colleagues rather than "instruction" by administrators.

A formal (one- or two-day) orientation before the start of the first semester. This orientation should go beyond the "nuts and bolts" of college life to include such items as an exchange of syllabi, discussion of grading and academic integrity policies, a review of student attendance policies, and a review of instructional technology and library resources.

Assignment of a faculty mentor from outside the department. A senior colleague from another department should join each new faculty member for occasional meals (paid for by the college), and the two should visit each others' classes and discuss their institution's norms, opportunities, and practices.

Continuing workshops for new faculty members in the first few months. These workshops should include discussions of tenure and promotion criteria, and review the relevant procedures and deadlines; offer various teaching techniques and ways to assess classes and students; discuss how to manage a classroom and handle disruptive student behavior; and cover opportunities and choices for departmental and university service. At midsemester, they should reflect on how courses are going.

An additional workshop for new faculty members at midyear. A period of reflection is appropriate to analyze how teaching went during the first semester, discuss syllabi revisions for fall courses and designs for spring courses, review assessments, and discuss how to interpret scores and comments on student evaluations.

Ph.D. completion working groups. Those faculty members who are ABD should have a support and working group in which they can read drafts, discuss roadblocks, and prepare for defenses. Such groups have been quite successful, particularly for those with lighter teaching loads and predoctoral scholarships or fellowships.

New faculty members should also be warmly welcomed into ongoing faculty development activities which are open to all faculty members. Successful such activities include *teaching roundtables*—lunchtime meetings for presentations on teaching innovations; *interdisciplinary reading groups* on topics of general interest, sometimes linked to the visit of a major speaker, sometimes leading toward development of a new program; and *research colloquia*—lunchtime or late afternoon presentations of scholarly and creative work open to the whole faculty (which supplement and complement departmental colloquia).

Faculty career development also requires us to be very intentional in recruiting new department heads and program chairs. We have found it most useful to acknowledge faculty members' major administrative contributions (chairing a department, heading an interdisciplinary program, taking a lead on a major project) by granting them reduced teaching loads. Time, more than money, forms the appropriate currency with which to compensate for time-consuming service or administrative tasks.

We have also found it important for faculty development to have informal topical support groups which often invite students and staff members to their program events. Most effective at DePauw are groups such as Women in Science (which holds lunches for speakers and discussions of research, organizes reunions with alumni, administers a summer high school science program, and arranges mentoring for undergraduate majors) and the Black Caucus (which offers mentoring and support for newcomers, a discussion forum for all, and close links to senior administrators).

These programs, which support new and continuing faculty members, also provide support to undergraduates. For example, at the monthly meetings of Women in Science, students and faculty members convivially discuss such topics as student research opportunities, a visiting graduate's graduate school work or job, career planning, and faculty research. A visiting recent graduate might also have leads for students about summer internships or contacts for faculty members for sabbatical research. Such meetings provide safe spaces outside of formal relationships, meetings where practical advice on how things work can be shared. They are also safe places for tough issues to be raised without embarrassment. And, perhaps most important, such meetings provide a real example of the advantages of cooperation rather than competition.

A coeducational institution such as DePauw has many strong support groups and programs. Perhaps it is because even after fifteen years of equity in hiring and a preponderance of women science majors overall (albeit with noticeable gender imbalances in mathematics, computer science, and physics), the senior faculty still show evidence of a highly sexist pattern of hiring and advancement for women, exemplified by the presence of just two women in math and science at the professorial rank, a rank at which men make up 80 percent of all the school's professors. The continued and steady advancement of women through the ranks and in the majors has drawn such strength from Women in Science programs that these programs are likely to continue for years to come. Why then were such meetings and support groups implicit rather than explicit at Bryn Mawr?

Perhaps it was founder M. Carey Thomas's determination that the education at a women's college must be just as "rigorous" as at the men's colleges, that women did not need a "different education," special programs, or special help. Or perhaps it was because the entire culture of a women's college was supportive of the advancement of women in the professions, and everywhere one looked women were playing key roles, from top administrators to faculty leaders, from the heads of student government and the editors of the student newspaper to the best student athletes and the most accomplished academic prizewinners, from the best research student to the one first with a hand up to answer questions in class, and from the laboratory teaching assistant to the tutors available to help (Michels 1948). Nonetheless, the number of science graduates waned in those years in which there were neither faculty nor student initiatives in each major subject to promote science careers and to encourage those in the entering classes, whereas intentional mentoring and deliberate curricular design got tangible results. Perhaps this means that both are needed: explicit admission of problems and concrete steps to address them, but also more subtle moves toward a student and faculty culture where it goes without saying that women can do anything, including science.

Students are often thought to be more at risk for dissuasion from continuing in science than faculty members, as students have less well formed career goals, more fragile egos, and more options. However, these case studies point to many similarities in strategies for success for both students and faculty members: creating supportive communities, providing individual attention, attending to the equitable outcomes of policies and procedures, and adapting good ideas from elsewhere to the local needs and contexts rather than simply adopting them.

We are reminded from examples such as those mentioned in this chapter that faculty members grow and develop from their experiences, and we have much to contribute to that growth and development. The common practice of hiring promising candidates who are then "left to prove they can succeed" in a manner reminiscent of the "weed-out" courses of traditional introductory science curricula is a waste of all sorts of resources, talents, and opportunities.

Sustained and consistent efforts to recruit, mentor, and retain faculty members lead to long-term satisfaction on all sides as well as to a stronger and more diverse population of students and faculty members. From this overview we see how important it is to sustained success across the country that the lessons learned be shared; far too often those things deemed daunting or impossible on one campus have proven to be quite possible at another. Those looking for new initiatives or greater success should turn to other institutions for ideas, as the evidence is clear that there are many successful programs to emulate.

Notes

I am pleased to acknowledge faculty members and faculty colleagues at Bryn Mawr College from whom I learned and with whom I worked for most of twenty-six years to facilitate the advancement of women, particularly in physics. Particular gratitude is due to Walter Michels, Rosalie Hoyt, Alfonso Albano, Peter Beckmann, and Elizabeth McCormack. Our work together earned the Presidential Award for Excellence in Science, Mathematics and Engineering Mentoring in fall 1998. The special efforts of Jeanne Narum and colleagues at Project Kaleidoscope helped me to broaden my understanding of the national dimensions of these issues. I am also grateful for the insights and energy of the leaders of the Women in Science program at DePauw University, Kathleen Jagger, Mary Kertzman, and Bridget Gourley, who nurtured this program from its inception. More recent energy and contributions from Terri Bonebright, Dana Dudle, Hilary Eppley, Jeane Jerz, Pam Propsom, Jackie Roberts, Jamie Stockton, and Janet Vaglia have been equally inspiring and encouraging as we have institutionalized some of the necessary changes across the sciences. Similarly, Gloria Townsend has provided exemplary leadership in the design and development of programs to support women in computer science at DePauw. Leadership from Meryl Altman, director of the Women's Studies Program, has been a source of inspiration and a constant reminder of how much better we can do.

1. See, for example, annual reports on employment and graduates in physics by the American Institute of Physics.
2. Project Kaleidoscope was commissioned by the National Science Foundation in 1989 to identify "what works" in undergraduate science education in the four-year colleges.
3. The results of these studies, the numbers and the success stories, are published in a two-volume report (PKAL 1991) which can be purchased from PKAL through the Independent Colleges Office in Washington, D.C.
4. For more details on a national program of predoctoral and postdoctoral minority fellowships, see the Web site of the Consortium for a Strong Minority Presence at http://www.grinnell.edu/offices/dean/csmp/.

References

Abraham, Neal B. 1999. "Mentoring the Whole Life of Emerging Scientists." In *Coming into Her Own: Educational Success in Girls and Women,* ed. Sara N. Davis, Mary Crawford, and Jadwiga Sebrechts, 211–28. San Francisco: Jossey-Bass.

Diversity Digest. 2002. "Faculty Leading Faculty: University of Colorado, Boulder Reflects on Good Teaching Across Difference." 6(3): 1.

Doyle, Michael P., ed. 2000. *Academic Excellence: The Role of Research in the Physical Sciences at Undergraduate Institutions.* Tucson, Ariz.: Research Corporation.

———. 2001. *Academic Excellence: The Sourcebook. A Study of the Role of Research in the Natural Sciences at Undergraduate Institutions.* Tucson, Ariz.: Research Corporation. http://www.rescorp.org/sourcebook/source_contents.htm. Accessed February 16, 2005.

Liberal Education. 2002. "Changing Course: Preparing Faculty for the Future." Special issue, 88(3).

Michels, Walter C. 1948. "Women in Physics." *Physics Today* 1: 16–19.

Nelson, Donna. 2002. "Faculty Diversity in Mathematics Departments at the 'Top 50' Research Universities." *AWIS Magazine* 31(3): 42–46.

Physics Today. 1998. News Briefs. 51(1): 49–50.

PKAL (Project Kaleidoscope). 1991. *What Works: Building Natural Science Communities.* 2 vols. Washington, D.C.: The Project.

Seife, Charles. 2003. "10 Years after the SSC: Physics Tries to Leave the Tunnel." *Science* 302(3): 36–38.

Wright, Dorothy. 2002. "Cultivating Success in Mathematics." *S&T: A Quarterly Newsletter on Research, Teaching, Management, Policy Making, and Leadership in Science and Technology* (Bryn Mawr College). July: 10–11.

17 Beyond Gender Schemas: Improving the Advancement of Women in Academia

Virginia Valian

What the 2002 Iowa State conference and the experiences of the first cohort of National Science Foundation ADVANCE Institutional Transformation awardees make clear is that we are fortunate to be at the inception of a new discipline—the discipline of gender equity. There is a distinct subject matter here. We see that our understanding of obstacles to and improvement in gender equity will draw on concepts and methods from women's studies, sociology, statistics, economics, psychology, mathematics, engineering, political science, and organizational structure and change, to name just a few fields.

We will need at least two types of explanation to account for the nature of equity problems and their ubiquity across the professions, and we will have to interweave these explanations and put them together in original ways. One type of explanation, which I will develop here, is a social-cognitive analysis; it brings to bear research and data on gender schemas and the accumulation of advantage. The other is an organizational analysis that brings to bear research and data on organizational structure and change. My particular target here is the paucity of women at the top of the professions, especially in academic science, engineering, and technology fields.

My explanation uses two key concepts: gender schemas and the accumulation of advantage. In brief, the application of gender schemas makes it more difficult for women than men to accumulate advantage.

Data Summary

Data from a range of sources[1] lead to the following conclusions. There has been progress in gender equity: men and women make roughly equal starting salaries at similar rank (but science and engineering salaries remain a problem and there are signs of early rank differences in science and associated fields). Nevertheless, problems remain in all fields: more women than men move into part-time positions; women advance more slowly than men; women earn less money than men except at entry level; women are particularly underrepresented at top-tier institutions; and women receive fewer national awards and prizes.

Gender Schemas

Schemas are hypotheses that we use to interpret social events (Fiske and Taylor 1991). Schemas are similar to stereotypes, but the term "schema" is more inclusive and more neutral, and a more appropriate term because it highlights the proto-scientific nature of our social hypotheses.

Gender schemas are hypotheses about what it means to be female or male, hypotheses that we all share, women and men alike. Schemas assign different psychological traits to males and females (Martin and Halverson 1987; Spence and Helmreich 1978; Spence and Sawin 1985). We think of males as capable of independent action, as oriented to the task at hand, and as doing things for a reason. We think of females as nurturing, expressive, and behaving communally. In brief: men act; women feel and express their feelings. Our beliefs have support. In questionnaires, men endorse more "instrumental" characteristics and women endorse more "expressive" characteristics.

The main answer to the question of why there are not more women at the top is that our gender schemas skew our perceptions and evaluations of men and women, causing us to overrate men and underrate women. Gender schemas affect our judgments of people's competence, ability, and worth.

Experimental Data on Perceptions of Sex Differences

Experimental data demonstrate that we do not see other people simply as people; we see them as males or females. Once gender schemas are invoked, they work to disadvantage women by directing and skewing our perception, even in the case of objective characteristics like height. In one example (Biernat, Manis, and Nelson 1991), the experimenters exploited the fact that our schemas include the—of course correct—information that men are on average taller than women. In this experiment, college students saw photographs of other students and estimated their height in feet and inches. The photos always contained a reference item, such as a desk or a doorway, so that height could be accurately estimated.

Unknown to the students who were doing the estimating, the experimenters had matched the photographs so that for every photograph of a male student of a given height there was a female student of the same height. But the students were affected by their knowledge that men are on average taller than women. They judged the women as shorter than they really were, and the men as taller.

In this experiment, as is typically the case, there were no differences in how male and female observers perceived the images; we all have nonconscious hypotheses about males and females and we all use those hypotheses in perceiving and evaluating others. The important point about this study is that a genuinely objective characteristic—height—is not immune from the effects of gender schemas.

In the case of professional competence, perceptions are similarly prone to error. We are likely to overvalue men and undervalue women. We can see why that would be the case: gender schemas play a large role in evaluations whenever (a) schemas make a clear differentiation between males and females, and they do for professional competence as much as for height; and (b) evidence is ambiguous and open to interpretation, as is the case with professional competence. It is tempting to think excellence is straightforward, but it is not.

Experimental Data on Perceptions of Women as Leaders

Not only do schemas affect perceptions of competence, they also make it difficult for women to reap the benefits of their achievements and be perceived as leaders. This is shown in the following three examples, in all of which there were no male-female differences among the observers.

In the head-of-the-table experiment (Porter and Geis 1981), college students were shown slides displaying five people seated around a table. The group was described as working together on a project. Two people sat at each side and one person sat at the head of the table. Sometimes all the people were male, sometimes they were all female, and sometimes the group included both males and females. The students were asked to identify the leader of the group. In same-sex groups, the man or woman sitting at the head of the table was always identified as the leader. In mixed-sex groups, a man at the head of the table was always identified as the leader. But if a woman was at the head, she was not reliably labeled as the leader; a man seated elsewhere at the table was labeled as the leader about equally as often.

As I indicated, there were no differences between male and female observers; both made the same judgments. There was no intention to discriminate. Nevertheless, a woman sitting at the head of the table is less likely to be considered a leader than is a man in that position. The symbolic position of leadership carries less weight for a woman than a man. Women are less likely to obtain the automatic deference that marks of leadership confer on men. Women are objectively hurt in such situations, even if observers intend no hurt. A woman has to work harder to demonstrate that her apparent position of leadership is a real position of leadership.

Not every person behaves on every occasion in accordance with gender schemas. Many different factors affect our evaluations and behaviors. But that variability should not distract us from what the odds are: the odds are that we will overrate men and underrate women, and we will see women as less capable of leadership than men.

Experiments investigating who looks at whom in a conversation show the effects of social dominance (Dovidio et al. 1988). When a subordinate and a superior are talking, the subordinate tends to look at his or her superior with the same frequency whether the subordinate is talking or listening. Looking at your interlocutor while listening is a sign of deference. The social superior in that situation tends to look at the inferior more while talking than listening, revealing and reinforcing a superior stance. When men and women talk (outside of a courtship setting), men look more while talking than listening, and women look the same amount whether talking or listening. The exception is if the topic is one that the woman has antecedently declared she knows a lot about and the man has antecedently declared he knows very little about. (Note how difficult it would be to recreate this laboratory situation in real life.) In that case, women look more when talking than listening.

Another set of experiments (Butler and Geis 1990) shows that women who adopt a friendly but assertive leadership role are responded to more negatively by both males and females than are men who adopt the same role. These experiments surreptitiously videotaped two naive participants reacting to two trained actors following a script. Men received more positive than negative facial expressions from the naive participants when they were leaders, but women received more negative than positive expressions. Again, there were no differences between male and female observers.

Accumulation of Advantage

Each example that I have discussed is a small thing. One might be tempted to dismiss concern about such imbalances as making a mountain out of a molehill. But mountains *are* molehills, piled one on top of another over time.

Small imbalances add up to disadvantage women. Success is largely the accumulation of advantage, exploiting small gains to obtain bigger ones (Merton 1968). A computer simulation (Martell, Lane, and Emrich 1996) showed the importance of very small amounts of bias. The researchers simulated an eight-level hierarchical institution with a pyramidal structure. They staffed this hypothetical institution with equal numbers of men and women at each level. The model assumed a tiny bias in favor of promoting men, a bias accounting for only 1 percent of the variability in promotion. After repeated iterations, the top level—which had been 50 percent male and 50 percent female at the beginning—was 65 percent male. Even very small amounts of disadvantage accumulate over time.

Summing Up the Causes of Women's Slow Advancement

What is responsible for women's lack of progress in the professions and in academia are the gender schemas through which we all—male and female alike—perceive and evaluate women. The small but systematic undervaluation of women culminates in women's smaller salaries and slower rates of promotion, compared to men.

We would like to think that our genuinely held egalitarian and meritocratic beliefs and ideals would buffer us from the effects of gender schemas (Lerner 1975). But our evaluations and reactions occur unintentionally and outside awareness. Indeed, our belief in our own goodwill can make it difficult for us to see what we are doing. That does not mean that we cannot institute remedies. We can, but we need to understand that good intentions are not enough. We need to understand how gender schemas work and the importance of the small daily inequities in our treatment of our colleagues.

Impact of Gender Schemas on Women's and Men's Self-Perceptions

Let us turn now to examples of the impact of gender schemas on a woman's perception of herself and a man's perception of himself. To be successful in academia, and in other areas, it is important to negotiate effectively. To do that, one must have a feeling of (at least moderate) *entitlement;* but women tend to be low in entitlement and men tend to be high. A number of experiments (e.g., Major 1987) show that women and men differ in how entitled they act: women work harder and more efficiently than men for the same pay, and accept as fair less pay for the same work.

An example from real life comes from tennis. In 1991, Monica Seles argued for equal prize money for men and women in tennis tournaments. Two other players responded publicly. Steffi Graf was quoted as saying, "We make enough, we don't need more," and Mary Joe Fernandez was quoted as saying, "I'm happy with what we have; I don't think we should be greedy" (Bailey 1991). In this example, equality is being perceived as greed. In 1995, Seles, Graf, and other top players wrote a letter to the Australian Open, protesting the decision to substantially increase the size of the men's purse for 1996, so that the men's purse was $390,000 more than the women's (Gallo 1996). But while the players protested, they also pledged not to boycott the tournament—for the good of the game. Not surprisingly, the Australian Open organizers saw no reason to equalize the prize, and the women played for less money.

One way that gender schemas affect women, then, is in women's perception of themselves as worth less and entitled to less; schemas conversely affect men, leading them to see themselves as worth more and entitled to more. Women also, through the chores they are given to do in childhood, become accustomed to acting for others' good, to laboring for love; men become accustomed to being recompensed for their labor.

Women's lack of entitlement, and people's expectations that women will not behave in an entitled manner, influence the jobs that women are called upon to do and accept doing: institutional "housework" and institutional "labors of love." These are usually low-visibility, low-power, low-reward, and labor-intensive tasks. Entitlement also plays a role in who teaches what. In one science department, a man and a woman in similar specialties had entered the department within a year of each other. They each taught two courses per term. The man taught the same introductory course in his specialty every term, plus a more advanced course in his specialty. There were many advantages to such an arrangement: the instructor needed minimal preparation time for the introductory course; he learned who the interested and talented undergraduates in his field were and could suggest to them that they work in his laboratory; in his seminars he taught the topics on which he was currently doing research. The woman taught many different introductory courses, not all in her specialty, and seldom taught an advanced course in her specialty. The disadvantages of such an arrangement were obvious: the instructor was always planning and developing a new course, frequently outside her specialty; she seldom met interested and talented undergraduates in her area; she seldom taught the topics on which she was currently doing research. A senior woman in the department spoke to the chair about the uneven division of labor between the two younger faculty. The chair said that the male faculty member would have put up a big fuss if he had tried to give him the same set of courses that he gave the woman. He also thought that the male faculty member would do a less conscientious job of teaching outside his specialty than the woman would.

It is not surprising—if lamentable in this case—that department chairs will take the path of least resistance. Entitled men will put up more resistance than unentitled women. It is difficult for others to take women seriously, and it is difficult for women to take themselves seriously. It is hard for everyone to see women as professionals who are entitled to a good salary and to a promotion, as people whose time is valuable. That makes it difficult for women to think they deserve, let alone negotiate successfully for, valuable resources such as time, space, and money.

Why Many Doubt There Is a Gender Equity Problem

The gender schemas analysis has implications for how to justify the need for remedies and for what remedies to propose. Schemas operate largely below the level of awareness; further, many people sincerely espouse meritocratic beliefs and perceive themselves as acting in concert with those beliefs. That makes it difficult for them to imagine that anything could be seriously wrong with their practices. In addition, people are distracted by exceptions: the fact that there are a few successful women misleads us into thinking that there is no problem. But an exception is just that: an exception to a general rule.

People rely on four common explanations for the gender disparities that exist in science, engineering, and technology fields: a) it is a pipeline problem; b) women's child-care responsibilities (which could be at best only ameliorated by providing day care) preclude their having enough time for research; c) women and men have different values and preferences (which cannot and should not be tampered with); and d) it is an acculturation problem, with women not being socialized to play by men's rules. Each of these explanations obviates the need for change. According to the first explanation, equity is a problem that will take care of itself once more women enter scientific fields. According to the second and third, the differences in men's and women's responsibilities and natures doom attempts to eliminate the disparities. According to the fourth, women simply need to learn the rules and then play by them.

Is it a *pipeline* problem? It is true that relatively few women obtain Ph.D.s in the natural sciences, computer science, mathematics, and engineering. But the science pipeline selectively leaks women, as the decline in the percentage of women from undergraduate to graduate to professorial status shows (see NSF 2000; MIT 1999). The problem is really a leaky-pipeline problem, and it is likely to be caused in part by gender imbalances in the professoriate and by the practices that produce these imbalances. An equally important point, however, is that even in the biological and social sciences, where women receive a large proportion of Ph.D.s, women fare worse than men. Numbers help, but they will not, by themselves, cause disparities to disappear.

Is it a *child-care* problem? Few working fathers do their share of child-care or housework and few institutions supply high-quality day care to their faculty. When child-care is seen as women's work rather than humans' work, there is a clear cost to women, to science, and to society. Women with children are much more likely to become part-time workers than are women without children and than men, in science and in other fields (Long 2001). We train and educate young people—an expensive undertaking—with the intention that they will increase the pool of those performing high-quality science. If we do not simultaneously keep those people in the full-time labor pool, we undercut our intentions.

But child-care is only part of the story. Women pay a price for simply being female: women without children do not progress at the same rate as their male peers. For example, men in the sciences are more likely than women to be tenured, even after controls are introduced for years since degree, discipline, parental status, and a host of other variables (Long 2001).[2] Institutions do better at developing their male faculty than their female faculty, even when both groups have much the same characteristics.

Is it a *values* problem? This is harder to evaluate. Survey data suggest that, by and large, men and women want the same things from their jobs. Yet it is probably true that men are more willing than women to forgo a balanced life in order to have a successful scientific career. What we need to question is whether it is a wise policy decision to have those who forgo a balanced life (regardless of their sex) to dominate science and other institutions. They are likely to, simply because they are more visible. But their domination may be an undesirable side effect of their visibility. Some data suggest that women emphasize quality in publishing over quantity, while men focus more on quantity than quality (Sonnert and Holton 1995, 1996). It may be a coincidence that the same people who focus on quality are leading a balanced life, but there may also be a causal relation. If we continue to emphasize and reward always being on the job, we will never find out whether leading a balanced life leads to equally good or better scientific work. In addition, of course, people who live a balanced life provide other benefits to an institution, benefits which add value but are insufficiently recognized and compensated.

The fourth common explanation for sex disparities is that it is an *acculturation* problem. If women only learned what was required for success and played by those rules, they would be successful. It is true that women receive less information about how to be successful than do men, especially the more informal information, and that it is important for everyone to have equal access to information about how success works. But the phrasing of the explanation presupposes that the rules and standards for success are good ones. What the fact of sex disparities offers us is the opportunity to question habits and practices that we have taken for granted. Speaking confidently, for example, is not the same as having something to say. We need to distinguish between someone who expresses a good point tentatively and someone who expresses a bad point confidently, listen to the former more than the latter, and reward the former more than the latter.

The first step in justifying attention to equity, then, is to neutralize the faulty reasoning behind reluctance to begin equity efforts. The second step is to show how the institution will benefit.

Why Gender Equity Is Desirable, above and beyond Fairness

Equity maximizes the chances of hiring the best new faculty by increasing the candidate pool. The larger the pool, the greater the choice and the higher the likelihood of finding well-qualified candidates. Also, women job candidates are likely to be slightly more talented than men, given their difficulties in accumulating advantage.

By modeling diversity, equity demonstrates to women and underrepresented minority students that they have a future—a good future—in academia and the professions. And if they do not have a future, why are we educating them? Students do not need to see people exactly like themselves among the faculty. But a faculty composed of a variety of social groups should have two effects. First, diversity suggests that there is room for the student: where there is a lot of variety it is plausible to think that there is room for more. Second, and relatedly, diversity will make the role of scientist one which is not sex- or race-specific (Heilman 1980). It will thus make it easier for everyone to accurately judge the qualifications and value of nontraditional scientists.

Equity increases the likelihood of innovations in teaching, scholarship, and research. Innovations arise from diverse groups of people with diverse perspectives. It is not that people reason differently as a function of their sex or race, but that they will have somewhat different interests and experiences, which in turn give rise to different ideas. (For example, as women and underrepresented minorities entered psychology, new areas of the discipline were developed.)

Further, innovations are more likely to be accepted by at least some members of a diverse group of people than by a homogeneous group.

Solving an equity problem can lead to solving a problem unrelated to equity. For example, a discovery that women receive computer support more slowly than men can lead to a more systematic and effective way of handling all computer help requests. Another example: the discovery that women receive less information about how to succeed can lead to better overall faculty development procedures. Thus, gender can be a window to institutional effectiveness.

Gender equity in salary, promotion, and access to resources maximizes the number of people who will receive the power and resources they need in order to do their best work. It also reduces the possibility that some people are prospering at the expense of others.

Equity creates a stronger and more viable institution by giving it a reputation for fairness. Demonstrations of fairness, and concern for fairness, build loyalty from within, attract interest from outside, and increase the attractiveness of the institution to underrepresented groups.

Equity improves students' experiences and leads to better job opportunities. Students leave college, in most cases, for the world of work. In that world, students will work for and with women and people of color (though fewer of each than we would like!). Students must learn, while they are still students, that authority figures and colleagues can be of any sex or race. Colleges and universities can reassure recruiters that their students have learned to accept and respect diversity.

What Institutions Can Do to Increase Gender Equity

In their efforts to improve the status and experiences of women, and to achieve gender equity, institutions of higher learning need to embrace several principles:

- Know the data; know the theory. In colleges and universities, everyone—students, faculty, staff, and administrators—should know how gender influences evaluations and rewards.
- Accept the fact that there are no one-time fixes. Equity requires consistent and constant effort.
- Install accountability from top to bottom.
- Take an experimental approach, in which failure leads to redesign. Relatively little is known at present about how to fine-tune equity efforts.
- Treat equity as a subject matter. Equity is not a matter of trying everything, but of trying strategies that are motivated by theory and past data.
- Choose a goal: meet the national average *or* be the best. The second is more likely to lead to a superior outcome.
- Make as many procedures as possible a matter of routine. If a routine tells people what to do, their unwitting biases have less room to take over.

The remaining portion of this chapter elaborates on two of the principles: increasing accountability and improving (search) procedures.[3]

Accountability

One way to achieve gender equity is to create accountability up and down the organizational ladder. That, in turn, requires creating a public measurement system. It should become part of standard practice to publish an annual review of equity benchmarks (tabulated by an institution's office of institutional research), such as salary, tenure rates, time in rank, and so on. In addition, departments should be rated by their current status with respect to gender equity (and diversity more generally) and by steps taken to improve gender equity.

A department's equity status can then be used as a criterion for allotting space and resources to departments and as a criterion for giving departments permission to search for new hires. The senior administration can reward departments that demonstrate equity in practice and allocate fewer resources to departments where credible evidence exists of bias, discrimination, harassment, or insufficient attention to gender equity. To increase people's willingness to work for gender equity, the institution can provide release time or other benefits to faculty who are doing so.

Leaders must *lead:* leaders have power. They must use it—and be seen to use it—to create equity. Leaders create other leaders by vouching for them. Leaders are responsible for placing other people into positions of power. By placing a diverse group of people into leadership positions, leaders show a commitment to equity.

Search Procedures

Improving hiring practices is an important step in creating equity. Institutions where women are underrepresented usually provide two reasons for this lack: there are no qualified women, and women choose not to come (or do not even apply).

Are there too few qualified women? What this claim frequently reduces to is that there are few women at the top-tier institutions from which the institution in question prefers to hire. Top-tier institutions in particular do not want to hire people from lower-tier institutions. Since women are overrepresented at lower-tier institutions, that reluctance reinforces the status quo. Institutions could instead use the knowledge that location creates productivity as much as or more than productivity creates location (Long and McGinnis 1981) and use an additional search strategy: identify women who are publishing more than is typical for their location. Such women are likely to do well at the new institution.

Do women choose not to come or fail to even apply? A paucity of female or minority candidates means that the institution has already failed. It is a sign that something is wrong with the institution, the search process, or both. If women and members of minority groups are not applying, the institution needs to make an extra effort to attract them. Women and minority group members do not want to apply for jobs they are certain they will not get or to be at places they are certain will not welcome them. Thus, institutions need to go out of their way and use different strategies to attract faculty from underrepresented groups. For example, institutions can search for couples: couples want to live together and are willing to give up other benefits in order to do so. Search committee members can personally contact people at schools that have graduated larger-than-average numbers of women or minority group members in order to make clear their institution's commitment to gender equity and diversity.

If underrepresented group members reject job offers, the institution must determine whether it has offered an insufficient start-up package or an unattractive teaching schedule. People usually accept good offers.

Institutions should make it clear that they are willing to entertain nontraditional candidates for senior positions; women and minority group members are less likely to fit the traditional profile of experience because they are less likely to have been chosen for leadership positions.

Search committees should be instructed on where they are likely to go wrong. For instance, women faculty candidates may be more likely than men to do interdisciplinary work. A narrow job description, based on replacing already existing faculty specialties, will inadvertently rule out those in interdisciplinary work. Further, people working in traditional areas may be unable to evaluate work in interdisciplinary areas, as new work is likely to be at variance with the methods and findings with which these people are most comfortable.

The person who writes the job description determines who is searched for. Typically, it is the people with power who write the job descriptions. The people with power are unlikely to be women and minority group members. Thus, spreading the power to write the job description and to create the short list will result in a wider range of applicants.

Finally, the interview is a crucible at both ends. Insufficient welcoming of women and minority group members as well as insensitive or inappropriate comments—no matter how few—will reduce a candidate's interest.

Gender schemas, guiding our perceptions and evaluations, make it harder for women than men to succeed. Since schemas operate covertly, it is difficult for people to see that they are putting women at a disadvantage. To be successful, gender equity strategies must take into account what we know about how gender schemas work.

Notes

1. For data sources, see Valian 1998, Long 2001, NSF 2000, the annotated bibliography at http://www.hunter.cuny.edu/genderequity (under the Resources link), and tutorial 1 at Tutorials for Change: Gender Schemas and Science Careers, http://www.hunter.cuny.edu/gendertutorial.

2. See also the slides on child-care in tutorial 1 at http://www.hunter.cuny.edu/gendertutorial.
3. Documents on analyzing and solving a wide range of visible and hidden gender equity problems can be found at the Web site of the Hunter College Gender Equity Project (http://www.hunter.cuny.edu/genderequity).

References

Bailey, Sandra. 1991. "Some Women Faulting Demand for Equal Pay." *Washington Post*, June 2: B9.

Biernat, Monica, Melvin Manis, and Thomas E. Nelson. 1991. "Stereotypes and Standards of Judgment." *Journal of Personality and Social Psychology* 60(4): 495–502.

Butler, Dore, and Florence L. Geis. 1990. "Nonverbal Affect Responses to Male and Female Leaders: Implications for Leadership Evaluations." *Journal of Personality and Social Psychology* 58(1): 48–59.

Dovidio, John F., Steve L. Ellyson, Caroline F. Keating, Karen Heltman, and Clifford E. Brown. 1988. "The Relationship of Social Power to Visual Displays of Dominance between Men and Women." *Journal of Personality and Social Psychology* 54(2): 233–42.

Fiske, Susan T., and Shelley E. Taylor. 1991. *Social Cognition*. 2nd ed. New York: McGraw-Hill.

Gallo, Bill. 1996. "A Grand Slam against Women." *Denver Westword*, January 31: 47.

Heilman, Madeline E. 1980. "The Impact of Situational Factors on Personnel Decisions concerning Women: Varying the Sex Composition of the Applicant Pool." *Organizational Behavior and Human Performance* 26: 386–95.

Lerner, Melvin J. 1975. "The Justice Motive in Social Behavior: An Introduction." *Journal of Social Issues* 31: 1–19.

Long, J. Scott, ed. 2001. *From Scarcity to Visibility: Gender Differences in the Careers of Doctoral Scientists and Engineers*. Washington, D.C.: National Academy Press.

Long, J. Scott, and Robert McGinnis. 1981. "Organizational Context and Scientific Productivity." *American Sociological Review* 46: 422–42.

Major, Brenda. 1987. "Gender, Justice, and the Psychology of Entitlement." In *Sex and Gender*, vol. 7 of *Review of Personality and Social Psychology*, ed. Phillip Shaver and Clyde Hendrick, 124–48. Newbury Park, Calif.: Sage.

Martell, Richard F., David M. Lane, and Cynthia Emrich. 1996. "Male-Female Differences: A Computer Simulation." *American Psychologist* 51: 157–58.

Martin, Carol Lynn, and Charles Halverson. 1987. "The Roles of Cognition in Sex Role Acquisition." In *Current Conceptions of Sex Roles and Sex Typing: Theory and Research*, ed. D. Bruce Carter, 123–37. New York: Praeger.

Merton, Robert K. 1968. "The Matthew Effect in Science." *Science* 159: 56–63.

MIT (Massachusetts Institute of Technology). 1999. *A Study of the Status of Women Faculty in Science at MIT*. Special issue of the *MIT Faculty Newsletter* 11(4). http://web.mit.edu/fnl/women/women.html. Accessed February 16, 2005.

NSF (National Science Foundation). 2000. *Women, Minorities, and Persons with Disabilities in Science and Engineering: 2000*. Arlington, Va.: National Science

Foundation (NSF 00-327). http://www.nsf.gov/sbe/srs/nsf00327. Accessed February 16, 2005.

Porter, Natalie, and Florence L. Geis. 1981. "Women and Nonverbal Leadership Cues: When Seeing Is Not Believing." In *Gender and Nonverbal Behavior,* ed. Clara Mayo and Nancy Henley, 45–63. New York: Springer-Verlag.

Sonnert, Gerhard, and Gerald Holton. 1995. *Gender Differences in Science Careers: The Project Access Study.* New Brunswick, N.J.: Rutgers University Press.

———. 1996. "Career Patterns of Women and Men in the Sciences." *American Scientist* 84: 63–71.

Spence, Janet T., and Robert L. Helmreich. 1978. *Masculinity and Femininity: Their Psychological Dimensions, Correlates, and Antecedents.* Austin: University of Texas Press.

Spence, Janet T., and Linda L. Sawin. 1985. "Images of Masculinity and Femininity: A Reconceptualization." In *Women, Gender, and Social Psychology,* ed. Virginia E. O'Leary, Rhoda K. Unger, and Barbara S. Wallston, 35–66. Hillsdale, N.J.: Erlbaum.

Valian, Virginia. 1998. *Why So Slow? The Advancement of Women.* Cambridge, Mass.: MIT Press.

Selected Readings

Ambrose, Susan A., ed. 1998. *Journeys of Women in Science and Engineering.* Philadelphia: Temple University Press.

Barr, Jean, and Lynda Birke. 1998. *Common Science? Women, Science, and Knowledge.* Bloomington: Indiana University Press.

Birke, Lynda, and Ruth Hubbard. 1995. *Reinventing Biology: Respect for Life and the Creation of Knowledge.* Bloomington: Indiana University Press.

Bleier, Ruth, ed. 1986. *Feminist Approaches to Science.* Oxford: Pergamon Press.

Bystydzienski, Jill M., ed. 2004. "(Re)Gendering Science Fields." Special issue, *NWSA Journal* 16(1).

Clewell, Beatriz Chu, ed. 2002. Special issue, *Journal of Women and Minorities in Science and Engineering* 8(3–4).

Duran, Jane. 1997. *Philosophies of Science: Feminist Theories.* Boulder, Colo.: Westview.

Eisenhart, Margaret A., and Elizabeth Finkel. 1998. *Women's Science: Learning and Succeeding from the Margins.* Chicago: University of Chicago Press.

Etzkowitz, Henry, Carol Kemelgor, and Brian Uzzi. 2000. *Athena Unbound: The Advancement of Women in Science and Technology.* New York: Cambridge University Press.

Fausto-Sterling, Anne. 1992. *Myths of Gender: Biological Theories about Women and Men.* 2nd ed. New York: Basic Books.

Griffin, Gabriele, Stevi Jackson, Sasha Roseneil, and Rita Felski, eds. 2004. "Feminist Theory and/of Science." Special issue, *Feminist Theory* 5(2).

Gumport, Patricia J. 2002. *Academic Pathfinders: Knowledge Creation and Feminist Scholarship.* Westport, Conn.: Greenwood.

Hanson, Sandra L. 1996. *Lost Talent: Women in the Sciences.* Philadelphia: Temple University Press.

Haraway, Donna. 1997. *Modest_Witness@Second_Millennium.FemaleMan_Meets_OncoMouse: Feminism and Technoscience.* New York: Routledge.

Harding, Sandra, ed. 1987. *Feminism and Methodology: Social Science Issues.* Bloomington: Indiana University Press.

———, ed. 1993. *The "Racial" Economy of Science: Toward a Democratic Future.* Bloomington: Indiana University Press.

———. 1998. *Is Science Multicultural? Postcolonialisms, Feminisms, and Epistemologies.* Bloomington: Indiana University Press.

Harding, Sandra, and Kathryn Norberg, eds. 2003. "Gender and Science: New Issues." Special issue, *Signs: Journal of Women in Culture and Society* 28(3).

Harding, Sandra, and Jean F. O'Barr, eds. 1987. *Sex and Scientific Inquiry.* Chicago: University of Chicago Press.

Hermann, Anne C., and Abigail J. Stewart, eds. 1994. *Theorizing Feminism: Parallel Trends in the Humanities and Social Sciences.* Boulder, Colo.: Westview.

Hubbard, Ruth. 1990. *The Politics of Women's Biology.* New Brunswick, N.J.: Rutgers University Press.

Keller, Evelyn Fox. 1995. *Reflections on Gender and Science.* 10th anniversary edition. New Haven, Conn.: Yale University Press.

———. 1995. *Refiguring Life: Metaphors of Twentieth-Century Biology.* New York: Columbia University Press.

Keller, Evelyn Fox, and Helen E. Longino, eds. 1996. *Feminism and Science.* Oxford: Oxford University Press.

Kohlstedt, Sally Gregory, and Helen Longino, eds. 1997. *Women, Gender, and Science: New Directions.* Chicago: University of Chicago Press.

Kourany, Janet A. 2002. *The Gender of Science.* Upper Saddle River, N.J.: Prentice Hall.

Lederman, Muriel, and Ingrid Bartsch, eds. 2001. *The Gender and Science Reader.* New York: Routledge.

Longino, Helen E. 1990. *Science as Social Knowledge: Values and Objectivity in Scientific Inquiry.* Princeton, N.J.: Princeton University Press.

Margolis, Jane, and Allan Fisher. 2002. *Unlocking the Clubhouse: Women in Computing.* Cambridge, Mass.: MIT Press.

Mayberry, Maralee, Banu Subramaniam, and Lisa Weasel, eds. 2001. *Feminist Science Studies: A New Generation.* New York: Routledge.

Pattatucci, Angela M., ed. 1998. *Women in Science: Meeting Career Challenges.* Thousand Oaks, Calif.: Sage.

Paxton, Nancy L. 1991. *George Eliot and Herbert Spencer: Feminism, Evolutionism, and the Reconstruction of Gender.* Princeton, N.J.: Princeton University Press.

Rose, Hillary. 1994. *Love, Power, and Knowledge: Towards a Feminist Transformation of the Sciences.* Bloomington: Indiana University Press.

Rosser, Sue V., ed. 1995. *Teaching the Majority: Breaking the Gender Barrier in Science, Mathematics, and Engineering.* New York: Teachers College Press.

———. 1997. *Re-engineering Female Friendly Science.* New York: Teachers College Press.

———. 2004. *The Science Glass Ceiling: Academic Women Scientists and the Struggle to Succeed.* New York: Routledge.

Rosser, Sue V., and Janet Zandy, eds. "Building Inclusive Science: Connecting Women's Studies and Women in Science and Engineering." Special issue, *Women's Studies Quarterly* 20(1–2).

Rossiter, Margaret W. 1982. *Women Scientists in America: Struggles and Strategies to 1940.* Baltimore, Md.: Johns Hopkins University Press.

Schiebinger, Londa. 1993. *Nature's Body: Gender in the Making of Modern Science.* Boston: Beacon.

Seymour, Elaine, and Nancy M. Hewitt. 1997. *Talking about Leaving: Why Undergraduates Leave the Sciences.* Boulder, Colo.: Westview.

Sonnert, Gerhard, and Gerald Holton. 1995. *Who Succeeds in Science? The Gender Dimension.* New Brunswick, N.J.: Rutgers University Press.

Spanier, Bonnie B. 1995. *Im/Partial Science: Gender Ideology in Molecular Biology.* Bloomington: Indiana University Press.

Tobach, Ethel, and Betty Rosoff, eds. 1994. *Challenging Racism and Sexism: Alternatives to Genetic Explanations.* New York: Feminist Press.

Tuana, Nancy, ed. 1989. *Feminism and Science.* Bloomington: Indiana University Press.

———. 1999. *The Less Noble Sex: Scientific, Religious, and Philosophical Conceptions of Woman's Nature.* Bloomington: Indiana University Press.

Valian, Virginia. 1998. *Why So Slow? The Advancement of Women.* Cambridge, Mass.: MIT Press.

van den Wijngaard, Marianne. 1997. *Reinventing the Sexes: The Biomedical Construction of Femininity and Masculinity.* Bloomington: Indiana University Press.

Wasserman, Elga R. 2000. *The Door in the Dream: Conversations with Eminent Women in Science.* Washington, D.C.: Joseph Henry.

Wyer, Mary, Mary Barbercheck, Donna Geisman, Hatice Orun Ozturk, and Marta Wayne, eds. 2001. *Women, Science, and Technology: A Reader in Feminist Science Studies.* New York: Routledge.

Contributors

Neal B. Abraham is currently Executive Vice President, Vice President for Academic Affairs, Dean of the Faculty, and professor of physics at DePauw University. Before coming to DePauw in 1998, he taught physics at Bryn Mawr College for eighteen years. He helped to found Project Kaleidoscope, served as an inaugural member of the Committee on Undergraduate Science Education of the National Research Council, and has played leadership roles in the Council on Undergraduate Research and the National Conferences on Undergraduate Research.

Meryl Altman is Director of Women's Studies at DePauw University, where she also teaches English and American literature. She has published articles about Djuna Barnes, H. D. Faulkner, William Carlos Williams, Simone de Beauvoir, metaphor, and the history of sexuality, and writes periodically for the *Women's Review of Books.*

Gilbert W. Bassett, Jr. is Department Head and professor of finance at the University of Illinois at Chicago.

Josephine Beoku-Betts is an associate professor of women's studies and sociology at Florida Atlantic University. She is co-editor of *Women and Education in Sub-Saharan Africa: Power, Opportunities, and Constraints* (1998). Her research focuses on the educational and employment experiences and perspectives of African and Caribbean women academic scientists.

Sharon R. Bird is an associate professor of sociology and an affiliate of the Women's Studies Program at Iowa State University. Her current research focuses on women and men academic engineers and scientists, and women and men small business owners: the barriers and supports to success they confront in their work environments, and how they balance work and family. She also examines in her research the gendering of work organizations and work spaces.

Amy Sue Bix is an associate professor in the Department of History at Iowa State University and teaches in ISU's Women's Studies Program. Author of work on the history of technological unemployment, on post–World War II physics and engineering, on breast cancer and AIDS research, on eugenics, home economics, and alternative medicine, she is currently finishing a book titled *Engineering Education for American Women: An Intellectual, Institutional, and Social History.*

Cynthia Burack is associate professor of women's studies at The Ohio State University. She is the author of *Healing Identities: Black Feminist Thought and the Politics of Groups* (2004) and co-editor (with Jyl J. Josephson) of *Fundamental Differences: Feminists Talk Back to Social Conservatives* (2003). She is currently writing about antigay discourse and politics of the Christian Right.

Jill M. Bystydzienski is director of the Women's Studies Program and professor of sociology at Iowa State University. She is the author of several books and numerous articles and book chapters on women in politics, women's movements in international perspective, and coalition politics. In 2004, she edited a special issue of the *NWSA Journal* (16, no. 1) titled "(Re)Gendering Science Fields."

Molly J. Dingel is a Ph.D. student at the University of Kansas. She is currently working on her dissertation, an analysis of the way the media represent the findings of genetics research.

Dana A. Dudle, a plant reproductive biologist, studies the interactions of genetics, environments, sex, and gender in flowering plants. She also teaches evolution, ecology, and plant biology (and the intersections of these) in the biology department of DePauw University.

Kirsten Smilla Ebeling is a professor of gender, technologies, and society and a member of the Center for Interdisciplinary Research on Women and Gender at the Carl-von-Ossietzky University of Oldenburg. She previously held a position as an assistant at the Center for Gender Studies at Basel University in Switzerland and worked as a visiting scholar at the Center for Feminist Studies at the University of Minnesota. Her current research interest is in the construction of sexual dimorphism in zoology and botany.

Carla Fehr is an associate professor of philosophy and an affiliate of Women's Studies at Iowa State University. Her work appears in *Philosophy of Science, Biology and Philosophy, The Minnesota Series in the Philosophy of Science, Molecular Ecology,* and the *NWSA Journal.*

Heidi Fencl is an assistant professor of physics at the University of Wisconsin–Green Bay, and past director of the University of Wisconsin's systemwide Women and Science Program. At UW-Green Bay, she continues to explore learning and retention issues for students in the introductory physics classroom.

Suzanne E. Franks has worked in the pharmaceutical industry and is the founding director of the Women in Engineering and Science Program at Kansas State University. She holds a Ph.D. in biomedical engineering and a graduate certificate in women's studies; her publications include works on gender and science as well as her scientific work.

Sandra L. Hanson is a professor and Chair of Sociology at Catholic University. She is the recipient of a number of grants from the National Science Foundation to study young women's experiences in science. She is the author of *Lost Talent: Women in the Sciences* (1996) and numerous journal articles on women in science in the U.S. and in a cross-national context. She is currently working on a project which uses new and mature data collection technologies to explore the factors that encourage and discourage young African American women in science.

Abbe H. Herzig is an assistant professor of mathematics education in the School of Education, University at Albany, State University of New York. Her primary research interests are diversity, equity, and social justice in mathematics education. She is currently researching the experiences of women and students of color in postgraduate mathematical sciences and in undergraduate engineering.

Sally Gregory Kohlstedt teaches the history of science and also served a term as chair of the Center for Advanced Feminist Studies at the University of Minnesota–Twin Cities. Her work concentrates on science in American culture, investigating the institutional context of scientific research, the mechanisms by which science becomes part of culture, and the pattern of participation especially by women in science and technology.

Anne J. MacLachlan is a senior researcher at the Center for Studies in Higher Education at the University of California, Berkeley. For the last eighteen years, she has conducted research and developed and presented programs on graduate education and academic careers, focusing on issues of women and minorities in science. She holds a Ph.D. in German economic history from the University of California, Irvine.

Anna M. Martinson is a doctoral candidate at Indiana University specializing in gender, discourse, and information technology. She has published in the areas of information technology and women's leisure, feminist science fiction, and the representation of gender on Web sites.

Sue V. Rosser is currently Dean of Ivan Allen College at the Georgia Institute of Technology, where she is a professor in the School of History, Technology, and Society. Before assuming her current position, she served as director of Women's Studies at the University of Florida, the University of South Carolina, and Mary Baldwin College for more than twenty years. She is the author of eight books and more than a hundred journal articles on the theoretical and applied issues of women, feminism, science, and health.

Karen R. Scheel received her Ph.D. in counseling psychology from the University of Iowa in 1999. She is an assistant professor in the Collaborative Program in Counseling Psychology in the Department of Counseling at the University of Akron. Her scholarly interests focus on college students in the areas of self-efficacy in the natural sciences, clinical training and supervision, and suicide assessment and prevention.

Mo-Yin S. Tam is professor of economics and Associate Vice Chancellor for Academic Affairs at the University of Illinois at Chicago.

Virginia Valian is Distinguished Professor of Psychology and Linguistics at Hunter College and the Graduate Center of the City University of New York (CUNY). She is a cognitive scientist whose research ranges from first- and second-language acquisition to gender differences and gender equity. She is the author of the landmark book *Why So Slow? The Advancement of Women* (1998).

Index

197, 216, 229; essentialism and, 217; plant biology and, 215–231
Fencl, Heidi, 8
Fernandez, Mary Joe, 324
Finland, women in science, 157
Fischer, Allen, 118
Fischer, Beth A., 249
Fiske, Susan T., 321
Fleming, Wilhamina, 27
Franks, Suzanne, 9, 10, 101
Friedan, Betty, 32

Gaiser, Megan, 276
Galassie, John P., 290
Galileo Galilei, 197
Gallo, Bill, 324
Gautreau, Ronald, 291
Geiss, Florence L., 322, 323
Geist, Valertus, 191
gender: accountability, 162; doing gender, 161–162, 173, 227; functional gender, 220–221; intersections with race, 147–149; multicultural approach to understanding, 124–125; pay differential and, 61; phenotypic, 220, 227; plant biology and functional gender, 227; plants and, 218, 226. See also plant "gender"
gender dualisms, perceptions of sex differences, 321–322
gender equity: in Science, Technology, Engineering and Math fields, 320; institutional remedies, 328–330
gender gap, undergraduate students in information technology, 110–113
gender schemas, 16, 320–330; defined, 321; perceptions of women as leaders and, 322; women's and men's self-perceptions and, 324; women's lack of entitlement and, 324; women's slow advancement and, 323–324
gender stereotyping, technology and, 271–272
gender theory: sexual difference and, 221; social constructionism and, 221
gendered language, 97–98
gendered metaphors, 16
Georgia Institute of Technology, 13, 46, 48, 49, 51, 89
Gilman, Charlotte Perkins, 189
"glass ceiling," 30
Glaubke, Christina, 278
Goldring, Winefred, 26
gonochorism. See reproduction
Graf, Steffi, 324
Greenberger, Marcia, 39

Greenfield, Patricia, 279
gynogenesis, 182, 183

Haag, Pamela S., 277
Hacket, Gail, 287, 289, 290
Hage, Ghassen, 149
Hall, Roberta M., 288, 297
Halliday, M. A. K., 222
Halverson, Charles, 321
Hammonds, Evelyn, 11
Hanson, Sandra, 9, 15
Haraway, Donna, 215, 226, 230n4
Harding, Sandra, 143, 200
Harris, Betty, 13
Hart, Lauri, 255
Harvard University, 69
Heller, Patricia, 291
Henrion, Claudia, 263
hermaphrodite: simultaneous, consecutive, sequential, 179, 180
Herzig, Abbe, 10, 14, 15
heterosexist biases in science, 11
Higginbothem, Elizabeth, 125
higher education: department culture, 12; enculturation and identity, 256–258; integration processes, 256–258; participation in a community of practice, 256–258; policies for changing institutions, 87–91; theoretical perspectives in, 256
Hispanic: students in computer science, 94; women, 1
Historically Black Colleges and Universities, 138, 304, 307
Holland, Dorothy, 118
Hopper, Grace Murray, 31
hormones, 206–209
Hubbard, Ruth, 191
Huff, Charles, 278
human capital, 244
human identity, 228
Huurre, Anne-Marie, 272, 277, 281

Indiana University South Bend, 90
information technology: computer game design, 15; computer science and, 10; employment statistics, 108–109; gender gap in performance, 113–116, gender gap in undergraduate majors, 110–113; women in, 108–120
Instedt, Karen Lafferty, 59
intelligibility, defined, 209–210
interactionist model, 208–209, 211–212, 228, 229

interdisciplinary collaboration, 215–218, 230

International Game Developers Association, 282

intersections of race and gender, 147–149, 228; "double jeopardy," 126. *See also individual race and ethnic groups*

intersex, plant biology and, 225

Iowa State University, 47, 48; Conference on Retaining Women in Science, 3

isolation of women in science, technology, engineering and math, 10

Ivey, Rachel, 251n2

Johnston, Betty, 260

Kafai, Yasmin, 276

Kahn, Jeffrey H., 289

Keith, Ronald, 291

Keller, Evelyn Fox, 201, 217, 223, 225–226, 230n2

Kemelgor, Carol, 12

Kenway, Jane, 149

Killian, James, 52

Kinsey, Alfred, 221, 226

Kinsman, Sharon, 221

Koblitz, Anne Hibner, 39

Kodak, 56

Kohlheb, Gerda, 54–55

Kohlstedt, Sally Gregory, 6

Kovalevskaia, Anna, 39

Ladson-Billings, Gloria, 258

Lane, David M., 323

language, gendered, 97–98

Lapan, Richard T., 289

Lapidus, Jules B., 251n1

Larkin, Kevin C., 289, 290, 298

Latina women, 1

Latino and Latina students in computer science, 94

Lave, Jean, 255

leadership, women, 97–99

learning communities, 15

Lent, Robert W., 289, 290, 298

Lerner, Melvin J., 324

liberal arts colleges, 14, 303–318; advancement of women and, 305; attracting and retaining faculty at, 312–316; graduate education, 310; increasing numbers of women in STEM, 304; liberal arts education, 303–318; preparing future faculty programs, 312, 315–316; support for undergraduate majors, 306–310; underrepresented groups in STEM, 304

Linn, Marcia, 273

Linnaeus, Carolus, 11, 218, 219, 224, 225

Lloyd, David, 220, 224, 225, 231nn8,9

Locke, John, 197

Logino, Helen, 198, 206–207

Long, J. Scott, 326, 329

Luttrell, Wendy, 149

Machamer, Peter, 205, 206, 208, 209, 210, 211

MacLachlan, Anne, 9, 10, 14, 15

MacLeaod, Margo, 288, 289

Major, Brenda, 324

Manis, Melvin, 321

Manley, Jason, 275

Marc, Mileva, 26

marginalization, being defined as "other" and, 169

Margolis, Jane, 118

Martell, Richard F., 323

Martin, Carol Lynn, 321

Martin, Emily, 203

Martin, J. R., 222

Martinsen, Anna, 10, 15, 118, 280

masculinism: in engineering, 46, 54; biological reproduction and, 3; male standard of science, 6; masculinist culture and ideology, 15

Massachusetts Institute of Technology, 38, 46, 51, 52, 53, 54, 55, 57, 58, 62, 69, 326; job preparation seminars for women, 55; Symposium on American Women in Science and Engineering, 53; 2001 Report on Women in Science, 88

mathematics, 10–11, 14; aesthetics of knowledge, 261; African American students in, 259; African American women's entry into, 126; faculty–graduate student relationships, 263–264; gender and, 113–116, 254–267; gendered and racialized beliefs about, 258–262; graduate students and, 254–267; mathematical autobiographies, 254; "mathematics for all," 255; mentoring, 266; obstacles to belonging, 262–265; recommendations for improving educational experience in, 266–267; retention of graduate students, 256; student relationships in, 264–265; students of color in, 265; theoretical model of belonging, 257; women's belonging in, 255

Mattfeld, Jacqueline, 53, 58

Mayer, Maria Goeppart, 31

McClintock, Barbara, 26, 31, 201, 229

McCloud, Margo, 297

McCormick, Katherine Dexter, 53

McDermott, Lillian C., 291

McDonough, Jerome, 278–279
McGinnis, Robert, 329
Mead, Margaret, 26
Meitner, Lisa, 29
Mendes, Pamela, 277
mentoring: African women in sciences, 151, 153; faculty development and, 315; liberal arts colleges' faculty and, 315; mentoring relationships, 15
Merchant, Carolyn, 197, 202–204
Merton, Robert K., 323
metaphor: across disciplines, 220–224; and anthropomorphizing, 223, 226; economic, 226; metaphor, functional agenda, 226; functional metaphor, 225, of gender, 223–224; plant biology and, 218–219, 222, 224–226
methodological pluralism, 197, 212
Meyerowitz, Joanne, 227
Michels, Walter C., 317
Miersch, Michael, 186
Millett, Catherine M., 251n1
Milliken, Robert A., 28
Minh-Ha, Trinh, 149
Mohanty, Chandra, 149
Mos, Toril, 217
Myrdal, Alva, 26
Myrdal, Gunnar, 26

Nagel, Ernest, 213n1
National Academy of Engineering, 93
National Academy of Sciences, 100–101
National Center for Educational Statistics, 115
National Educational Longitudinal Study (NELS), 124, 128–132, 137
National Institutes of Health, 244
National Organization for Women, 55
National Research Council, 35, 237
National Science Board, 123
National Science Foundation, 34, 35, 61, 69, 115, 118, 123, 126, 244, 307, 326; ADVANCE Institutional grant (see ADVANCE Institutional Transformation program of the National Science Foundation)
National Society of Black Physicists, 247
Native American students in computer science, 94
Nauta, Margaret, 289
Nelson, Thomas E., 321
Nettles, Michael T., 251n1
Newton, Carol, 247
Newton, Sir Isaac, 197
Noddings, Nel, 264

normative career path, 7
Novesky, Lisa, 291

O'Bannon, Helen, 51
Ohio State University, 59
Olive, Dennis, 47
ontological reductionism, defined, 198–199
ontology, 198
organizational culture, "relational department," 12–13
Owen, Richard, 182

Palmer, Craig, 218
Parker, Chris, 275
pedagogy: science education and, 14; teaching methods and, 8
Peden, Irene Carswell, 50
Pennsylvania State University, 47, 51, 56
"performance pressure," 163
physics: Bryn Mawr introductory courses, 308; classroom climate and class size, 288–289; retention of women, 287, 288–300; teaching styles, 287; women's course participation, 287–300
pipeline theory of women's advancement in Science, Technology, Engineering and Math fields, 4, 34, 326
plant biology: intersex, 225; metaphor and, 218–219; plant identities and, 228
plant "gender," 218, 226; biological sex and, 227. See also gender
pleiotropy, 228
Porter, Natalie, 322
Pratt, M. L., 150
Princeton University, 69
Pritchard, Alice, 118
Professional Opportunities for Women in Research and Education (POWRE), 69, 70–87, 90
Project Kaleidoscope, 307; PKAL study, 309, 310
Prosser, Jay, 230n1
psychodynamic group theory, 93–96
Purdue University, 47, 51, 55

race, intersections with gender. See intersections of race and gender; individual race and ethnic groups
racism in science, 137–138
Raven, Peter H., 231n7
recruitment of women in engineering, 55–56
Reimers, Tekla, 189
Rensselaer Polytechnic Institute, 46, 47
reproduction: asexual, 180; gonochorism, 179,

181; heterosexual, 11; parthenogenesis, 182–192, 185, 191

retention, pedagogy and, 289–291; women's self-efficacy, 289–291

retention of women, cooperative learning strategies, 291

Rice, A. K., 95

Rich, Adrienne, 227, 231n10

Richards, Ellen Swallow, 26, 27

Richmond, Geraldine, 39

Riewenherm, Sabine, 179

"Rosie the Riveter," 47

Rosiland, Franklin, 31

Rosser, Sue, 5, 7, 10, 13, 37, 144, 202

Rossi, Alice, 53

Rossiter, Margaret, 23, 25, 26, 28

Rubin, Gayle, 220, 229, 230nn1,6

Russo, Jan, 276

Sakai, Anne, 223

Salmon, Wesley, 205, 206, 208, 211

Sandler, Bernice R., 288, 297

Sawin, Linda L., 321

Schaefers, Kathleen G., 289

Scheel, Karen, 8, 298

Schiebinger, Londa, 38, 197

Schifano, Ross S., 299

Schlupp, Ingo, 183–184, 185

Schwartz, Nancy, 280

science: as a gendered institution, 6; courses, 8; Finland women in, 157; third world women in, 142

Science Citation Index, 226

science education: mixed methods, 292–298; nontraditional, 292–298; retention and, 291–299; science education, traditional methods, 292–298

Science, Technology, Engineering and Mathematics disciplines (STEM): acculturation problem for women, 327; applicant pool, 304; gender equity in, 320; gender equity problem in, 325–328

sciences: decreased funding for science research, 90–91; sexism in, 11, 137–138; women's reasons for leaving, 287–289

second-wave feminism and feminist theory, 215

Sedgwick, Eve, 217, 222, 228

Seife, Charles, 304

Seles, Monica, 324

sex-gender binary, 215–216, 217, 219–222. See also gender dualisms, perceptions of sex differences

sex-gender system, 229

sexual difference, gender theory and, 221. See

also gender dualisms, perceptions of sex differences

sexual dimorphism, 11, 179

Seymour, Elaine, 287, 288, 289, 290, 296, 297

Shaffer, Peter S., 291

Shaughnassey, Peter, 289

Shneiderman, Ben, 281

Siegal, Ross G., 289

Silverberg, Lisa A., 288, 297

Silverman, Suzanne, 118

single-sex colleges, 24

Skerry, Peter, 104

social constructionism, 162, 221; gender theory and, 217

Society for the Advancement of Chicanos and Native Americans in Science (SACNAS), 247

Society for Women Engineers, 49, 50, 51, 55, 56, 59, 60; climate of activism, 59–61; promoting women in science at high schools, 60; support for junior women professionals in engineering, 59

Society of Women Geographers, 30

Spanier, Bonnie, 191

Spence, Janet T., 321

"status shield," 162, 172

Steinem, Gloria, 58

stereotyping, women in science and engineering, 88–89

Stiles, Florence, 52

Stone, Allucquère Rosanne, 272

Stone, Paula, 57

Stowe, Katie, 251n2

Stratton-Porter, Gene, 38

Subrahmanyam, Kaveri, 279

Subramaniam, Banu, 11

Syracuse University, 49

Tam, Mo-Ying, 10

Tate, William, 258

Taylor, Shelley E., 321

technology, gender stereotyping and, 271–282

Texas A&M University, 15

Thacker, Beth, 291

"third-wave" feminism, 216–217, 227

Thomas, M. Carey, 316

Thompson, E. P., 227

Thornhill, Randy, 218

Tinto, Vincent, 255

Title IX, Education Amendments of 1972, 33, 39, 94

Tobias, Sheila, 35

tokenism. See "performance pressure"

Trower, Cathy, 100

underrepresented groups, 14; climate for women of color in education, 9; in computer science, 94; data on gender distribution of undergraduate students of, 111–112; "double jeopardy" and, 126; women in science and, 1, 15, 123–138. *See also individual race and ethnic groups*

University of Arizona, 88, 89

University of California, 87

University of California at Berkeley, 69, 247

University of Californa at Los Angeles, 59, 247; computer use survey among students, 112–113; math survey of female and male students' confidence, 118

University of Colorado, 51

University of Florida, 60

University of Illinois at Chicago, 110

University of Illinois at Urbana-Champaign, 60

University of Iowa, 247

University of Maryland, 265

University of Michigan, 69, 277

University of Minnesota, 38, 47, 55

University of Missouri, 51

University of Pennsylvania, 69

University of Pittsburg, 249

University of South Florida, 60

University of Texas, 47

University of Washington, 59

Uzzi, Brian, 12

Valian, Virginia, 16, 163

Van Leer, Blake, 48–49

Vaughan, Misha, 280

Vetter, Betty, 33, 35

Voytin, Karla Klein, 289

Walker, Alice: space for people of color, 102; "womanism," 102

Walker, Eric, 51

Ware, William B., 290

Webber, Lynn, 125

Weiner, Kimberly, 118

Weller, Stephen, 223

Wenger, Etienne, 255

West, Candice, and Sarah Fenstermaker, 162

West, Candice, and Don Zimmerman, 161, 162

Wick, Emily, 58

Wickler, Wolfgang, 191

Widnall, Sheila, 55, 60

Wiesner, Jerome, 58

Wilcox, Clyde, 125

Wiles, Peter, 256

Wineman, Janice, 108

Wittgenstein, Ludwig, 218

women: competition for resources, 78; cumulative disadvantage and, 16; as deficient, 3, 4; discrimination and harassment, 78–79; isolation, 79; lack of camaraderie, 79; leadership, 97–99; liberation movement and, 183, 186; mentoring, 79; scientists, work environment, 82; students' self-esteem in science, technology, engineering and math, 289; tokenism in science and engineering and, 76–78; underrepresentation in science and engineering, 76–78, 79; "womanism," 102

Women's Anthropological Society of America, 29

Women's Bureau of the Department of Labor, 33

women's colleges, 26, 138

Women's International Science Collaboration Program, 39

Women in Science and Engineering (WISE), 271; DePauw University's WISE program, 307; discrimination and harassment and, 89–90; low numbers, 88–89; new approaches to science and engineering and, 86; statistics on, 1–2, 35–37, 71–72; stereotyping and, 88–89

Women in Technology (WIT), 271

Womengamers.com, 274

Woodward, Jim, 205–206

work-life balance, 7, 74–76, 82, 87–88, 326; African women in graduate school, 154–156; extension of tenure clock, 13; family-friendly policies and practices and, 13, flexible work loads and, 13

Wright, Dorothy, 305

Wulf, William A., 93, 98

Yale University, 69

Zigmund, Michael J., 249

Zita, Jacquelyn, 202

Zuk, Marlene, 223, 226, 229